Mobile Crane Operations

Level Three

Trainee Guide

Pearson

Boston Columbus Indianapolis New York San Francisco Amsterdam
Cape Town Dubai London Madrid Milan Munich Paris Montreal Toronto Delhi
Mexico City Sao Paulo Sydney Hong Kong Seoul Singapore Taipei Tokyo

NCCER

President: Don Whyte
Vice President: Steve Greene
Chief Operations Officer: Katrina Kersch
Mobile Crane Operations Curriculum Project Manager: Chris Wilson
Senior Development Manager: Mark Thomas
Senior Production Manager: Tim Davis
Quality Assurance Coordinator: Karyn Payne
Desktop Publishing Coordinator: James McKay
Permissions Specialists: Kelly Sadler
Production Specialist: Kelly Sadler
Editors: Graham Hack, Debie Hicks

Writing and development services provided by Topaz Publications, Liverpool, NY

Lead Writer/Project Manager: Troy Staton
Desktop Publisher: Joanne Hart
Art Director: Alison Richmond
Permissions Editor: Andrea LaBarge
Writers: Troy Staton, Terry Egolf

Pearson

Director of Alliance/Partnership Management: Andrew Taylor
Editorial Assistant: Collin Lamothe
Program Manager: Alexandrina B. Wolf
Assistant Content Producer: Alma Dabral
Digital Content Producer: Jose Carchi
Director of Marketing: Leigh Ann Simms
Senior Marketing Manager: Brian Hoehl
Composition: NCCER
Printer/Binder: LSC Communications
Cover Printer: LSC Communications
Text Fonts: Palatino and Univers

Credits and acknowledgments for content borrowed from other sources and reproduced, with permission, in this textbook appear at the end of each module.

Pearson

ISBN-13: 978-0-13-518502-5
ISBN-10: 0-13-518502-5

ScoutAutomatedPrintCode

Preface

To the Trainee

Mobile cranes are found on almost every construction job site and in every facet of the construction industry. Residential and commercial construction companies, industrial and petrochemical plants, shipbuilders, and highway and bridge construction companies all use mobile cranes to do the heavy lifting.

Mobile crane operators use cranes to move heavy loads around a job site with precision. They are required to inspect and make adjustments to the crane. They use gauges and monitors to understand the stresses on the machinery and to maintain safety during a lift. Of course, safety is of paramount importance when using such large machines to move heavy loads. Training as a mobile crane operator, therefore, is extremely important to the safety of those who will work around a lifting site.

The need for mobile crane operators will continue to increase in all industries as experienced operators retire. Those who meet the qualifications under OSHA 29 *CFR* Part 1926 Subpart CC, *Cranes and Derricks in Construction* will continue to be in high demand in this field that the US Department of Labor expects to experience growth over the next several years.

New with *Mobile Crane Operations*

NCCER is pleased to release *Mobile Crane Operations* in full color, with new photographs and figures. This version is now presented in NCCER's improved instructional systems design, in which the sections of each module are directly tied to learning objectives.

This version of *Mobile Crane Operations Level Three* features a great deal of updated material. The module "Load Charts" offers additional examples and study questions on load calculations. "Advanced Operational Techniques" covers a majority of the advanced techniques listed in the technical knowledge criteria found under OSHA 29 *CFR* Part 1926 Subpart CC Appendix C. The module "Lift Planning" covers the most important sections of ASME P30.1, *Planning for Load Handling Activities*. The role of the A/D director is covered in both the "Telescopic Boom Attachment Setup and Assembly" and "Lattice Boom Assembly and Disassembly" modules.

We wish you success as you progress through this training program. If you have any comments on how NCCER might improve upon this textbook, please complete the User Update form located at the back of each module and send it to us. We will always consider and respond to input from our customers.

We invite you to visit the NCCER website at **www.nccer.org** for information on the latest product releases and training, as well as online versions of the *Cornerstone* magazine and Pearson's NCCER product catalog.

Your feedback is welcome. You may email your comments to **curriculum@nccer.org** or send general comments and inquiries to **info@nccer.org**.

NCCER Standardized Curricula

NCCER is a not-for-profit 501(c)(3) education foundation established in 1996 by the world's largest and most progressive construction companies and national construction associations. It was founded to address the severe workforce shortage facing the industry and to develop a standardized training process and curricula. Today, NCCER is supported by hundreds of leading construction and maintenance companies, manufacturers, and national associations. The NCCER Standardized Curricula was developed by NCCER in partnership with Pearson, the world's largest educational publisher.

Some features of the NCCER Standardized Curricula are as follows:

- An industry-proven record of success
- Curricula developed by the industry, for the industry
- National standardization providing portability of learned job skills and educational credits
- Compliance with the Office of Apprenticeship requirements for related classroom training (*CFR 29:29*)
- Well-illustrated, up-to-date, and practical information

NCCER also maintains the NCCER Registry, which provides transcripts, certificates, and wallet cards to individuals who have successfully completed a level of training within a craft in NCCER's Curricula. *Training programs must be delivered by an NCCER Accredited Training Sponsor in order to receive these credentials.*

Special Features

In an effort to provide a comprehensive and user-friendly training resource, this curriculum showcases several informative features. Whether you are a visual or hands-on learner, these features are intended to enhance your knowledge of the construction industry as you progress in your training. Some of the features you may find in the curriculum are explained below.

Introduction

This introductory page, found at the beginning of each module, lists the module Objectives, Performance Tasks, and Trade Terms. The Objectives list the knowledge you will acquire after successfully completing the module. The Performance Tasks give you an opportunity to apply your knowledge to real-world tasks. The Trade Terms are industry-specific vocabulary that you will learn as you study this module.

00104-15
Introduction to Power Tools

Objectives

When you have completed this module, you will be able to do the following:

1. Identify and explain how to use various types of power drills and impact wrenches.
   a. Identify and explain how to use common power drills and bits.
   b. Identify and explain how to use a hammer drill.
   c. Identify and explain how to use pneumatic drills and impact wrenches.
2. Identify and explain how to use various types of power saws.
   a. Identify and explain how to use a circular saw.
   b. Identify and explain how to use saber and reciprocating saws.
   c. Identify and explain how to use a portable band saw.
   d. Identify and explain how to use miter and cutoff saws.
3. Identify and explain how to use various grinders and grinder attachments.
   a. Identify and explain how to use various types of grinders.
   b. Identify and explain how to use various grinder accessories and attachments.
4. Identify and explain how to use miscellaneous power tools.
   a. Identify and explain how to use pneumatic and powder-actuated fastening tools.
   b. Identify and explain how to use pavement breakers.
   c. Identify and explain the uses of hydraulic jacks.

Performance Tasks

Under supervision of your instructor, you should be able to do the following:

1. Safely and properly demonstrate the use of three of the following tools:
   - Electric drill
   - Hammer drill or rotary hammer
   - Circular saw
   - Reciprocating saw
   - Portable band saw
   - Miter or cutoff saw
   - Portable or bench grinder
   - Pneumatic nail gun
   - Pavement breaker

Trade Terms

Abrasive	Chuck key	Ground fault circuit interrupter (GFCI)	Revolutions per minute (rpm)
Alternating current (AC)	Countersink	Ground fault protection	Ring test
Arbor	Direct current (DC)	Kerf	Shank
Auger bit	Forstner bit	Masonry bit	Trigger lock
Carbide	Grit	Reciprocating	
Chuck			

Industry Recognized Credentials

If you are training through an NCCER-accredited sponsor, you may be eligible for credentials from NCCER's Registry. The ID number for this module is 00104-15. Note that this module may have been used in other NCCER curricula and may apply to other level completions. Contact NCCER's Registry at 888.622.3720 or go to **www.nccer.org** for more information.

Trade Features

Trade features present technical tips and professional practices based on real-life scenarios similar to those you might encounter on the job site.

Bowline Trivia

Some people use this saying to help them remember how to tie a bowline: "The rabbit comes out of his hole, around a tree, and back into the hole."

Figures and Tables

Photographs, drawings, diagrams, and tables are used throughout each module to illustrate important concepts and provide clarity for complex instructions. Text references to figures and tables are emphasized with *italic* type.

Notes, Cautions, and Warnings

Safety features are set off from the main text in highlighted boxes and categorized according to the potential danger involved. Notes simply provide additional information. Cautions flag a hazardous issue that could cause damage to materials or equipment. Warnings stress a potentially dangerous situation that could result in injury or death to workers.

NOTE

Nameplates must be posted on each material handling device. The nameplate must indicate

CAUTION

It is essential to note the revision designation on a construction drawing and to use only the latest

WARNING!

Saw teeth are very sharp. Use gloves and do not handle the saw teeth with bare hands. When cutting with a saw, ensure that your fingers remain clear of the teeth at all times.

Case History

Case History features emphasize the importance of safety by citing examples of the costly (and often devastating) consequences of ignoring best practices or OSHA regulations.

Case History

Requesting an Outage

An electrical contractor requested an outage when asked to install two bolt-in, 240V breakers in panels in a data processing room. It was denied due to the 24/7 worldwide information processing hosted by the facility. The contractor agreed to proceed only if the client would sign a letter agreeing not to hold them responsible if an event occurred that damaged computers or resulted in loss of data. No member of upper management would accept liability for this possibility, and the outage was scheduled.

The Bottom Line: If you can communicate the liability associated with an electrical event, you can influence management's decision to work energized.

Going Green

Going Green features present steps being taken within the construction industry to protect the environment and save energy, emphasizing choices that can be made on the job to preserve the health of the planet.

GOING GREEN

Reducing Your Carbon Footprint

Many companies are taking part in the paperless movement. They reduce their environmental impact by reducing the amount of paper they use. Using email helps to reduce the amount of paper used,

Did You Know

Did You Know features introduce historical tidbits or interesting and sometimes surprising facts about the trade.

Did You Know?

Safety First

Safety training is required for all activities. Never operate tools, machinery, or equipment without prior training. Always refer to the manufacturer's instructions.

Step-by-Step Instructions

Step-by-step instructions are used throughout to guide you through technical procedures and tasks from start to finish. These steps show you how to perform a task safely and efficiently.

Perform the following steps to erect this system area scaffold:

Step 1 Gather and inspect all scaffold equipment for the scaffold arrangement.

Step 2 Place appropriate mudsills in their approximate locations.

Step 3 Attach the screw jacks to the mudsills.

Trade Terms

Each module presents a list of Trade Terms that are discussed within the text and defined in the Glossary at the end of the module. These terms are presented in the text with **bold, blue** type upon their first occurrence. To make searches for key information easier, a comprehensive Glossary of Trade Terms from all modules is located at the back of this book.

During a rigging operation, the **load** being lifted or moved must be connected to the apparatus, such as a crane, that will provide the power for movement. The connector—the link between the load and the apparatus—is often a sling made of synthetic, chain, or **wire rope** materials. This section focuses on three types of slings:

Section Review

Each section of the module wraps up with a list of Additional Resources for further study and Section Review questions designed to test your knowledge of the Objectives for that section.

Additional Resources

Materials Handling Handbook, The American Society of Mechanical Engineers (ASME) and The International Material Management Society (IMMS), Raymond A. Kulwiec, Editor-in-Chief. 1985. New York, NY: Wiley-Interscience.

Manufacturing Facilities Design & Material Handling, Matthew P. Stevens, Fred E. Meyers. 2013. West Lafayette, IN: Purdue University Press.

1.0.0 Section Review

1. For material handling tasks, it is just as important to be mentally fit as it is to be _____.
   a. physically fit
   b. physically aggressive
   c. closely supervised
   d. over 200 pounds

2. Which of the following is a type of knot that is often used to join the ends of two ropes in noncritical, low-strain applications?
   a. Bowline
   b. Clove hitch
   c. Half hitch
   d. Square knot

Review Questions

The end-of-module Review Questions can be used to measure and reinforce your knowledge of the module's content.

Review Questions

1. Identification tags for slings must include the _____.
   a. type of protective pads to use
   b. type of damage sustained during use
   c. color of the tattle-tail
   d. manufacturer's name or trademark

2. The type of wire rope core that is susceptible to heat damage at relatively low temperatures is the _____.
   a. fiber core
   b. strand core
   c. independent wire rope core
   d. metallic link supporting core

3. Synthetic slings must be inspected _____.
   a. once every month
   b. visually at the start of each work week
   c. before every use
   d. once wear or damage becomes apparent

4. An alloy steel chain sling must be removed from service if there is evidence that _____.
   a. the sling has been used in different hitch configurations
   b. replacement links have been used to repair the chain
   c. the sling has been used for more than one year
   d. strands in the supporting core have weakened

5. A piece of rigging hardware used to couple the end of a wire rope to eye fittings, hooks, or other connections is a(n) _____.
   a. eyebolt
   b. hitch
   c. shackle
   d. U-bolt

6. A lifting clamp is most likely to be used to move loads such as _____.
   a. steel plates
   b. piping bundles
   c. concrete blocks
   d. plastic tubing

7. Chain hoists are able to lift heavy loads by utilizing a _____.
   a. rope and pulley system
   b. rigger's strength
   c. stationary counterweight
   d. gear system

8. Before attempting to lift a load with a chain hoist, make sure that the _____.
   a. hoist is secured to a come-along
   b. load is properly balanced
   c. tag lines are properly anchored
   d. tackle is connected to its power source

9. A hitch configuration that allows slings to be connected to the same load without using a spreader beam is a _____.
   a. double-wrap hitch
   b. choker hitch
   c. bridle hitch
   d. basket hitch

10. To make the emergency stop signal that is used by riggers, extend both arms _____.
    a. horizontally with palms down and quickly move both arms back and forth
    b. directly in front and then move both arms up and down repeatedly
    c. vertically above the head and wave both arms back and forth
    d. horizontally with clenched fists and move both arms up and down

00106-15 Introduction to Basic Rigging Module Six 31

NCCER Standardized Curricula

NCCER's training programs comprise more than 80 construction, maintenance, pipeline, and utility areas and include skills assessments, safety training, and management education.

Boilermaking
Cabinetmaking
Carpentry
Concrete Finishing
Construction Craft Laborer
Construction Technology
Core Curriculum: Introductory Craft Skills
Drywall
Electrical
Electronic Systems Technician
Heating, Ventilating, and Air Conditioning
Heavy Equipment Operations
Heavy Highway Construction
Hydroblasting
Industrial Coating and Lining Application Specialist
Industrial Maintenance Electrical and Instrumentation Technician
Industrial Maintenance Mechanic
Instrumentation
Ironworking
Manufactured Construction Technology
Masonry
Mechanical Insulating
Millwright
Mobile Crane Operations
Painting
Painting, Industrial
Pipefitting
Pipelayer
Plumbing
Reinforcing Ironwork
Rigging
Scaffolding
Sheet Metal
Signal Person
Site Layout
Sprinkler Fitting
Tower Crane Operator
Welding

Maritime

Maritime Industry Fundamentals
Maritime Pipefitting
Maritime Structural Fitter

Green/Sustainable Construction

Building Auditor
Fundamentals of Weatherization
Introduction to Weatherization
Sustainable Construction Supervisor
Weatherization Crew Chief
Weatherization Technician
Your Role in the Green Environment

Energy

Alternative Energy
Introduction to the Power Industry
Introduction to Solar Photovoltaics
Power Generation Maintenance Electrician
Power Generation I&C Maintenance Technician
Power Generation Maintenance Mechanic
Power Line Worker
Power Line Worker: Distribution
Power Line Worker: Substation
Power Line Worker: Transmission
Solar Photovoltaic Systems Installer
Wind Energy
Wind Turbine Maintenance Technician

Pipeline

Abnormal Operating Conditions, Control Center
Abnormal Operating Conditions, Field and Gas
Corrosion Control
Electrical and Instrumentation
Field and Control Center Operations
Introduction to the Pipeline Industry
Maintenance
Mechanical

Safety

Field Safety
Safety Orientation
Safety Technology

Supplemental Titles

Applied Construction Math
Tools for Success

Management

Construction Workforce Development Professional
Fundamentals of Crew Leadership
Mentoring for Craft Professionals
Project Management
Project Supervision

Spanish Titles

Acabado de concreto: nivel uno (*Concrete Finishing Level One*)
Aislamiento: nivel uno (*Insulating Level One*)
Albañilería: nivel uno (*Masonry Level One*)
Andamios (*Scaffolding*)
Carpintería: Formas para carpintería, nivel tres (*Carpentry: Carpentry Forms, Level Three*)
Currículo básico: habilidades introductorias del oficio (*Core Curriculum: Introductory Craft Skills*)
Electricidad: nivel uno (*Electrical Level One*)
Herrería: nivel uno (*Ironworking Level One*)
Herrería de refuerzo: nivel uno (*Reinforcing Ironwork Level One*)
Instalación de rociadores: nivel uno (*Sprinkler Fitting Level One*)
Instalación de tuberías: nivel uno (*Pipefitting Level One*)
Instrumentación: nivel uno, nivel dos, nivel tres, nivel cuatro (*Instrumentation Levels One through Four*)
Orientación de seguridad (*Safety Orientation*)
Paneles de yeso: nivel uno (*Drywall Level One*)
Seguridad de campo (*Field Safety*)

Acknowledgments

This curriculum was revised as a result of the farsightedness and leadership of the following sponsors:

ABC Pelican Chapter
Bay Ltd.
Bo-Mac Contractors, Ltd.
Cowboyscranes.com
Exelon Generation
Jacobs Field Services
KBR, Inc.
Kelley Construction
Mammoet USA
North American Crane Bureau
Orion Marine Group
Southland Safety

This curriculum would not exist were it not for the dedication and unselfish energy of those volunteers who served on the Authoring Team. A sincere thanks is extended to the following:

Ed Burke
Robert Capelli
Anthony Johnson
Richard Laird
Don McDonald
Timothy Prakop
Larry "Cowboy" Proemsey
Joseph Watts
Harold Williamson

A sincere thanks is also extended to the dedication and assistance provided by the following technical advisors:

Grant Chutz
Keith Denham
Frank Jones

NCCER Partners

American Council for Construction Education
American Fire Sprinkler Association
Associated Builders and Contractors, Inc.
Associated General Contractors of America
Association for Career and Technical Education
Association for Skilled and Technical Sciences
Construction Industry Institute
Construction Users Roundtable
Design Build Institute of America
GSSC – Gulf States Shipbuilders Consortium
ISN
Manufacturing Institute
Mason Contractors Association of America
Merit Contractors Association of Canada
NACE International
National Association of Women in Construction
National Insulation Association
National Technical Honor Society
National Utility Contractors Association
NAWIC Education Foundation
North American Crane Bureau
North American Technician Excellence
Pearson
Prov
SkillsUSA®
Steel Erectors Association of America
U.S. Army Corps of Engineers
University of Florida, M. E. Rinker Sr., School of Construction Management
Women Construction Owners & Executives, USA

NCCER Business Partners

Contents

Module One

Load Charts

Explains how mobile crane operators use load charts to ensure safe lifting operations. Operators must understand the forces acting on the crane at any given time during a lift. Prior to the lift, operators must know the limitations of the crane based on the load weight and the crane configuration. This information is obtained primarily from load charts. This module explains how to read and use the important information in load charts for various types of cranes. (Module ID 21301; 35 Hours)

Module Two

Lift Planning

Supports industrial and safety standards requiring some form of lift planning. The level of detail required for a lift plan depends on the complexity of the lift, and the real and potential hazards involved. The more hazardous or sensitive the lift, the more detailed and formal the plan must be. A lift plan contains information relating to the crane(s) used, the load and its rigging, and all the site coordination issues. This module explains what lift planning involves and describes how to prepare a lift plan. (Module ID 21304; 30 Hours)

Module Three

Telescopic-Boom Attachment Setup and Assembly

Covers the setup and assembly of telescopic-boom attachments for mobile cranes. Manufacturers build and sell cranes in various models with specific options. However, operators can greatly increase the utility of cranes by installing add-on components provided by manufacturers. This module describes the assembly and disassembly of several types of telescopic-boom extensions and accessories. (Module ID 21302; 20 Hours)

Module Four

Lattice Boom Assembly and Disassembly

Presents common boom assembly and disassembly procedures for lattice-boom cranes. The long length of some lattice booms makes it impossible for them to remain assembled for transport. Each brand and model of crane have their differences, however, and boom assembly and disassembly can be a hazardous task. Due to the weight and size of boom components, assembly and disassembly requires careful thought as well as careful actions. (Module ID 21306; 20 Hours)

Module Five

Hoisting Personnel

Presents the hazardous task of hoisting personnel using a crane. Loss of control of the crane or the failure of a hoisting component will likely result in the serious injury or death of the persons involved. It was once common practice for workers to ride a ball as a convenient means of moving between the ground and a high point. However, the potential for catastrophic injuries was extreme, and falls and injuries were common. For this reason, the government enacted strict safety regulations pertaining to lifting personnel by crane. This module addresses these requirements in detail. (Module ID 21305; 20 Hours)

Module Six

Advanced Operational Techniques

Presents a wide variety of unique mobile-crane applications and circumstances. Up to this point, training has focused on operating cranes safely in common lifting operations. In this module, trainees will learn about several advanced mobile-crane operations that many operators will eventually experience during their professional career. (Module ID 21303; 20 Hours)

Glossary

Index

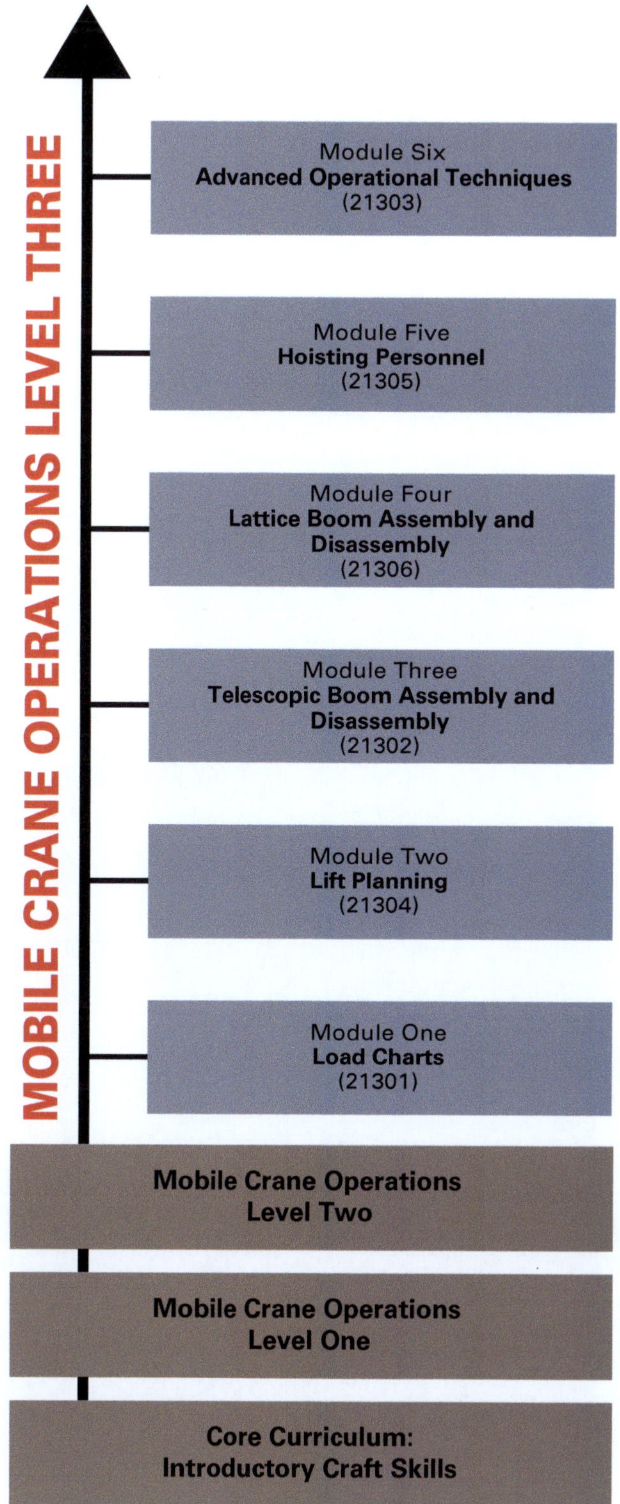

This course map shows all of the modules in *Mobile Crane Operations Level Three*. The suggested training order begins at the bottom and proceeds up. Skill levels increase as you advance on the course map. The local Training Program Sponsor may adjust the training order.

Load Charts

Overview

A lift can be accomplished safely as long as the operator has a working knowledge of the forces acting on the crane at any given time during the operation. Prior to the lift, operators must know the limitations of the crane based on the load weight and the crane configuration. This important information is obtained primarily from load charts. This module explains how the information on load charts for various types of cranes is used to ensure safe lifts.

Module 21301

Trainees with successful module completions may be eligible for credentialing through the NCCER Registry. To learn more, go to **www.nccer.org** or contact us at 1.888.622.3720. Our website has information on the latest product releases and training, as well as online versions of our *Cornerstone* magazine and Pearson's product catalog.

Your feedback is welcome. You may email your comments to **curriculum@nccer.org**, send general comments and inquiries to **info@nccer.org**, or fill in the User Update form at the back of this module.

This information is general in nature and intended for training purposes only. Actual performance of activities described in this manual requires compliance with all applicable operating, service, maintenance, and safety procedures under the direction of qualified personnel. References in this manual to patented or proprietary devices do not constitute a recommendation of their use.

21301 V3

21301
LOAD CHARTS

Objectives

When you have completed this module, you will be able to do the following:

1. Explain how center of gravity and leverage affects crane capacity.
   a. Explain the concepts of center of gravity and leverage.
   b. Describe the content of typical load charts.
2. Describe the various crane configurations that affect load chart selection.
   a. Define and describe the four standard mobile crane configurations.
   b. Explain the significance of boom length, boom angle, and load radius.
   c. Identify the quadrants of operation and their relationship to load charts.
   d. Explain the significance of the crane's base and its configuration.
3. Describe how to apply load charts and deductions from capacity.
   a. Describe the use of jib charts.
   b. Describe how the parts of line can affect capacity.
   c. Describe how to read a range diagram.
   d. Describe how to calculate a crane's capacity from its load charts.

Performance Tasks

Under the supervision of your instructor, you should be able to do the following:

1. Identify boom angle, boom length, and load radius on a load chart.
2. Identify the requirements of the on-rubber load chart.
3. Identify the requirements of the on-outrigger load chart.
4. Properly identify load charts that are used in different configurations.
5. Identify parts of line and counterweight considerations in load chart information.
6. Calculate minimum parts of line required.

Trade Terms

Boom angle
Boom length
Boom-point elevation
Configuration
Effective weight
Gross capacity
Gross load
Interpolate
Line pull
Load operating radius
Net capacity
Offset
Ringer crane
Tipping axis

Industry Recognized Credentials

If you are training through an NCCER-accredited sponsor, you may be eligible for credentials from NCCER's Registry. The ID number for this module is 21301. Note that this module may have been used in other NCCER curricula and may apply to other level completions. Contact NCCER's Registry at 888.622.3720 or go to **www.nccer.org** for more information.

Contents

1.0.0 Gravity, Leverage, and Load Charts 1
1.1.0 Center of Gravity and Leverage 1
1.1.1 Center of Gravity 1
1.1.2 Leverage 2
1.2.0 Crane Load Charts 2
2.0.0 Crane Configuration 9
2.1.0 Types of Cranes and Hardware 9
2.1.1 Industrial/All-Purpose Cranes 9
2.1.2 Lattice-Boom Cranes 10
2.1.3 Telescopic-Boom Cranes 10
2.1.4 Boom Trucks 10
2.1.5 Crane Configuration Considerations 10
2.2.0 Boom Length, Boom Angle, and Load Radius 11
2.3.0 Quadrants of Operation 14
2.4.0 Configuration of the Crane Base 15
2.4.1 On Rubber 15
2.4.2 On Outriggers 15
2.4.3 On Crawlers 18
2.4.4 Tower and Ring Attachments 23
2.4.5 Counterweight Configurations 23
3.0.0 Working with Load Charts 25
3.1.0 Jib Charts 25
3.2.0 Line Pull and Parts of Line 26
3.2.1 Hoist Rope Properties 26
3.2.2 Line Pull 28
3.2.3 Parts of Line 28
3.3.0 Range Diagram 29
3.4.0 Calculating Crane Capacity 30
3.4.1 Example One: Industrial/All-Purpose Crane 32
3.4.2 Example Two: Telescopic-Boom Crane 33
3.4.3 Example Three: Lattice-Boom Crane 33
3.4.4 Example Four: Boom Truck 38

Appendix A Terex RT555-1 Load Chart 50
Appendix B Manitowoc 4100W-2 Load Chart 52
Appendix C Grove TM1500 Load Chart 56

Figures and Tables

Figure 1 Typical center of gravity and leverage distances for each position of a truck-mounted crane 2
Figure 2 Relative tipping moment 3
Figure 3 Typical deductions for a telescopic-boom crane 4
Figure 4 Typical deductions for a lattice-boom crane 4
Figure 5 Boom length, boom angle, load operating radius, and boom-point elevation 5
Figure 6A Structural strength and stability charts (1 of 2) 6
Figure 6B Structural strength and stability charts (2 of 2) 7
Figure 7 Industrial/all-purpose crane 9
Figure 8 Truck-mounted and crawler-mounted lattice-boom cranes 10
Figure 9 Telescopic-boom crane with rotating control station 11
Figure 10 Boom truck 11
Figure 11 Telescopic-boom crane load chart 12
Figure 12 Lattice-boom crawler crane load chart 12
Figure 13 Example load chart 1 13
Figure 14 Example load chart 2 14
Figure 15 Example load chart 3 25
Figure 16 Telescopic-boom extension mode chart 16
Figure 17 Quadrant definition drawings for rough-terrain, truck, and crawler cranes 17
Figure 18 Capacity differences 18
Figure 19 On-rubber load chart 19
Figure 20 Capacities on outriggers at no extension (0 percent) 20
Figure 21 Capacities on outriggers at intermediate extension (50 percent) 21
Figure 22 Capacities on outriggers at full extension (100 percent) 22
Figure 23 Crane with tower attachment 23
Figure 24 3,000 metric-ton ringer crane 23
Figure 25 Crawler crane counterweights 23
Figure 26 Counterweights on an auxiliary carrier 24
Figure 27 Deductions chart 26
Figure 28 Jib/boom extension chart 27
Figure 29 Reeving for various parts of line 29
Figure 30 Wire rope capacity chart 30
Figure 31 Range diagram 31
Figure 32 Grove YB4409 industrial crane 32
Figure 33 Graphic illustration of YB4409 configuration 32
Figure 34 Grove YB4409 load chart 34
Figure 35 Grove TM-1500 crane 35
Figure 36 Graphic illustration of the TM-1500 configuration 35
Figure 37 TM-1500 load chart notes 35
Figure 38 Grove TM-1500 load chart 36
Figure 39 Grove TM-1500 weight deduction table 37
Figure 40 Manitowoc 4100W-2 lattice-boom crawler crane 37
Figure 41 Graphic illustration of the Manitowoc 4100-W2 configuration 38

Figure 42A Manitowoc 4100 W-2 load chart (1 of 4) 39
Figure 42B Manitowoc 4100 W-2 load chart (2 of 4) 40
Figure 42C Manitowoc 4100 W-2 load chart (3 of 4) 41
Figure 42D Manitowoc 4100 W-2 load chart (4 of 4) 42
Figure 43 NC 1800 Series boom truck 43
Figure 44 Graphic illustration of the National 18103 boom truck configuration 43
Figure 45 NC 1800 Series load chart 44
Figure 46 NC 1800 Series boom truck deductions 45

Table 1 Selected Wire Rope Properties 26
Table 2 Sheave Ratio Factors for Parts of Line 30

SECTION ONE

1.0.0 GRAVITY, LEVERAGE, AND LOAD CHARTS

Objective

Explain how center of gravity and leverage affects crane capacity.

a. Explain the concepts of center of gravity and leverage.
b. Describe the contents of typical load charts.

Performance Tasks

1. Identify boom angle, boom length, and load radius on a load chart.

Trade Terms

Boom angle: In load charts, the angle between the boom relative to the horizontal plane. On telescopic-boom cranes, boom angle instruments indicate the angle between the bottom of the base section and the horizontal with the boom loaded. On lattice-boom cranes, it is the angle between the center line of the boom and a horizontal line.

Boom length: Measured from the boom foot pins to the center of the boom-point sheaves. It does not include the length of an attached jib.

Boom-point elevation: Distance from the ground to the center of the boom-point sheaves.

Configuration: A term that addresses all aspects of a crane's condition for a given lift that can include boom assembly and position, hoisting line and accessories, base arrangement, counterweights, and other factors addressed in a load chart.

Gross capacity: A crane capacity value as displayed in the crane's load chart for a specified configuration.

Gross load: The total weight of all the items lifted by the crane. The gross capacity of the crane minus any deductions for the crane configuration results in the gross load that can be lifted. Gross load is directly related to net capacity.

Load operating radius: The horizontal distance from a vertical line through the axis of crane rotation to the center of the vertical hoist line or tackle, with a load applied. Also called load radius, lifting radius, or operating radius.

Net capacity: The crane's hoisting force available after taking into account the crane's and/or load's configurations relating to a lift. Equal to load chart gross capacity minus weight deductions.

Tipping axis: In crane operations, an imaginary horizontal line around which a crane's center of gravity would rotate as it tips.

Long ago, crane operators relied on the feel of the crane as it began to tip as a capacity indicator. They would feel the crane begin to tip, and then react the best they could to keep it from overturning. Older cranes were built using components that were heavier than necessary in order to ensure safe lifts. These overbuilt cranes would tip before there was any chance of structural failure.

Today, there are much better methods available to ensure safe lifts, due to technological developments in crane construction. Cranes are now built with lighter boom components; however, this means that they are as likely to suffer structural failure as they are to tip. While relying on the feel of a crane was never a safe way to avoid tipping, an operator doing so today could experience structural failure of the crane before tipping occurs.

It is extremely important for crane operators to understand the information provided on load charts. Thousands of man-hours go into developing load charts for cranes to ensure safety. Operators must be able to distinguish between the different tables and countless numbers found on load charts, and know how to apply the information correctly when planning a lift.

This module provides information related to the use of load charts. It also describes methods used to determine a crane's **gross capacity** and **net capacity**.

1.1.0 Center of Gravity and Leverage

The principles of center of gravity and leverage play a critical role in crane operations. An understanding of these principles is essential to every lifting operation.

1.1.1 Center of Gravity

A key concept associated with lifting ability is center of gravity. The crane's center of gravity (CG) is a conceptual point where crane designers calculate the crane's weight can be located. The actual position of a crane's CG depends on the locations of the weights of all its attached components taken together. The components that contribute most to the crane's CG include the boom, carrier, upperworks, and counterweight. The load also has a CG, which riggers must know when preparing a load in order to ensure stability when it is hoisted.

The location of the crane's CG can change depending on its **configuration**, as shown in *Figure 1*. When the crane is in over-the-side, over-the-front, or over-the rear configurations, the center of gravity is shifted either closer to or farther away from the associated **tipping axis**. This, in turn, may decrease or increase the crane's leverage.

1.1.2 Leverage

Leverage is another important concept associated with crane lifting operations. The crane has leverage, and the load also exerts leverage. The load's leverage is the product of the load's weight and the shortest horizontal distance between the load's

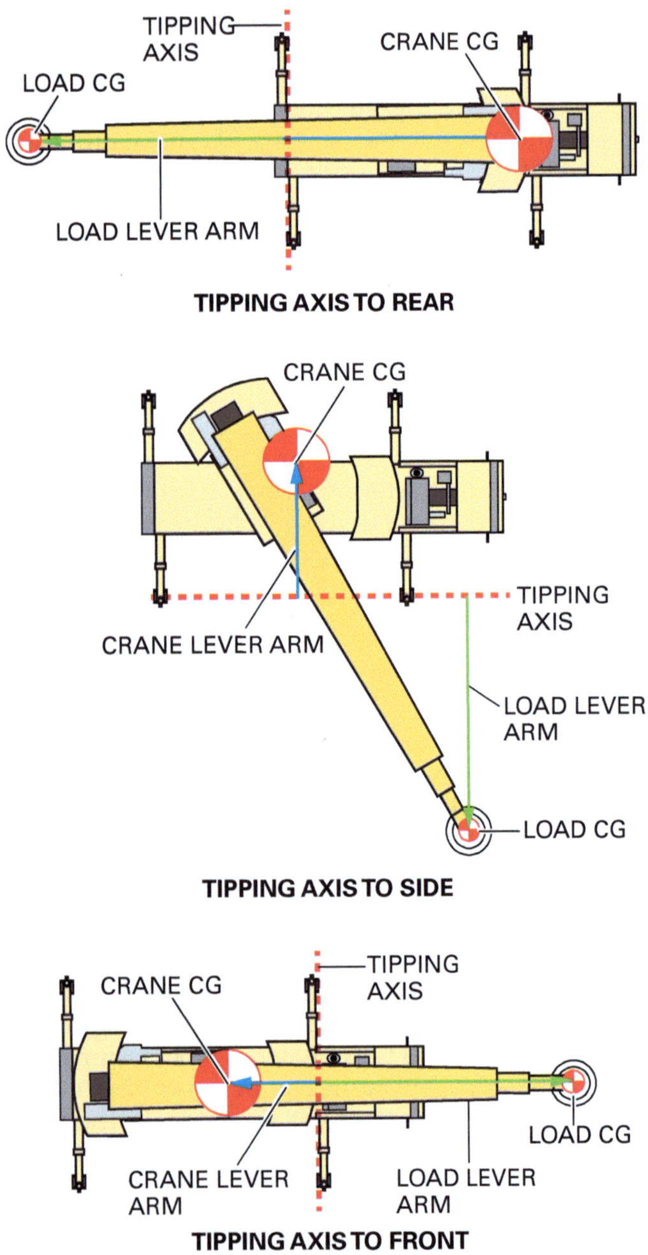

Figure 1 Typical center of gravity and leverage distances for each position of a truck-mounted crane.

CG and the tipping axis. To make a safe, stable lift, the crane's leverage has to be greater than the load's leverage (*Figure 2*). If the load's leverage is roughly equal to or greater than the crane's leverage, a tipping situation results. In addition to the configuration of the crane, booming down or telescoping out increases the **load operating radius** and moves the load's center of gravity further from the tipping axis. This, in turn, increases the load's leverage.

Leverage is addressed in load charts, represented by the crane's gross capacity. This is the balance of leverage remaining after subtracting the crane's weight from the leverage moment.

For a practical example of this concept, pick up a heavy book and place it directly over your head with your arms straight up. Then, keeping your arms straight, begin lowering the book from above your head until it is directly in front of you. You can quickly feel the leverage of the book. Notice how much more difficult it is to hold the book in this position due to the leverage applied by the book.

Crane and load leverages are the fundamental quantities that load charts take into account when establishing limits for safe crane operations. When you have a thorough understanding of these principles, you will see why it is essential for operators to refer to load charts before planning any lift.

1.2.0 Crane Load Charts

The crane manufacturer designs a load chart for each crane model. The charts may even be serial-number specific, meaning a unique load chart is assigned to each individual crane. This may be done if there is a slight difference between cranes due to design changes, manufacturing techniques, materials used, or permissible boom configurations. While some load charts may show the crane serial number, this isn't always the case, since some load charts apply to more than one crane. A load chart applies to only to the crane(s) specifically identified on the chart.

> **WARNING!**
> Load charts found online are not typically crane-specific. As a result, online charts should never be relied upon. Always locate and use crane-specific charts.

If the applicable load chart for a crane becomes lost or illegible, only the manufacturer can replace it. An individual can't legally redraw or recreate a crane's load chart for lifting purposes.

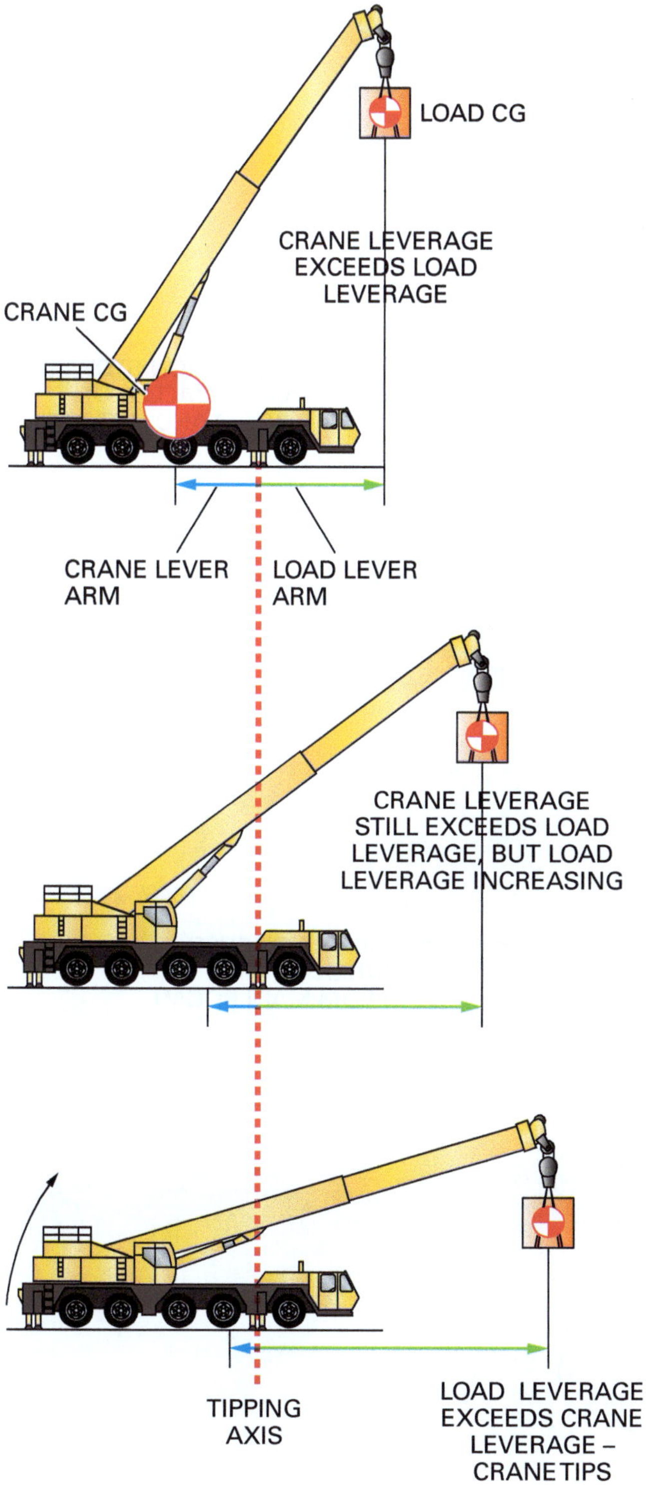

Figure 2 Relative tipping moment.

A crane cannot be operated legally without its load chart.

According to *ASMEB30.5, Mobile and Locomotive Cranes*, manufacturers are required to provide specific information in load charts for a crane. Furthermore, 29 *CFR* 1926.1417(c) requires that owners post load charts in a location continuously accessible to the operator at the controls. For situations where the operator can access only an electronic load chart, any loss of the electronic display requires that all crane operations must cease and be secured until access to the crane's load chart is restored.

Each load chart must display identifying information associating it with a specific crane(s). The complete set of load charts for a crane should generally include the following information:

- The complete range of loads for all operating configurations of the crane, including the use and non-use of any optional equipment that may be installed and that may affect the ratings
- A work area chart, which is often referred to as a *quadrant-of-operation diagram*, corresponding to the load chart in use
- Component weights, such as jibs and similar accessories
- Work areas and loads where no loads are permitted to be handled
- Specifications for lifting directions other than the most stable, if applicable
- Recommended wire rope dimension and weight limitations
- Recommended parts of line for reeving
- Load limits and precautions during boom extension and retraction, if applicable
- Tire pressures, if applicable
- Cautions, warnings, and notes relating to load limitations

NOTE

Standards no longer require load limitations based on structural and/or mechanical strength of materials, stability, hydraulics, and/or other factors. However, load charts for older cranes may include this information.

Load charts indicate the gross lifting capacity of the crane under stated conditions. This is not the weight of the load alone suspended from the crane hook. The **gross load** represents the total weight of all the items lifted by the crane. Net capacity, the crane's actual capacity available for a safe lift, is the gross capacity minus weight deductions related to the various crane configurations. This also represents the gross load the crane can safely lift. Refer to *Figure 3* and *Figure 4*.

The manufacturer bases load chart capacities on the maximum allowable freely suspended loads at a given **boom length**, **boom angle**, and load operating radius, as shown in *Figure 5*. **Boom-point elevation** relates directly to boom length and boom angle. Boom length is the length of the main boom and boom extensions, if applicable, as defined by standards. It can vary with different crane models and configurations.

Figure 3 Typical deductions for a telescopic-boom crane.

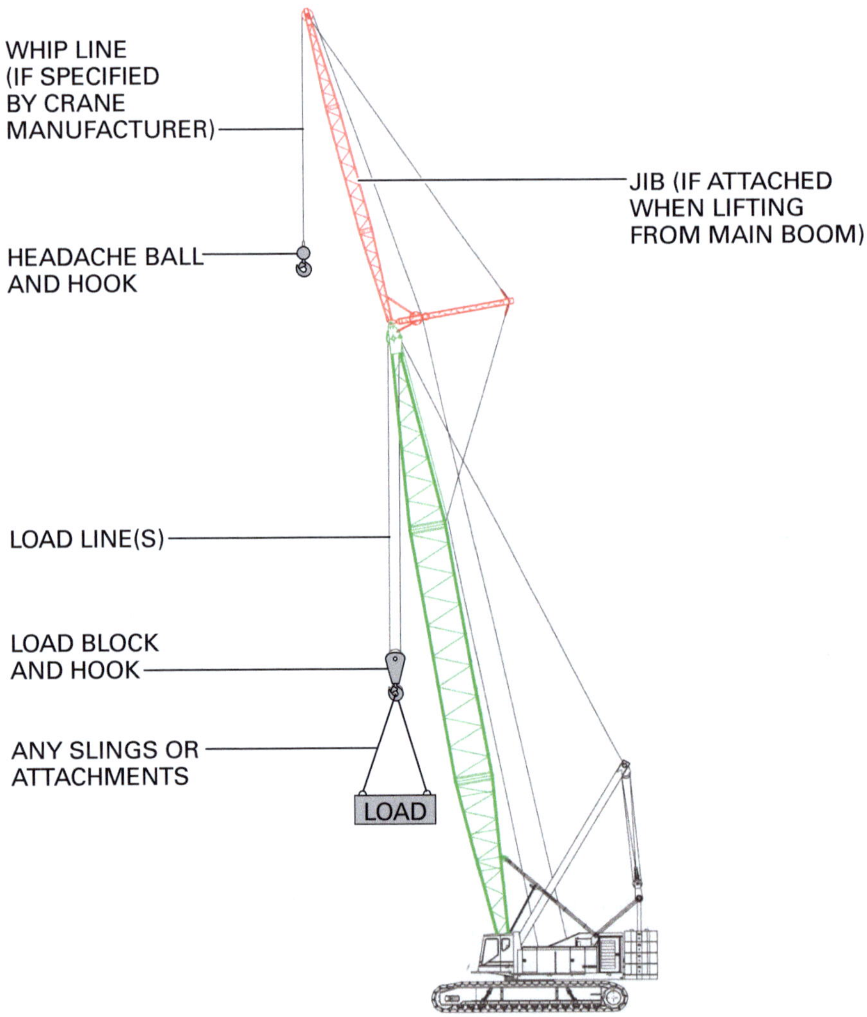

Figure 4 Typical deductions for a lattice-boom crane.

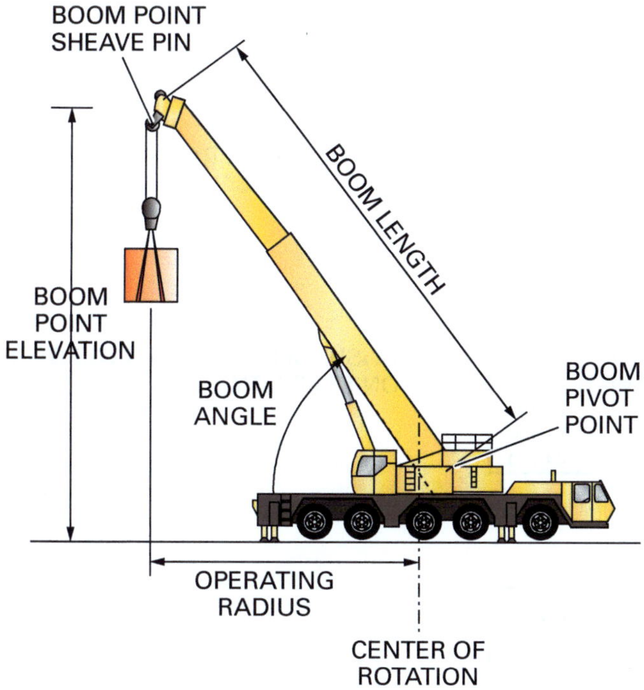

Figure 5 Boom length, boom angle, load operating radius, and boom-point elevation.

Load charts are determined for a machine standing level on a firm, supporting surface under the following ideal operating conditions:

- Equipment in original operating condition
- Equipment operated by an experienced, qualified operator
- No inclement weather during operation or with a suspended load
- Known load weight
- Correct rigging and counterweights used
- Boom length, operating radius, and boom angle known
- Load block or hook over the center of gravity of the load

Refer to the manufacturer's load charts for notes concerning allowances for conditions other than ideal. The operator must limit the load to allow for adverse conditions. Also note that load chart capacities do not exceed the allowable percentage of tipping loads. OSHA standards do not permit on-rubber or crawler capacities to exceed 75 percent of the tipping capacity as actually measured by testing. The limit is 85 percent when on outriggers.

CAUTION

Crane operators working in countries other than the United States should be aware that many countries abide by ISO standards for determining crane stability. In most cases, these standards meet or exceed the 75% / 85% tipping capacity limits specified by ASME and OSHA standards. However, this is not consistent, and operators should never assume that the same standards have been met. They must carefully study the load charts provided to understand the basis for the crane's stability limits and operate accordingly.

When operators are unable to interpret a load chart correctly, they may be tempted to rely on guesswork or the sensation of tipping to determine the limits of a crane. These practices have resulted in serious accidents and fatalities. Using the onset of tipping as an indication that the crane has reached capacity is never acceptable. Structural overloading of a crane can occur before any signs of tipping are evident, resulting in a catastrophic failure. While older, overbuilt cranes are likely to tip before a structural failure occurs, there is no reason to operate them differently. The load chart must be followed regardless.

If there is no capacity listing for a particular boom length or radius, standards prohibit the operator from lifting in that configuration. Also, be aware that load charts limit lowering a boom below a minimum angle in some quadrants, because the weight of the boom and attachments may cause tipping, even without a load.

Operators must understand that load chart capacities reflect operating limits imposed by either the structural strength of the crane or its stability. The use of shading, bold lines, or asterisks in a load chart generally indicate the operating conditions allowed within these limitations. *Figure 6A* and*Figure 6B* show examples of such load charts.

Main Boom Capacities* – 110 Ft. – Open Throat Boom

Load Radius (ft)	Boom Angle (deg)	360 Rotation				Over End Blocked	
		135K + LWR (lb)	135K + 0 (lb)	83K + 0 (lb)	31K + 0 (lb)	135K + LWR (lb)	135K + 0 (lb)
18.5	82.0	346,500	346,500	329,000	283,900	346,500	346,500
19	81.7	344,400	344,400	320,000	276,000	344,400	344,400
20	81.2	341,100	341,100	303,900	246,900	341,100	341,100
25	78.6	276,100	276,100	233,200	158,900	276,100	276,100
30	75.9	229,100	226,500	171,200	116,000	229,100	229,100
35	73.2	192,600	178,500	134,500	90,500	195,200	195,200
40	70.4	158,500	146,800	110,300	73,700	169,700	169,700
50	64.8	116,200	107,400	80,100	52,900	133,700	133,000
60	58.9	90,900	83,900	62,200	40,400	109,700	102,800
70	52.6	74,100	68,300	50,200	32,100	90,700	83,100
80	45.7	62,100	57,200	41,600	26,100	75,800	69,300
90	37.9	53,100	48,800	35,200	21,700	64,700	59,100
100	28.3	46,100	42,200	30,200	18,100	56,100	51,100
110	13.8	40,400	36,900	26,100	15,200	49,200	44,700

*Ratings are not more than 75% of tipping loads

SHADED AREA RATINGS BASED ON STRUCTURAL STRENGTH

UNSHADED AREA RATINGS BASED ON STABILITY

(A) EXAMPLE FROM LINK-BELT CRANE

Boom and Jib Length	Jib Radius (ft)	5.0 Offset		15.0 Offset		25.0 Offset	
		Boom Angle	Ratings (lbs)	Boom Angle	Ratings (lbs)	Boom Angle	Ratings (lbs)
40' (12.2M) JIB & 200' (61.0M) BOOM	45	81.0	19,380*				
	50	79.8	18,840*				
	60	77.3	17,740*	78.9	16,740*	80.4	15,650*
	70	74.9	16,670*	76.5	15,890*	77.9	15,050*
	80	72.4	15,640*	74.0	15,050*	75.4	14,440*
	90	69.9	13,330	71.5	13,330	72.9	13,330
	100	67.4	11,190	68.9	11,200	70.3	11,200
	110	64.7	9,480	66.3	9,480	67.6	9,490
	120	62.1	8,070	63.6	8,070	64.9	8,070
	130	59.3	6,900	60.8	6,900	62.1	6,900
	140	56.5	5,900	58.0	5,900	59.3	5,900
	150	53.6	5,040	55.1	5,040	56.3	5,040
	160	50.6	4,290	52.0	4,290	53.2	4,290
	170	47.4	3,640	48.8	3,640	49.9	3,650
	180	44.0	3,070	45.4	3,080	46.5	3,080
	190	40.4	2,560	41.8	2,560	42.8	2,570
	200	36.6	2,110	37.9	2,110	38.8	2,120

RATINGS WITH AN ASTERISK BASED ON STRUCTURAL STRENGTH

RATINGS WITHOUT AN ASTERISK BASED ON STABILITY

(B) EXAMPLE FROM TEREX CRANE

Figure 6A Structural strength and stability charts (1 of 2).

Rated Lifting Capacities in Pounds (35 Ft – 110 Ft Boom)
On Outriggers Fully Extended – 360

Radius in Feet	#0001								
	Main Boom Length in Feet								
	35	40	50	*60	70	80	90	100	110
10	100,000 (63.5)	80,400 (66.5)	74,400 (71.5)	44,600 (75.5)					
12	88,050 (60)	79,050 (63.5)	70,900 (69)	44,600 (74)	@35,600 (75.5)				
15	74,500 (54)	67,450 (59)	63,350 (65.5)	44,600 (71)	35,600 (74)	@33,000 (75.5)			
20	54,700 (43)	53,850 (50.5)	50,900 (59)	44,600 (66)	35,600 (70)	33,000 (72.5)	25,500 (75)	@23,300 (75.5)	
25	41,450 (29)	41,150 (40.5)	40,700 (52.5)	40,350 (60.5)	35,550 (65.5)	33,000 (69)	25,500 (71.5)	23,300 (74)	@18,500 (75.5)
30		32,450 (28)	32,050 (45)	31,750 (55)	30,550 (61)	28,950 (65)	25,500 (68)	23,300 (71)	18,500 (73)
35			25,950 (36.5)	25,650 (48.5)	26,500 (56.5)	24,900 (61)	23,000 (64.5)	21,200 (68)	18,500 (70.5)
40	See Note 16		21,400 (25)	21,150 (41.5)	22,000 (51.5)	21,750 (57)	20,000 (61)	18,450 (65)	18,000 (67.5)
45				17,600 (33.5)	18,500 (46)	19,100 (53)	17,600 (57.5)	16,300 (61.5)	15,750 (65)
50				14,600 (23)	15,250 (39.5)	15,700 (48)	15,650 (53.5)	14,400 (58)	13,950 (62)
55					12,650 (32.5)	13,100 (43)	13,550 (49.5)	12,850 (54.5)	12,450 (59)
60					10,500 (23)	11,000 (37.5)	11,450 (45)	11,550 (51)	11,150 (55.5)
65						9,350 (31)	9,780 (40.5)	10,200 (47)	10,050 (52.5)
70						8,870 (22)	8,370 (35)	8,780 (43)	9,090 (49)
75							7,180 (28.5)	7,590 (38.5)	7,980 (45)
80							6,120 (20)	6,560 (33)	6,950 (41)
85								5,680 (27)	6,060 (37)
90								4,910 (19)	5,280 (32)
95									4,600 (26)
100									3,990 (18.5)
Minimum boom angle (deg.) for indicated length									0
Maximum boom length (ft.) at 0 deg. boom angle (no load)									110

Note: ( ) Boom angles are in degrees.

RATINGS ABOVE BOLD LINE IN EACH COLUMN BASED ON STRUCTURAL STRENGTH

RATINGS BELOW BOLD LINE IN EACH COLUMN BASED ON STABILITY

(C) EXAMPLE FROM MANITOWOC CRANE

Figure 6B Structural strength and stability charts (2 of 2).

Additional Resources

ASME Standard B30.5, Mobile and Locomotive Cranes. Current edition. New York, NY: American Society of Mechanical Engineers.

29 *CFR* 1926.1417(c), *Cranes and Derricks in Construction – Operation.* **www.ecfr.gov**

1.0.0 Section Review

1. The leverage of a load is found by _____.
   a. multiplying the load's weight by the shortest horizontal distance between the load's CG and the tipping axis
   b. multiplying the crane's weight by the horizontal distance from the load's CG to the tipping axis
   c. dividing the load's weight by the shortest horizontal distance between the load's CG and the tipping axis
   d. dividing the crane's weight by the horizontal distance from the load's CG to the tipping axis

2. A given load chart applies to _____.
   a. all models built by a crane manufacturer
   b. all cranes of the same model built by a manufacturer
   c. all cranes that have a common design
   d. only the cranes specified by the chart

Section Two

2.0.0 Crane Configuration

Objective

Describe the various crane configurations that affect load chart selection.

a. Define and describe the four standard mobile crane configurations.
b. Explain the significance of boom length, boom angle, and load radius.
c. Identify the quadrants of operation and their relationship to load charts.
d. Explain the significance of the crane's base and its configuration.

Performance Tasks

2. Identify the requirements of the on-rubber load chart.
3. Identify the requirements of the on-outrigger load chart.

Trade Terms

Interpolate: To estimate or calculate an unknown intermediate value that falls between two known values.

Ringer crane: A large, high-capacity crane erected on a circular track (the ring, usually installed directly on the ground) that supports the entire crane base.

The capacity of a given crane depends on many factors, such as the type of crane, the type of boom and its attachments, the configuration of the crane base, and other specifics. Lift planning determines the likely boom length, boom angle, and load radius that are going to be employed during a lift. Load charts provide consideration of all of these factors, including whether the crane is lifting on rubber or on outriggers. This section addresses each of these items.

2.1.0 Types of Cranes and Hardware

There are four categories of mobile cranes for which NCCER certification is available:

- Industrial/all-purpose cranes
- Lattice-boom cranes
- Telescopic-boom cranes
- Boom trucks

These designated categories are based on the crane application, boom type, and/or vehicle type.

2.1.1 Industrial/All-Purpose Cranes

Industrial/all-purpose cranes are relatively small, lightweight, four-wheeled machines designed for industrial yard operations (*Figure 7*). They are not intended for highway operation. Manufacturers often sell this type of crane, especially the larger versions, with four-wheel crab and coordinated steering as well as the normal two-wheel front or rear capabilities. They are operated from the driver's seat, and have a telescoping-boom crane mounted on the chassis. They are suitable for on-rubber operations, and some are designed for pick-and-carry work, as well as lifts on outriggers. The boom upperworks can usually rotate a full 360 degrees.

Many industrial/all-purpose cranes include a flat, load-carrying deck in their design. Note that the term *pick-and-carry* refers to moving a load on the hook, not on the deck. Those that can only carry a load on the deck and not on the hook are often referred to as *carry-deck cranes*.

Figure 7 Industrial/all-purpose crane.

2.1.2 Lattice-Boom Cranes

Lattice-boom cranes can be truck-mounted or crawler-mounted (*Figure 8*). Because lattice booms are strong but relatively lightweight, they can be constructed or extended to lengths up to several hundred feet—longer than equal-weight telescopic booms. Lattice-boom cranes also vary in the types of load control systems used and the kind of work the crane does. Most modern models and those used for straight lifting work use hydraulic motors and controls. Older model cranes and those used for large dynamic loads, such as pile-pulling and driving, use friction clutches and brakes. These are known as *friction cranes*. Site preparation and erection of lattice-boom cranes may take time—from a day for smaller rigs, to weeks for larger, heavy-lift units with their various attachments.

2.1.3 Telescopic-Boom Cranes

Telescopic-boom cranes are categorized for NCCER certifications by carrier type—rough/all-terrain (RT/AT) cranes (*Figure 9*), tire-mounted truck cranes, and crawler cranes. It is common to further categorize RT/AT types by whether the control station rotates with the boom or is stationary. For larger cranes, the operator's cab usually rotates with the upperworks supporting the boom. Hydraulic components operate and extend all telescopic booms. Because of their construction, they are generally shorter for a given boom weight, compared to lattice-boom cranes.

2.1.4 Boom Trucks

Boom trucks (*Figure 10*) consist of a telescopic- or articulating-boom upperworks placed on a tire-mounted carrier designed and engineered for highway travel. Certifications in this category designate fixed or rotating control stations, and whether the boom is of the telescoping or articulating type. Boom trucks are used in many applications where highly mobile units with limited capacity are needed. Models with special insulation systems are regularly used to service power distribution systems.

2.1.5 Crane Configuration Considerations

Manufacturers base load charts of their mobile cranes on the configuration of the crane at the time of the lift. Some examples of configurations are:

- Jibs and boom extensions not installed
- Jib and/or boom extensions installed but load lifted from main boom

Figure 8 Truck-mounted and crawler-mounted lattice-boom cranes

- Jibs and/or boom extensions installed and load lifted from jib or boom extension

The principal factors influencing the machine's capacity and correct interpretation of the chart include these considerations:

- Boom length, boom angle, and load radius

Figure 9 Telescopic-boom crane with rotating control station.

Figure 10 Boom truck.

- Type of boom
  - Telescopic
  - Lattice
- Operational quadrant in which the superstructure is working
- Configuration of crane base
  - On rubber
  - On outriggers (stabilizers)
  - On crawlers
- Telescopic boom modes (A or B)
- Effective/stowed weight of jib or boom extension (or other attachments)
- Ring or tower attachment
- Type, size, and weight of load block and headache ball
- Weight of all required rigging
- Weight of rope per manufacturer's requirement
- Counterweight configuration
- Parts of line

2.2.0 Boom Length, Boom Angle, and Load Radius

The load radius for telescopic- and lattice-boom cranes can be found in various locations in a set of load charts. The location of the boom angle information can also vary. Understanding where these resources are on a load chart is key to reading and properly interpreting a load chart. Operators must demonstrate their ability to read, comprehend, and write in the language of the crane manufacturer's operation and maintenance documentation. They must also exhibit arithmetic skills and the ability to use the manufacturer's load charts.

Most telescopic-boom crane load charts (*Figure 11*) start with the shortest boom length at the upper left region and read to the right and down for progressively longer booms. Start by locating the number that corresponds to the length (in feet) of the boom in the Boom Length heading. Then, locate the operating radius of the crane in the left- or right-hand column. The capacity where the operating radius row and an operating quadrant under the Boom Length heading intersect is the gross capacity with the indicated counterweight configuration.

Lattice-boom crawler crane load charts may show their information differently, depending on the manufacturer. In the example chart shown in *Figure 12*, the boom length is in the chart heading, and the load radius is in the left-hand column.

The operator must first know what boom length the crane is operating at, determine the required radius for the lift, and then read across to determine the gross capacity, depending on the counterweight configuration. Some manufacturers may also include the boom elevation point in their charts for reference, especially if they require consideration of the weight of suspended hoist cable in the load calculations.

Note that the load charts shown in this module are examples from certain manufacturers. The standards do not specify a mandatory, universal format for presenting this information. Load chart layouts are as diverse as the number of manufacturers that build cranes. They can even vary greatly between different crane models from the same manufacturer. Some have illustrations and counterweight requirements at the top; some express their load radius or boom lengths in feet and/or meters (*Figure 13*, *Figure 14*, and *Figure 15*), and some list their capacities in pounds, kilopound-force (kip), or kilograms (*Figure 13* and *Figure 15*). This variety presents a challenge to operators; they must pay close attention to load charts and the operating notes when switching between cranes to avoid making mistakes on load chart calculations and interpretations.

REQUIRED COUNTERWEIGHT CONFIGURATION

BOOM LENGTH HEADING (1 OF 4)

BOOM MODE

LOADED BOOM ANGLE COLUMN

LOAD RADIUS COLUMN

CAPACITY IN ANY QUADRANT

CAPACITY IN REAR QUADRANT ONLY

Rated Lifting Capacities In Pounds
Fully Extended Outriggers
See Set Up Note 2

4,700 lbs.

FULL EXTENSION

MAIN BOOM "A"

Load Radius (Ft.)	33 Ft.			40 Ft.			Load Radius (Ft.)
	∡°	360°	Over Rear	∡°	360°	Over Rear	
9	68.0	80,000	80,000				9
10	66.0	72,300	72,300	70.5	72,300	72,300	10
12	62.0	65,500	65,500	67.5	65,200	65,200	12
15	55.5	55,600	55,600	62.5	55,100	55,100	15
20	43.5	38,400	38,500	54.0	37,900	37,900	20
25	26.5	25,600	25,600	43.5	25,300	25,300	25
30				31.0	18,100	18,100	30
Min.Bm Ang/Cap	0 (27.5)	18,400	18,400	0 (34.5)	13,700	13,700	Min.Bm Ang/Cap

Load Radius (Ft.)	50 Ft.			57 Ft.			Load Radius (Ft.)
	∡°	360°	Over Rear	∡°	360°	Over Rear	
10	75.0	67,500	67,500	77.0	43,800	43,800	10
12	73.0	61,200	61,200	75.0	43,800	43,800	12
15	69.0	53,400	53,400	72.0	42,100	42,100	15
20	62.5	37,300	37,300	66.5	34,300	34,300	20
25	55.5	24,900	24,900	60.5	24,600	24,600	25
30	48.0	17,900	17,900	54.5	17,600	17,600	30
35	39.0	13,200	13,200	47.5	13,100	13,100	35
40	27.5	9,800	10,000	40.0	9,600	9,900	40
45				30.5	7,200	7,600	45
50				16.0	5,200	5,700	50
Min.Bm Ang/Cap	0 (44.5)	7,400	7,800	0 (51.5)	4,700	5,200	Min.Bm Ang/Cap

Note: Refer to page 5 for (Capacity Deductions for Auxiliary Load Handling Equipment).

∡° Loaded Boom Angle in Degrees

( ) Reference radius for minimum boom angle capacities (shown in parenthesis) are in feet.

Figure 11 Telescopic-boom crane load chart.

BOOM LENGTH (CHART HEADING)

LOAD RATINGS COLUMN

CAPACITY IN ANY QUADRANT

LOADED BOOM ANGLE COLUMN

CAPACITY IN EITHER END QUADRANT ONLY

COUNTERWEIGHT CONFIGURATIONS (LWR = BOTH LOWER CRAWLER FRAME COUNTERWEIGHTS)

Main Boom Capacities* – 110 Ft. — Open Throat Boom

Load Radius (ft)	Boom Angle (deg)	360° Rotation				Over End Blocked	
		135K + LWR (lb)	135K + 0 (lb)	83K + 0 (lb)	31K + 0 (lb)	135K + LWR(lb)	135K + 0 (lb)
18.5	82.0	346,500	346,500	329,000	283,900	346,500	346,500
19	81.7	344,400	344,400	320,000	276,000	344,400	344,400
20	81.2	341,100	341,100	303,900	246,900	341,100	341,100
25	78.6	276,100	276,100	233,200	158,900	276,100	276,100
30	75.9	229,100	226,500	171,200	116,000	229,100	229,100
35	73.2	192,600	178,500	134,500	90,500	195,200	195,200
40	70.4	158,500	146,800	110,300	73,700	169,700	169,700
50	64.8	116,200	107,400	80,100	52,900	133,700	133,000
60	58.9	90,900	83,900	62,200	40,400	109,700	102,800
70	52.6	74,100	68,300	50,200	32,100	90,700	83,100
80	45.7	62,100	57,200	41,600	26,100	75,800	69,300
90	37.9	53,100	48,800	35,200	21,700	64,700	59,100
100	28.3	46,100	42,200	30,200	18,100	56,100	51,100
110	13.8	40,400	36,900	26,100	15,200	49,200	□□□□00

***Ratings are not more than 75% of tipping loads**

Figure 12 Lattice-boom crawler crane load chart.

Outriggers Fully Extended 25' - 43' *(7.62m - 13.11m)* Offset Telescoping Fly + 70.25' *(21.41m)* Main Boom with Manual Retracted													
	25' *(7.62m) Fly*						43' *(13.11m) Fly*						
	2 OFFSET		15 OFFSET		30 OFFSET		2 OFFSET		15 OFFSET		30 OFFSET		
LOAD RADIUS (ft/m)	LOADED BOOM ANGLE (deg.)	LOAD (lbs/kgs)	LOADED BOOM ANGLE (deg.)	LOAD (lbs/kgs)	LOADED BOOM ANGLE (deg.)	LOAD (lbs/kgs)	LOADED BOOM ANGLE (deg.)	LOAD (lbs/kgs)	LOADED BOOM ANGLE (deg.)	LOAD (lbs/kgs)	LOADED BOOM ANGLE (deg.)	LOAD (lbs/kgs)	LOAD RADIUS (ft/m)
20 *6.10*	78.5	12,700 *5,761*											20 *6.10*
25 *7.62*	75.5	11,500 *5,216*	79.0	8,800 *3,992*			77.5	4,900 *2,223*					25 *7.62*
30 *9.14*	72.5	10,600 *4,808*	76.0	8,200 *3,720*	79.0	6,100 *2,767*	75.5	4,700 *2,132*					30 *9.14*
35 *10.67*	69.5	9,600 *4,355*	73.0	7,500 *3,402*	76.0	5,700 *2,586*	73.0	4,500 *2,041*	78.0	4,000 *1,614*			35 *10.67*
40 *12.19*	66.0	8,700 *3,946*	69.5	6,900 *3,130*	72.5	5,400 *2,449*	70.0	4,300 *1,950*	75.5	3,900 *1,769*			40 *12.19*
45 *13.72*	63.0	8,000 *3,629*	66.0	6,400 *2,903*	69.0	5,100 *2,313*	67.5	4,100 *1,860*	72.5	3,700 *1,678*	77.5	3,000 *1,361*	45 *13.72*
50 *15.24*	59.5	7,400 *3,357*	62.5	5,900 *2,676*	65.5	4,800 *2,177*	65.0	3,900 *1,769*	70.0	3,500 *1,588*	74.5	2,900 *1,315*	50 *15.24*
55 *16.76*	56.0	6,500 *2,948*	59.0	5,300 *2,404*	62.0	4,500 *2,041*	62.0	3,800 *1,724*	67.0	3,300 *1,497*	71.5	2,700 *1,225*	55 *16.76*
60 *18.29*	52.0	5,400 *2,449*	55.0	4,900 *2,223*	58.0	4,200 *1,905*	59.5	3,700 *1,678*	64.0	3,100 *1,406*	68.5	2,600 *1,179*	60 *18.29*
65 *19.81*	48.0	4,500 *2,041*	51.0	4,500 *2,041*	53.5	3,900 *1,769*	56.5	3,500 *1,588*	61.0	2,900 *1,315*	65.5	2,500 *1,134*	65 *19.81*
70 *21.34*	43.5	3,700 *1,678*	46.5	3,900 *1,769*	49.0	3,600 *1,633*	53.0	3,200 *1,452*	58.0	2,800 *1,270*	62.5	2,400 *1,089*	70 *21.34*
75 *22.86*	38.5	3,000 *1,361*	41.5	3,200 *1,452*	44.0	3,400 *1,542*	50.5	3,000 *1,361*	54.5	2,600 *1,179*	59.0	2,300 *1,043*	75 *22.86*
80 *24.38*	33.0	2,500 *1,134*	36.0	2,600 *1,179*	37.5	2,700 *1,225*	46.5	2,800 *1,270*	51.0	2,500 *1,134*	55.0	2,200 *998*	80 *24.38*
85 *25.91*	26.5	2,000 *907*	29.0	2,100 *953*	30.0	2,100 *953*	43.0	2,700 *1,225*	47.5	2,400 *1,089*	51.0	2,100 *953*	85 *25.91*
90 *27.43*	17.5	1,600 *726*	19.5	1,600 *726*	17.0	1,600 *726*	38.5	2,300 *1,043*	43.0	2,300 *1,043*	46.5	2,000 *907*	90 *27.43*
95 *28.96*							34.0	2,000 *907*	38.5	2,200 *998*	41.5	2,000 *907*	95 *28.96*
100 *30.48*							28.5	1,600 *726*	33.0	1,800 *816*	35.5	1,900 *862*	100 *30.48*
105 *32.00*									25.5	1,400 *635*	26.0	1,400 *635*	105 *32.00*
Warning: Do not lower 25' offset telescoping fly in working position below 12 unless main boom is 68' or less, since loss of stability will occur causing a tipping condition.							**Warning:** Do not lower 43' offset telescoping fly in working position below 20 unless main boom is 65' or less, since loss of stability will occur causing a tipping condition.						

Figure 13 Example load chart 1.

> **NOTE**
>
> European crane manufacturers often follow the European practice of using commas in the place of decimal points in numbers for their documentation.

If you are not using a load moment indicator (LMI) but operating from load charts alone, enter the chart at the next highest load radius or next lowest boom angle; do not **interpolate** chart data. LMIs may mathematically interpolate the gross capacity in the charts as programmed by the manufacturer, but interpolation must not be done by the operator. Always use the lower value on a chart when a value falls between chart values.

If a load radius, boom length, or boom angle falls between values listed on the load chart, the operator cannot interpolate the crane's gross capacity. When either boom length or radius, or both, are between the values listed, the operator shall use the smallest load value shown at either the next larger radius or next longer or shorter boom length.

> **NOTE**
>
> Some manufacturers direct load chart users to check both the next longer and the next shorter boom length for the same radius. In some cases, operators find that the longer boom provides the best advantage, which may be counter to what seems logical.

Some telescoping cranes are equipped with multiple boom-extension modes. The operator selects the mode with a switch. Depending on the manufacturer, these modes may be identified by letters, numbers, or Roman numerals (*Figure 16*). The first extension mode generally involves all sections of the boom extended equally to the boom's maximum length. Other modes may either extend all sections at smaller percentages equally, or involve extending one or more of the sections by certain percentages. By delaying the

8,40 m	9.8 m/s	360°	ISO											
	SH_1													
	24 m	30 m	36 m	42 m	48 m	54 m	60 m	66 m	72 m	78 m	84 m	90 m	96 m	
m	t	t	t	t	t	t	t	t	t	t	t	t	t	m
5,5	625,0	-	-	-	-	-	-	-	-	-	-	-	-	5,5
6	574,0	577,0	-	-	-	-	-	-	-	-	-	-	-	6
7	512,0	463,0	423,0	388,0	-	-	-	-	-	-	-	-	-	7
8	420,0	386,0	357,0	331,0	308,0	288,0	-	-	-	-	-	-	-	8
9	356,0	331,0	308,0	288,0	270,0	254,0	239,0	-	-	-	-	-	-	9
10	308,0	288,0	270,0	254,0	239,0	226,0	213,0	202,0	192,0	-	-	-	-	10
11	267,5	257,5	243,0	229,5	216,5	205,0	194,0	184,5	175,5	166,0	158,0	-	-	11
12	227,0	227,0	216,0	205,0	194,0	184,0	175,0	167,0	159,0	152,0	144,0	139,0	132,0	12
14	177,0	176,0	175,0	170,0	162,0	155,0	147,0	141,0	134,0	129,0	123,0	118,0	112,0	14
16	143,0	142,0	142,0	140,0	138,0	132,0	126,0	121,0	115,0	111,0	106,0	102,0	97,5	16
18	120,0	119,0	118,0	117,0	116,0	115,0	109,0	105,0	100,0	97,0	92,0	89,5	85,0	18
20	103,0	101,0	100,0	99,5	98,5	97,5	96,5	92,5	88,0	85,0	81,0	78,5	74,5	20
22	89,5	88,5	87,0	85,5	85,0	84,0	83,0	82,0	78,0	75,5	71,5	69,5	65,5	22
23	84,0	83,0	81,7	80,2	79,5	78,5	77,5	76,7	73,7	71,2	67,5	65,7	62,0	23
24	-	77,5	76,5	75,0	74,0	73,0	72,0	71,5	69,5	67,0	63,5	62,0	58,5	24
26	-	69,0	67,5	66,0	65,0	64,0	63,0	62,5	61,0	60,0	56,5	55,0	52,0	26
28	-	62,0	60,5	59,0	58,0	56,5	55,5	55,0	53,5	53,0	50,5	49,6	46,4	28
30	-	-	54,5	53,0	51,5	50,5	49,4	48,9	47,6	47,1	45,7	44,5	41,5	30
33	-	-	47,3	45,8	44,4	43,2	41,9	41,2	39,5	38,9	37,6	37,2	35,2	33
34	-	-	-	43,4	42,1	40,8	39,4	38,7	37,2	36,5	35,0	34,8	33,1	34
38	-	-	-	36,3	34,6	33,0	31,4	30,7	29,1	28,4	26,9	26,6	25,3	38
42	-	-	-	-	28,7	26,9	25,2	24,4	22,8	22,0	20,5	20,2	18,8	42
44	-	-	-	-	26,2	24,5	22,6	21,8	20,2	19,4	17,9	17,6	16,2	44
46	-	-	-	-	-	22,1	20,3	19,4	17,7	16,9	15,4	15,1	13,7	46
49	-	-	-	-	-	19,2	17,3	16,4	14,6	13,8	12,2	11,9	10,4	49
50	-	-	-	-	-	-	16,3	15,4	13,6	12,8	11,2	10,9	9,4	50
54	-	-	-	-	-	-	13,1	12,0	10,1	9,3	7,7	7,3	5,9	54
55	-	-	-	-	-	-	-	11,3	9,4	8,5	6,9	6,5	5,1	55
57	-	-	-	-	-	-	-	9,9	8,0	7,1	5,5	5,1	-	57
58	-	-	-	-	-	-	-	9,3	7,3	6,4	-	-	-	58
59	-	-	-	-	-	-	-	8,7	6,7	5,8	-	-	-	59
60	-	-	-	-	-	-	-	-	6,1	5,2	-	-	-	60
61	-	-	-	-	-	-	-	-	5,5	-	-	-	-	61

125 t 165 t

NOTE: Data published herein is intended as a guide only and shall not be construed to warrant applicability for lifting purposes. Crane operation is subject to the computer charts and operation manual both supplied with the crane.

Figure 14 Example load chart 2.

extension of the outermost sections, the boom has a greater capacity at shorter boom lengths than it would if all sections are extended the same percentage. The different configurations will result in different lifting capacities, which, in turn, necessitate different load charts for each mode.

2.3.0 Quadrants of Operation

Crane operational quadrants are horizontal sectors surrounding the crane, defined by the tipping axes for a given configuration and the direction of the crane boom. The quadrants represent variations of crane stability for positions of the boom for which the load chart must account. As a boom swings to various quadrants with no change in boom angle, the distance between the crane's or load's CGs and the nearest tipping axis changes dramatically. This is because mobile cranes, especially wheeled ones, are typically longer than they are wide. This rectangular shape tends to result in more stability of the crane from front-to-rear than from side-to-side, though the locations of tires, outriggers, and stabilizers affect the shape of operational quadrants. It is not unusual for a crane to have different capacities from one quadrant to another.

The operational quadrants are over-the-front, over-the-rear, and over-the-side (right or left) (*Figure 17*). Note that manufacturers may differ on the range of the defined arc, although 6 degrees in each direction is common.

GR-800XL-1 RATED LIFTING CAPACITIES (IN POUNDS)

ON OUTRIGGERS FULLY EXTENDED 23' 7-1/2" (7.2 M) SPREAD
360° ROTATION

A	37.7'		51'		64.4' (19.62 m)				91' (27.75 m)				117.7' (35.87 m)				131'		144.4'	
B	C	(11.5 m)	C	(15.56 m)	C		C		C		C		C		C		C	(39.93 m)	C	(44.0 m)
10	68	160,000	75	103,600	78	88,100	78	44,000												
12	65	125,000	72	103,600	76	88,100	76	44,000												
15	60	108,000	69	103,600	73	88,100	73	44,000	79	44,000	79	30,800								
20	50	78,400	63	77,800	69	71,900	69	44,000	76	44,000	76	30,800	79	30,800	79	17,600				
25	38	59,400	56	59,000	64	56,100	64	44,000	73	44,000	73	30,800	77	30,800	77	17,600	79	17,600	80	17,600
30	21	45,900	48	44,600	59	42,600	59	44,000	70	39,000	70	26,700	75	30,800	75	17,600	77	17,600	78	17,600
35			39	33,800	53	33,000	53	39,900	66	34,000	66	23,200	72	28,200	72	17,600	75	17,600	76	17,600
40			28	26,300	47	25,500	47	32,300	63	28,700	63	20,400	70	24,700	70	17,600	73	17,600	74	17,600
45			5	20,900	40	20,000	40	26,400	59	23,600	59	18,200	67	21,800	67	16,400	71	17,600	73	17,600
50					32	15,900	32	21,900	55	19,300	56	16,400	65	19,500	64	14,700	68	16,200	71	17,100
60									46	13,200	47	14,500	59	14,800	59	11,900	63	13,300	66	13,900
70									36	9,000	37	11,400	53	10,700	53	9,900	58	11,100	62	10,900
80									22	6,100	24	9,500	46	7,600	46	8,400	53	9,000	57	8,200
90													38	5,300	38	7,200	46	6,900	51	6,100
100													27	3,500	28	5,900	39	5,100	45	4,300
110													13	2,100	12	4,600	31	3,600	39	2,900
120																	19	2,600	32	1,800
																	18°		32°	

...ions (%)

	I	II	II	

Figure 15 Example load chart 3.

Many load charts cover 360 degrees of operation. However, crane manufacturers sometimes provide separate load charts for certain quadrants due to variances in capacity. For example, notice the capacity differences in the load charts in *Figure 18* for a telescopic-boom crane.

With a boom length of 36 feet, the gross capacity remains the same in both configurations at an operating radius of 10 feet. However, at all other boom lengths and operating radiuses, the capacity for 360 degrees is less than that for over-the-front operation. This is because the crane leverage is much larger in this quadrant alone than can be allowed for in 360-degree operation.

2.4.0 Configuration of the Crane Base

The crane can be set up on rubber, outriggers, or stabilizers, and with crawlers extended or retracted. Each configuration affects capacity, requiring a different load chart for each base configuration.

2.4.1 On Rubber

Figure 19 shows on-rubber load charts for various operations. Load charts assume specified tire pressures for determining gross capacities. Manufacturers must identify the required tire pressures in notes on the load charts or in a separate table.

There are several requirements for on-rubber lifting. Capacities for on-rubber lifting printed on the load chart do not exceed 75 percent of tipping loads as determined by test. This percentage is strictly an engineering limit established by American safety standards, used by the manufacturer in creating load-chart capacities. It is important to emphasize that this does not mean that the operator has a 25-percent cushion to work with. The operator must not exceed published load chart capacities under any circumstances.

> **CAUTION**
>
> Crane operators working in countries other than the United States should be aware that many countries abide by ISO standards for determining crane stability. In most cases, these standards meet or exceed the 75% / 85% tipping capacity limits specified by ASME and OSHA standards. Operators must carefully study the load charts provided to understand the basis for the crane's stability limits and operate accordingly.

2.4.2 On Outriggers

Load chart capacities for on-outrigger lifting do not exceed 85 percent of tipping loads. As with on-rubber lifting, this is an engineering safety limit. It does not mean the operator has a 15-percent cushion to work within. Depending on the design

Boom Extend Modes

Boom Length		Section Length		
ft	m	T3	T2	T1
50	15.2	32%		
60	18.3	68%		
70	21.3	100%	4%	
80	24.4	100%	39%	
90	27.4	100%	74%	
100	30.5	100%	100%	8%
110	33.5	100%	100%	42%
120	36.6	100%	100%	76%
127	38.71	100%	100%	100%

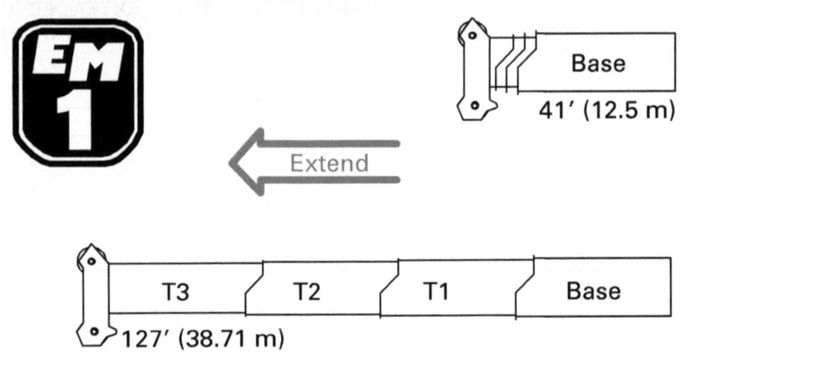

Boom Length		Section Length		
ft	m	T3	T2	T1
50	15.2	32%		
60	18.3	50%	18%	
70	21.3	50%	53%	
80	24.4	50%	87%	
90	27.4	50%	100%	22%
100	30.5	50%	100%	56%
113.1	34.5	50%	100%	100%

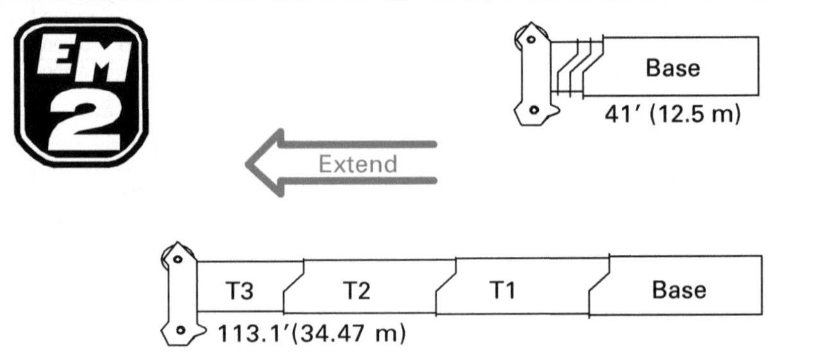

Boom Length		Section Length		
ft	m	T3	T2	T1
50	15.2	32%		
60	18.3	50%	18%	
70	21.3	50%	50%	3%
80	24.4	50%	50%	37%
90	27.4	50%	50%	71%
98.7	30.1	50%	50%	100%

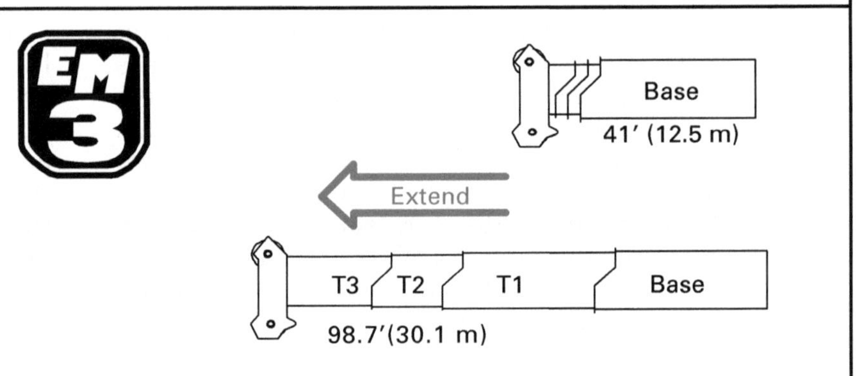

Boom Length		Section Length		
ft	m	T3	T2	T1
47	14.3	22%		
61.3	18.7	22%	50%	
76	23.2	22%	50%	50%

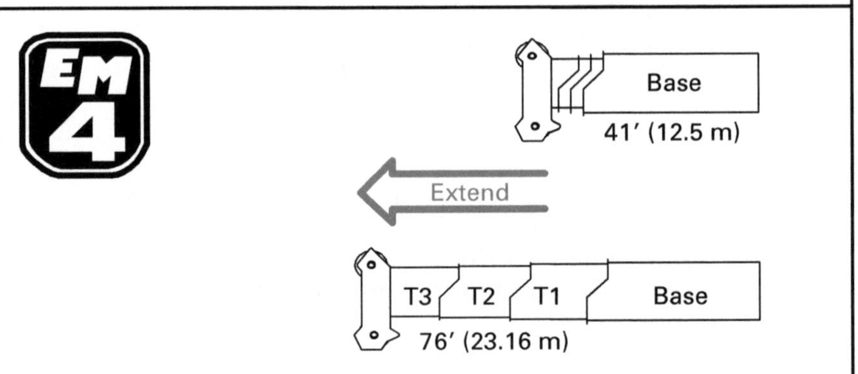

Figure 16 Telescopic-boom extension mode chart.

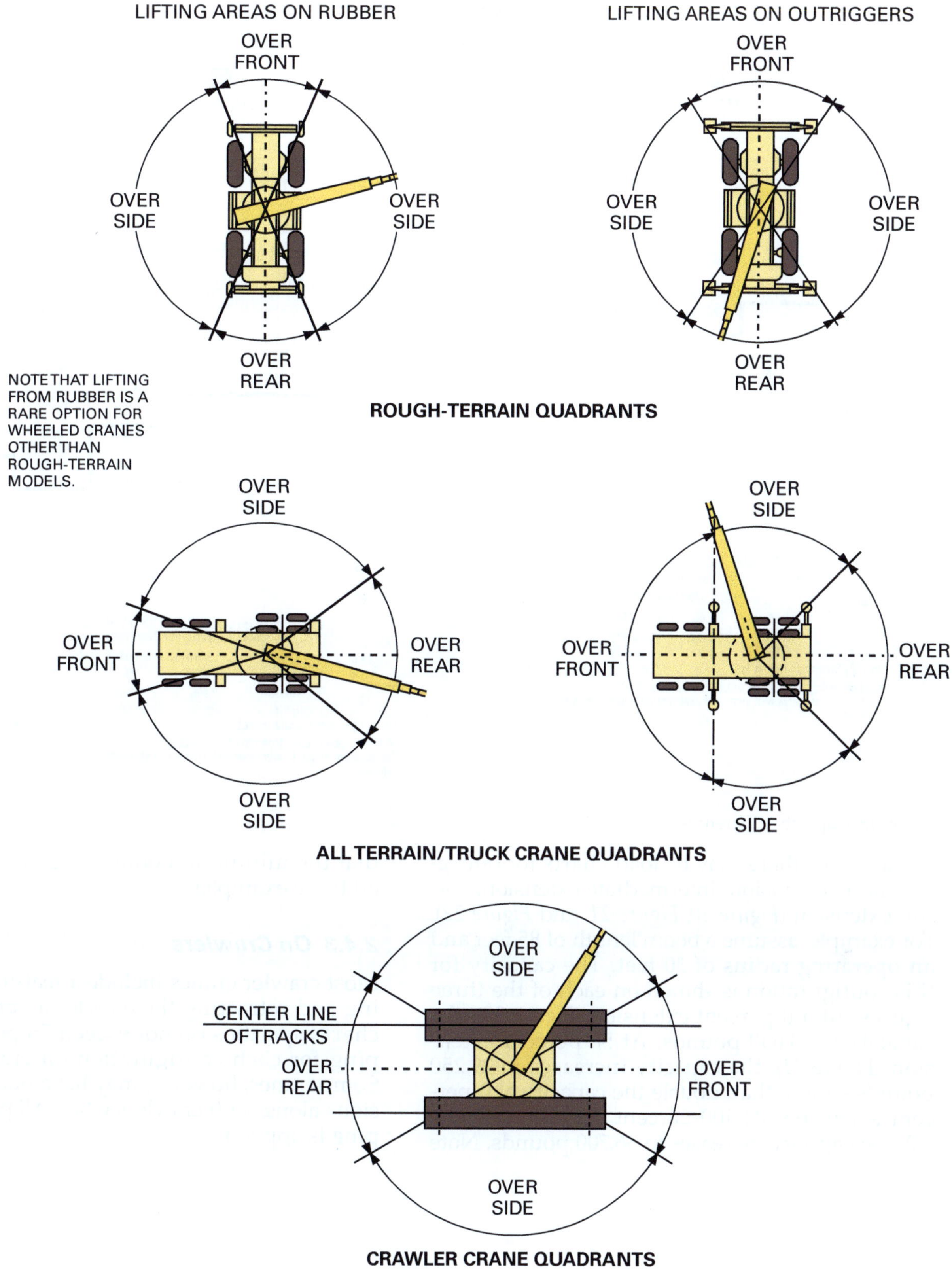

Figure 17 Quadrant definition drawings for rough-terrain, truck, and crawler cranes.

QUADRANT

36-70 ft. | 12,150 lbs | Stationary | 360°

Pounds

#9005

Feet	Main Boom Length in Feet 36	40	50	*60	70
10	45,300 (69)	39,700 (72)			
12	41,750 (65.5)	39,700 (68.5)	29,600 (73.5)		
15	29,350 (59.5)	26,450 (63.5)	26,450 (70)	20,900 (74)	
20	17,800 (49)	17,650 (55)	17,050 (63.5)	16,250 (69)	16,250 (72)
25	11,750 (36)	11,700 (45)	11,350 (56.5)	10,850 (63.5)	10,850 (68)
30		8,040 (31.5)	7,820 (48.5)	7,470 (57.5)	7,470 (63)
35			5,400 (40)	5,120 (51.5)	5,120 (58)
40			3,660 (28)	3,200 (45)	3,430 (53)
45				1,000 (37)	2,150 (47.5)
50					1,150 (41.0)

Lifting Capacities at Zero Degree Boom Angle On Rubber - Stationary 360

Boom Angle	Main Boom Length in Feet 36	40	50
0°	8,180 (29.7)	5,890 (34.2)	2,170 (44.2)

Note: ( ) Reference radii in feet. A6-829-101048A

#LMI operating code. Refer to LMI manual for instructions.
*60 ft. boom length is with inner-mid extended and outer-mid & fly retracted.

QUADRANT

36-70 ft. | 12,150 lbs | Stationary | Defined Arc Over Front

Pounds

#9005

Feet	Main Boom Length in Feet 36	40	50	*60	70
10	45,300 (69)	42,850 (72)	29,600 (76)		
12	43,650 (65.5)	41,350 (68.5)	29,600 (73.5)		
15	38,300 (59.5)	36,300 (63.5)	29,600 (70)	20,900 (74)	17,300 (76.5)
20	31,150 (49)	29,550 (55)	25,900 (63.5)	20,900 (69)	17,300 (72)
25	24,100 (36)	24,150 (45)	21,800 (56.5)	18,800 (63.5)	17,300 (68)
30		17,400 (31.5)	17,200 (48.5)	15,300 (57.5)	15,300 (63)
35			12,800 (40)	12,500 (51.5)	11,000 (58)
40			9,720 (28)	9,390 (45)	9,390 (53)
45				7,090 (37)	7,090 (47.5)
50				5,310 (26.5)	5,310 (41)
55					3,870 (33.5)
60					3,170 (24)

Lifting Capacities at Zero Degree Boom Angle On Rubber - Defined Arc Over Front

Boom Angle	Main Boom Length in Feet 36	40	50	*60	70
0°	17,600 (29.7)	13,600 (34.2)	7,750 (44.2)	4,010 (54.6)	2,670 (64.2)

Note: ( ) Reference radii in feet. A6-829-101047

#LMI operating code. Refer to LMI manual for instructions.
*60 ft. boom length is with inner-mid extended and outer-mid & fly retracted.

Figure 18 Capacity differences.

of the crane, there may be load charts for outriggers at no extension, intermediate extension, and full extension (*Figure 20*, *Figure 21*, and *Figure 22*). For example, assume a boom length of 85 feet and an operating radius of 30 feet. The capacity for this configuration is shown on each of the three figures. At a 0-percent extension (*Figure 20*), the capacity is 13,650 pounds. At 50-percent extension (*Figure 21*), the capacity increases to 29,250 pounds—more than double the capacity at 0-percent extension. At 100-percent extension (*Figure 22*), the capacity increases to 36,700 pounds. Note that the minimum boom angle is unchanged in all three examples.

2.4.3 On Crawlers

Most crawler cranes include a feature for extending and retracting the crawler assemblies. Load chart capacities do not exceed 75 percent of tipping for each configuration on crawler cranes. Some cranes, however, may have outrigger extensions along with crawlers where 85 percent of tipping is applied.

36-70 ft. | 12,150 lbs | Stationary | 360°

Pounds

#9005

Main Boom Length in Feet

Feet	36	40	50	*60	70
10	45,300 (69)	39,700 (72)			
12	41,750 (65.5)	39,700 (68.5)	29,600 (73.5)		
15	29,350 (59.5)	26,450 (63.5)	26,450 (70)	20,900 (74)	
20	17,800 (49)	17,650 (55)	17,050 (63.5)	16,250 (69)	16,250 (72)
25	11,750 (36)	11,700 (45)	11,350 (56.5)	10,850 (63.5)	10,850 (68)
30		8,040 (31.5)	7,820 (48.5)	7,470 (57.5)	7,470 (63)
35			5,400 (40)	5,120 (51.5)	5,120 (58)
40			3,660 (28)	3,200 (45)	3,430 (53)
45				1,000 (37)	2,150 (47.5)
50					1,150 (41.0)

Lifting Capacities at Zero Degree Boom Angle
On Rubber - Stationary 360

Boom Angle	36	40	50
0°	8,180 (29.7)	5,890 (34.2)	2,170 (44.2)

Note: ( ) Reference radii in feet. A6-829-101048A

#LMI operating code. Refer to LMI manual for instructions.
*60 ft. boom length is with inner-mid extended and outer-mid & fly retracted.

36-70 ft. | 12,150 lbs | Stationary | Defined Arc Over Front

Pounds

#9005

Main Boom Length in Feet

Feet	36	40	50	*60	70
10	45,300 (69)	42,850 (72)	29,600 (76)		
12	43,650 (65.5)	41,350 (68.5)	29,600 (73.5)		
15	38,300 (59.5)	36,300 (63.5)	29,600 (70)	20,900 (74)	17,300 (76.5)
20	31,150 (49)	29,550 (55)	25,900 (63.5)	20,900 (69)	17,300 (72)
25	24,100 (36)	24,150 (45)	21,800 (56.5)	18,800 (63.5)	17,300 (68)
30		17,400 (31.5)	17,200 (48.5)	15,300 (57.5)	15,300 (63)
35			12,800 (40)	12,500 (51.5)	11,000 (58)
40			9,720 (28)	9,390 (45)	9,390 (53)
45				7,090 (37)	7,090 (47.5)
50				5,310 (26.5)	5,310 (41)
55					3,870 (33.5)
60					3,170 (24)

Lifting Capacities at Zero Degree Boom Angle
On Rubber - Defined Arc Over Front

Boom Angle	36	40	50	*60	70
0°	17,600 (29.7)	13,600 (34.2)	7,750 (44.2)	4,010 (54.6)	2,670 (64.2)

Note: ( ) Reference radii in feet. A6-829-101047

#LMI operating code. Refer to LMI manual for instructions.
*60 ft. boom length is with inner-mid extended and outer-mid & fly retracted.

36-70 ft. | 12,150 lbs | Pick & Carry Up to 2.5 mph | Boom Center Over Front

Pounds

#9006

Main Boom Length in Feet

Feet	36	40	50	*60	70
10	42,500 (69)	37,600 (72)	30,700 (76)		
12	42,500 (65.5)	37,600 (68.5)	30,700 (73.5)	23,450 (77)	
15	37,600 (59.5)	37,600 (63.5)	30,700 (70)	19,700 (74)	16,250 (76.5)
20	32,950 (49)	32,900 (55)	30,700 (63.5)	19,700 (69)	16,250 (72)
25	24,100 (36)	24,150 (45)	23,850 (56.5)	19,700 (63.5)	16,250 (68)
30		17,400 (31.5)	17,200 (48.5)	16,800 (57.5)	16,250 (63)
35			12,800 (40)	12,500 (51.5)	12,500 (58)
40			9,720 (28)	9,390 (45)	9,390 (53)
45				7,090 (37)	7,090 (47.5)
50				5,310 (26.5)	5,310 (41)
55					3,870 (33.5)
60					3,090 (24)

Lifting Capacities at Zero Degree Boom Angle
On Rubber - Pick & Carry

Boom Angle	36	40	50	*60	70
0°	17,600 (29.7)	13,600 (34.2)	7,750 (44.2)	4,010 (54.6)	2,670 (64.2)

Note: ( ) Reference radii in feet. A6-829-101049

#LMI operating code. Refer to LMI manual for instructions.
*60 ft. boom length is with inner-mid extended and outer-mid & fly retracted.

NOTES:

1. Capacities are in pounds and do not exceed 75% of tipping loads as determined by test in accordance with SAE J765.
2. Capacities are applicable to machines equipped with 29.6x25 (28 ply) tires at 65 psi cold inflation pressure.
3. Defined Arc – Over front includes 6' on either side of longitudinal centerline of machine (ref. drawing C6-829-003529).
4. Capacities appearing above the bold line are based on structural strength and tipping should not be relied upon as a capacity limitation.
5. Capacities are applicable only with machine on firm level surface.
6. On rubber lifting with boom extensions not permitted.
7. For pick and carry operation, boom must be centered over front of machine, mechanical swing lock engaged and load restrained from swinging. When handling loads in the structural range with capacities close to maximum ratings, travel should be reduced to creep speeds.
8. Axle lockouts must be functioning when lifting on rubber.
9. All lifting depends on proper tire inflation, capacity and condition. Capacities must be reduced for lower tire inflation pressures. See lifting capacity chart for tire used. Damaged tires are hazardous to safe operation of crane.
10. Creep – Not over 200 ft. of movement in any 30 minute period and not exceeding 1 mph.

Figure 19 On-rubber load chart.

40 - 125 ft.
(12 - 38 m)

11,500 lbs.
(5216 kg)

0%
10' 4" Spread

360

Pounds

Feet	40	45	55	65	75	85	95	105	115	125
10	91,900 (70)	86,850 (72.5)								
12	70,100 (67)	66,600 (70)	61,200 (74)							
15	50,500 (61.5)	48,150 (65.5)	44,550 (70.5)	41,550 (74)						
20	31,200 (52.5)	30,950 (58)	28,900 (65)	26,950 (69)	26,600 (72.5)	26,600 (75)				
25	20,850 (41.5)	20,550 (49.5)	19,900 (58.5)	18,500 (64.5)	18,500 (68.5)	18,800 (71.5)	18,950 (73.5)	19,050 (75.5)		
30	14,650 (26)	14,350 (39.5)	13,850 (52)	12,950 (59)	13,850 (64)	13,650 (67.5)	13,950 (70.5)	14,200 (72.5)	14,200 (75)	*14,200 (76.5)
35		10,200 (26)	9,750 (44.5)	9,090 (53.5)	9,370 (59.5)	9,970 (64)	10,350 (67)	10,700 (69.5)	10,750 (72)	10,850 (74)
40			6,810 (35.5)	6,190 (47.5)	6,550 (54.5)	7,200 (60)	7,660 (63.5)	8,070 (66.5)	8,200 (69.5)	8,350 (71.5)
45			4,600 (23)	3,950 (40.5)	4,350 (49.5)	5,040 (55.5)	5,540 (60)	5,990 (63.5)	6,170 (66.5)	6,360 (69)
50				2,170 (32.5)	2,610 (44)	3,320 (51.5)	3,850 (56.5)	4,330 (60.5)	4,540 (64)	4,750 (66.5)
60							1,300 (48.5)	1,820 (53.5)	2,070 (58)	2,320 (61.5)
Minimum boom angle (deg.) for indicated length (no load)										51
Maximum boom length (ft.) at 0 degree boom angle (no load)										65

NOTE: ( ) Boom angles are in degrees.
*Based on maximum obtainable boom angle.

A6-829-011927A

Boom Angle	40	45	55
0	12,550 (32.3)	8,440 (37.8)	3,610 (47.8)

NOTE: ( ) Reference radii are in feet.

A6-829-012282

Figure 20 Capacities on outriggers at no extension (0 percent).

40 - 125 ft. (12 - 38 m) | 11,500 lbs. (5216 kg) | 50% 17' 4" Spread | 360

Pounds

Feet	40	45	55	65	75	85	95	105	115	125
10	116,000 (70)	105,000 (72.5)								
12	102,500 (67)	102,000 (70)	94,600 (74)							
15	86,800 (61.5)	86,350 (65.5)	85,800 (70.5)	71,050 (74)						
20	67,250 (52.5)	64,850 (58)	60,550 (65)	56,950 (69)	55,250 (72.5)	48,150 (75)				
25	45,750 (41.5)	44,950 (49.5)	42,050 (58.5)	39,700 (64.5)	39,100 (68.5)	38,900 (71.5)	38,000 (73.5)	33,350 (75.5)		
30	32,500 (26)	32,050 (39.5)	31,000 (52)	29,250 (59)	29,050 (64)	29,250 (67.5)	29,300 (70.5)	29,250 (72.5)	24,550 (75)	*23,700 (76.5)
35		23,900 (26)	23,350 (44.5)	22,250 (53.5)	22,300 (59.5)	22,700 (64)	22,900 (67)	23,050 (69.5)	21,700 (72)	21,900 (74)
40			17,650 (35.5)	17,200 (47.5)	17,400 (54.5)	17,900 (60)	18,250 (63.5)	18,550 (66.5)	18,550 (69.5)	18,550 (71.5)
45			13,400 (23)	13,000 (40.5)	13,600 (49.5)	14,300 (55.5)	14,700 (60)	15,050 (63.5)	15,150 (66.5)	15,250 (69)
50				9,850 (32.5)	10,450 (44)	11,450 (51.5)	11,900 (56.5)	12,350 (60.5)	12,450 (64)	12,600 (66.5)
60					5,970 (30)	6,940 (41.5)	7,840 (48.5)	8,310 (53.5)	8,500 (58)	8,700 (61.5)
70						4,020 (28.5)	4,840 (39)	5,410 (46)	5,700 (51.5)	5,930 (55.5)
80							2,630 (27)	3,170 (37)	3,620 (44.5)	3,870 (49.5)
90								1,490 (25.5)	2,000 (36)	2,280 (42.5)
100										1,020 (34.5)
Minimum boom angle (deg.) for indicated length (no load)										27.5
Maximum boom length (ft.) at 0 degree boom angle (no load)										105

NOTE: ( ) Boom angles are in degrees.
*Based on maximum obtainable boom angle.

A6-829-011802A

Boom Angle	40	45	55	65	75	85	95
0	22,800 (32.3)	18,250 (37.8)	11,500 (47.8)	6,220 (57.8)	3,740 (67.8)	2,350 (77.8)	1,300 (87.8)

NOTE: ( ) Reference radii are in feet.

A6-829-012282

Figure 21 Capacities on outriggers at intermediate extension (50 percent).

40 - 125 ft. (12 - 38 m)	11,500 lbs. (5216 kg)	100%	360

Pounds

Feet	40	45	55	65	75	85	95	105	115	125
10	130,000 (70)	105,000 (72.5)								
12	111,000 (67)	105,000 (70)	94,600 (74)							
15	91,450 (61.5)	91,000 (65.5)	88,250 (70.5)	71,050 (74)						
20	69,550 (52.5)	69,050 (58)	68,400 (65)	60,400 (69)	55,250 (72.5)	48,150 (75)				
25	55,050 (41.5)	54,600 (49.5)	53,950 (58.5)	53,450 (64.5)	47,950 (68.5)	41,700 (71.5)	38,000 (73.5)	33,350 (75.5)		
30	42,950 (26)	42,450 (39.5)	41,700 (52)	41,200 (59)	41,950 (64)	36,700 (67.5)	33,300 (70.5)	30,750 (72.5)	24,550 (75)	*23,700 (76.5)
35		33,700 (26)	33,000 (44.5)	32,500 (53.5)	33,250 (59.5)	32,[illegible]00 (64)	29,550 (67)	27,300 (69.5)	21,700 (72)	21,900 (74)
40			26,650 (35.5)	26,150 (47.5)	26,900 (54.5)	27,850 (60)	26,450 (63.5)	24,450 (66.5)	19,350 (69.5)	20,300 (71.5)
45			21,750 (23)	21,300 (40.5)	22,050 (49.5)	23,000 (55.5)	23,700 (60)	22,000 (63.5)	17,450 (66.5)	18,800 (69)
50				17,500 (32.5)	18,250 (44)	19,150 (51.5)	19,900 (56.5)	19,850 (60.5)	15,800 (64)	17,050 (66.5)
60					12,400 (30)	13,250 (41.5)	14,100 (48.5)	14,650 (53.5)	13,250 (58)	14,150 (61.5)
70						9,190 (28.5)	9,910 (39)	10,400 (46)	10,850 (51.5)	11,350 (55.5)
80							6,930 (27)	6,740 (37)	7,850 (44.5)	8,290 (49.5)
90								5,170 (25.5)	5,600 (36)	6,010 (42.5)
100									3,880 (25)	4,250 (34.5)
110										2,840 (24)
Minimum boom angle (deg.) for indicated length (no load)										0
Maximum boom length (ft.) at 0 degree boom angle (no load)										125

NOTE: ( ) Boom angles are in degrees.
*Based on maximum obtainable boom angle.

A6-829-011801A

Boom Angle	40	45	55	65	75	85	95	105	115
0	22,800 (32.3)	18,250 (37.8)	12,200 (47.8)	7,990 (57.8)	5,720 (67.8)	4,320 (77.8)	3,210 (87.8)	2,380 (97.8)	1,570 (107.8)

NOTE: ( ) Reference radii are in feet.

A6-829-012282

Figure 22 Capacities on outriggers at full extension (100 percent).

2.4.4 Tower and Ring Attachments

A tower luffing-jib attachment (*Figure 23*) is an alternative to a freestanding tower crane where a lower erection height allows its use. It can also increase the load radius of a standard crane where reach is the primary concern. As with all attachments, the operator should follow the crane manufacturer's and/or attachment manufacturer's recommendations for lifting.

When a work site requires a stationary crane with major capacity, some manufacturers offer **ringer crane** (*Figure 24*) packages. A ring attachment installed on the ground greatly increases the crane's capacity and stability, permitting weight transfer to the ring foundation in all positions. The capacity remains the same regardless of the operating quadrant.

2.4.5 Counterweight Configurations

Counterweights give extra stability to the crane during assembly and lifting operations. A crane can have front, rear, side, or upper counterweights. Each crane has counterweight specifications, and manufacturers explain in the operational notes that the capacities listed are with the specified counterweights in place. The assembled counterweight sections have to match the configurations specified in the load charts. If a counterweight is extendable, its extension must be according to the manufacturer's recommendations. *Figure 25* and *Figure 26* show examples of counterweight configurations.

It's important to note that no form of counterweight can be used other than the counterweight specified by the manufacturer. For example, placing or connecting another heavy piece of equipment onto the crane is prohibited.

Figure 23 Crane with tower attachment.

Figure 24 3,000 metric-ton ringer crane.

Figure 25 Crawler crane counterweights.

Figure 26 Counterweights on an auxiliary carrier.

Additional Resources

ASME Standard B30.5, Mobile and Locomotive Cranes. Current edition. New York, NY: American Society of Mechanical Engineers.

Cranes: Design, Practice, and Maintenance, Ing J. Verschoof. 2002. Hoboken, NJ: John Wiley and Sons, Inc.

29 *CFR* 1926, Subpart CC,*Cranes and Derricks in Construction.* **www.ecfr.gov**

2.0.0 Section Review

1. Relatively small four-wheeled cranes typically operated by the driver are _____.
   a. industrial/all-purpose cranes
   b. crawler cranes
   c. lattice-boom cranes
   d. articulated boom cranes

2. The arrangement of information on a crane load chart _____.
   a. is specified by safety standards
   b. varies according to the manufacturer and crane type
   c. always shows boom height along the top row
   d. always shows gross capacities in the right column

3. Most mobile cranes are more stable front-to-rear because _____.
   a. the crane upperworks are centered on the carrier chassis
   b. they are typically wider than they are long
   c. the carrier's drive engine lowers the center of gravity
   d. they are typically longer than they are wide

4. All the items listed here relate specifically to the crane base configuration for load chart purposes *except* _____.
   a. fuel-tank fill level
   b. outriggers
   c. tires
   d. crawlers

SECTION THREE

3.0.0 WORKING WITH LOAD CHARTS

Objective

Describe how to apply load charts and deductions from capacity.

a. Describe the use of jib charts.
b. Describe how the parts of line can affect capacity.
c. Describe how to read a range diagram.
d. Describe how to calculate a crane's capacity from its load charts.

Performance Tasks

4. Properly identify load charts that are used in different configurations.
5. Identify parts of line and counterweight considerations in load chart information.
6. Calculate minimum parts of line required.

Trade Terms

Effective weight: With reference to jib/boom extensions, the weight calculated by the manufacturer which, when applied at the boom tip, has the same effect on the crane capacity as the jib and/or boom extension itself.

Offset: The angle formed by an erected boom, measured from an imaginary line extending through the center line of the main boom. The crane's load chart provides capacities for allowable jib or boom-extension offset angles.

Line pull: The maximum pulling force a crane hoist drum can apply to a hoist cable, accounted for by the torque that the motor can apply to the drum and the working radius of the rope as it leaves the drum.

Load charts establish the gross weight that a crane can lift in a specific configuration. Extra weight, such as the load block, headache ball, wire rope, and parts of hoist line, reduce the primary load weight that the crane can safely hoist. These weights must be deducted from the gross capacity of the crane unless they are specifically included in the load chart configuration. The weight of the main boom is configured and included.

Load charts usually have a separate list of items with their weight deductions (*Figure 27*). Also refer to the load charts in *Appendix B* for examples of these lists. Operators should always review the notes section of load charts as well. The notes provide valuable information that must often be applied.

Some load charts require the deduction of the suspended hoist line, while others include the rope weight in the load chart calculations. If you must calculate a rope deduction, find the product of (multiply) the full boom-tip height from which the load is suspended, the parts of line, and the line's weight per unit. This is illustrated by the following equation:

Suspended line weight (lb) =
boom-tip height (ft) × parts of line × unit weight (lb/ft)

Assume that the boom-tip height of a crane is 45 feet, there are 3 parts of line, and the ¾" wire rope weighs 1.15 lb/ft:

Suspended line weight (lb) =
45 ft × 3 parts of line × 1.15 lb/ft = **156 lb**

The exact result of this equation is actually 155.25 lb. However, when determining the weight of variable loads like rope, these calculations are usually rounded up to the nearest whole number (in this case, 156 lb). This calculation can also be used with metric units, substituting kilograms and meters, for example. It provides a conservative value for the weight of suspended cable, which can amount to thousands of pounds for larger cranes.

3.1.0 Jib Charts

Jib load charts, such as the one shown in *Figure 28*, show the capacities in pounds, kips, or kilograms for the luffing or fixed boom extensions at different degrees of jib **offset**.

> **NOTE**
> Jib chart styles vary widely from manufacturer to manufacturer.

The following factors affect the capacity of a crane equipped with a jib boom extension:

- Load radius
- Length of the main boom
- Main boom angle
- Jib length
- Jib angle with respect to the boom (offset)
- Weight of the rigging and hooks on both the jib and main boom

CAPACITY DEDUCTIONS FOR AUXILIARY LOAD HANDLING EQUIPMENT

Load Handling Equipment:	**(lbs)**
Auxiliary Head Attached	100
25-ton quick reeve 3 sheave hook block (see hook block for actual weight)	670
40-ton quick reeve 4 sheave hook block (see hook block for actual weight)	780
8.5-ton hook ball (see hook ball for actual weight)	360
Lifting From Main Boom With:	**(lbs)**
28.5 ft or 51 ft fly stowed on base (see operation note 4)	0
28.5 ft. offset fly erected but not used	2600
51 ft. offset fly erected but not used	4800
Lifting From 28.5 ft Offset Fly With:	
22.5 ft fly tip erected but not used	**PROHIBITED**
22.5 ft. fly tip stowed on 28.5 ft. offset fly	**PROHIBITED**
Note: Capacity deductions are for Link-Belt supplied equipment only.	

Figure 27 Deductions chart.

When not in use, some jibs and/or boom extensions are stowed on the main boom. In some cases, the operator must deduct the extension's weight from the crane's gross capacity in this configuration. The operator must consult the manufacturer's documentation to determine the effect of a stowed jib/boom extension on the crane's net capacity.

When mounted on the boom tip, the operator must deduct the **effective weight** of the jib/extension from the crane's gross capacity. This deduction is made only when hoisting with the main boom. When lifting from an erected jib, do not count the weight of the jib as part of the load; the appropriate load chart already compensates for the specified jib configuration.

3.2.0 Line Pull and Parts of Line

Determination of a crane's gross and net capacity using a load chart assures the operator that the crane configuration will allow hoisting a given load. However, to physically lift a load requires pulling on the hoist line. How do you know if the line is strong enough to lift the weight of the load, and whether the crane can apply enough tension to the hoist line?

3.2.1 Hoist Rope Properties

Crane hoist rope must be strong enough for its intended purpose, flexible enough to run smoothly through sheaves, and have minimal tendency to rotate or twist under load. Unfortunately, these properties can work against each other. Rope manufacturers compromise to obtain the optimum rope qualities for an intended purpose. Strength and flexibility are obtained by using the finest wire strands possible and plying them in various ways to maximize flexibility at a given strength.

Rotation-resistant wire rope is manufactured by applying multiple strand layers in opposite directions. Thus, when the rope is under load, it resists untwisting and spinning the load. While rotation-resistant rope provides this desirable characteristic, it is not as durable as its standard counterpart.

For safety purposes, a wire rope's breaking strength must be much larger than any tension it will experience during operation. *Table 1* lists typical wire rope sizes, associated breaking strengths, and their weights for a given type of rotation-resistant rope.

Table 1 Selected Wire Rope Properties

Wire Rope Diameter (in)	Minimum Breaking Strength (lb)	Unit Weight (lb/ft)
5/8	47,400	0.78
3/4	69,400	1.15
1	122,800	2.09
1 1/2	269,000	4.56

33-56 ft. luffing folding boom extension (mode B) (fixed offset angles)

37.3-141.7 ft.	33 - 56 ft.	22,000 lbs	100% 34'-6" Spread	360°

Pounds

Feet	33 ft. LENGTH 5° OFFSET #0091	33 ft. LENGTH 20° OFFSET #0091	33 ft. LENGTH 40° OFFSET #0091	56 ft. LENGTH 5° OFFSET #0092	56 ft. LENGTH 20° OFFSET #0092	56 ft. LENGTH 40° OFFSET #0092
40	*13,700 (78)					
45	13,700 (77)					
50	13,700 (75)	13,700 (77.5)		*8,200 (78)		
55	13,700 (73.5)	13,700 (75.5)	*11,000 (78)	8,200 (77.5)		
60	13,700 (71.5)	13,700 (74)	11,000 (76)	8,200 (76)		
65	13,700 (70)	12,850 (72)	10,950 (74.5)	8,200 (74.5)	8,200 (77.5)	
70	12,500 (68)	12,000 (70)	10,350 (72.5)	8,200 (73)	8,200 (76)	
75	11,350 (66)	11,200 (68)	9,830 (70.5)	8,200 (71.5)	8,100 (74)	6,400 (77.5)
80	9,730 (64.5)	10,450 (66.5)	9,330 (68.5)	8,200 (69.5)	7,600 (72.5)	6,400 (76)
85	8,300 (62.5)	8,980 (64.5)	8,860 (66.5)	8,200 (68)	7,150 (71)	6,230 (74)
90	7,060 (60.5)	7,660 (62.5)	8,210 (64.5)	7,740 (66.5)	6,730 (69)	5,920 (72.5)
95	5,960 (58.5)	6,500 (60.5)	6,980 (62)	7,130 (64.5)	6,350 (67.5)	5,640 (70.5)
100	4,990 (56.5)	5,470 (58)	5,880 (60)	6,130 (63)	6,000 (65.5)	5,380 (68.5)
105	4,120 (54)	4,560 (56)	4,900 (58)	5,230 (61)	5,690 (64)	5,140 (67)
110	3,340 (52)	3,730 (54)	4,020 (55.5)	4,430 (59.5)	5,290 (62)	4,900 (65)
115	2,640 (49.5)	2,990 (51.5)	3,230 (53)	3,700 (57.5)	4,490 (60)	4,690 (63)
120	2,000 (47.5)	2,320 (49)	2,510 (50.5)	3,040 (55.5)	3,760 (58.5)	4,470 (61)
125	1,420 (45)	1,700 (46.5)	1,850 (47.5)	2,440 (53.5)	3,100 (56.5)	3,710 (58.5)
130		1,140 (44)	1,250 (45)	1,900 (51.5)	2,500 (54.5)	3,030 (56.5)
135				1,390 (49.5)	1,940 (52)	2,390 (54)
140					1,420 (50)	1,810 (52)
145						1,270 (49)
Minimum boom angle (°) for indicated length (no load)	42	43	43	48	48	47
Maximum boom length (ft.) at 0° boom angle (no load)		89.8			76.7	

NOTE: ( ) Boom angles are in degrees.

A6-829-103522

#LMI operating code. Refer to LMI manual for operating instructions.
*This capacity is based upon maximum boom angle.

NOTES:

1. All capacities above the bold line are based on structural strength of boom extension and do not exceed 85% of tipping loads, in accordance with SAE J-765.
2. The 33 ft. luffing folding boom extension may be used for single or double line lifting service. The 56 ft. luffing folding boom extension may be used for single line lifting service only.
   WARNING: Lifting with the 33 ft. extension base, with the 23 ft. extension fly either erected or folded along side of extension base, is strictly prohibited.
3. WARNING: Operation of this machine with heavier loads than the capacities listed is strictly prohibited. Machine tipping with boom extension occurs rapidly and without advance warning.
4. Boom angle is the angle above or below horizontal of the longitudinal axis of the boom base section after lifting rated load.
5. Capacities listed are with outriggers properly extended and vertical jacks set only.
6. For main boom lengths less than 141.7 ft. with the boom extension erected, the rated loads are determined by boom angle. Use only the column which corresponds to the boom extension length and offset for which the machine is set up. For boom angles not shown, use rating of the next lower boom angle.
7. When lifting over the main boom nose with 33 ft. or 56 ft. extension erected, the outriggers must be fully extended or 50% extended (17.3 ft. spread).

Figure 28 Jib/boom extension chart.

Did You Know?

Synthetic Hoist Rope

Steel wire rope has been in use since the late 1800s. It has many advantages in terms of low cost, ease of manufacture, and strength. But wire rope also has many disadvantages. Twisted wire rope, which is the most common form, tends to kink and birdcage, and the metal strands fatigue and break. Steel wire rope also corrodes, shortening its lifespan. Metal rope is also difficult to handle when reeving, adding hardware, and repairing.

In recent years, advances in materials technology have led to the development of nonmetallic, synthetic fiber ropes suitable for crane hoisting. These new rope products offer breaking strengths comparable to steel rope at the same diameters. Additionally, the ropes are about 80 percent lighter, can achieve the same line pull, have low bending fatigue, and are easier and safer to handle. However, it will likely take some time for synthetic fiber rope to become widely accepted and applied in the field.

3.2.2 Line Pull

Manufacturers describe a crane's ability to apply tension to the hoist cable as its **line pull**. This is the maximum amount of force the hoist drum can apply to the hoist cable for a given layer on the drum. A hoist drum motor can apply only a certain amount of torque to the drum. As the drum wraps more layers of rope, its diameter increases, and the maximum line pull it can exert decreases proportionately. It is an industry standard that the ratio of the minimum breaking strength of the specified hoist cable to the crane's maximum line pull should be at least 5:1. Therefore, the limiting factor is usually the crane's line pull, not the minimum breaking strength of the wire rope.

Another factor that affects line pull is the type of hoist sheave bearings the crane has. Cranes have sheaves with either bronze-bushed bearings or low-friction roller bearings. The type of bearings affects the amount of friction force that the line pull must overcome during the lift. Also, the number of sheaves and their location in the system affect how fast they are moving and the amount of friction they generate. To lift a certain load, the line pull must be greater than the suspended load weight, plus the additional resistance created by the friction force of the rotating sheaves.

3.2.3 Parts of Line

Maximum crane capacities may far exceed the breaking strength of a single hoist line. So how do cranes manage to lift such loads? By using multiple lines to support a load block, the weight of the load is divided among the number of line parts reeved through the block. Two parts of line reduce the weight each line supports by half. Each of five lines supports only one-fifth of the suspended load. Before making a lift, the operator must ensure the crane has the required minimum parts of line (*Figure 29*) to lift the load.

NOTE

The rope dead end of an odd number of parts attaches to the load block. The dead end of an even number of parts attaches to the boom head/sheave housing.

There are several methods the operator can use to determine the parts of line needed for a lift. Some crane manufacturers provide this information along with the load charts (*Figure 30*). This is the easiest method, and it is safe and accurate if the crane has the manufacturer's specified wire rope.

Without this resource, use the following procedure to determine the minimum number of parts of line required for a given lift:

Step 1 Determine the suspended weight of the load, rigging, and other hoisting accessories.

Step 2 Determine the crane's minimum single line pull available during the lift operation.

Step 3 Determine the type of hoist sheave bearings in the hoist system.

Step 4 Divide the suspended weight found in Step 1 by the crane's single line pull.

Step 5 Compare the value found in Step 4 to the listed ratios found in *Table 2* for either the bronze-bushed sheaves or the anti-friction or roller bearing sheaves. Always round up for intermediate ratios.

Step 6 In the Number of Line Parts column, read the number corresponding to the resulting ratio found in Step 5. This is the minimum required parts of line for the lift.

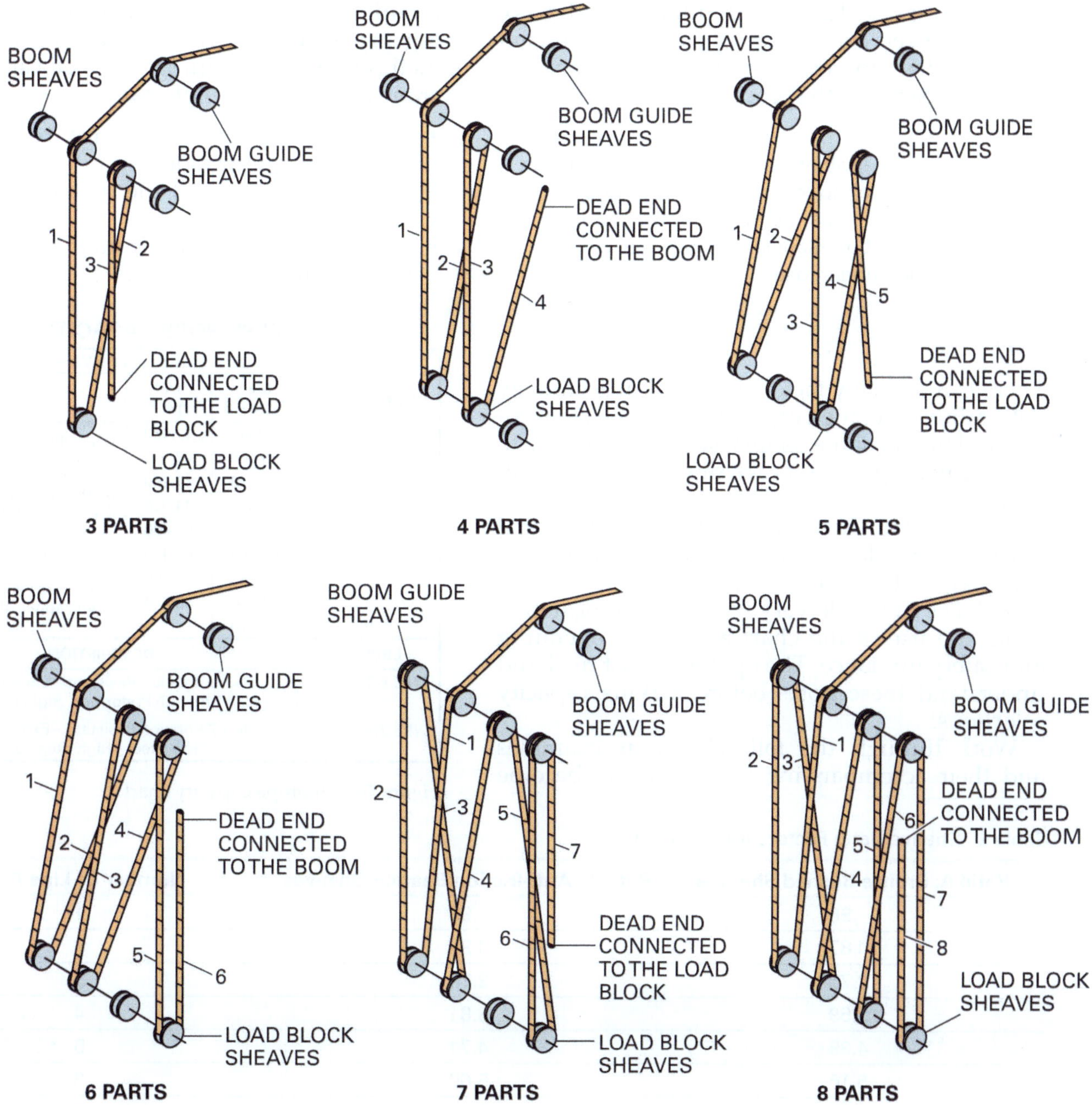

Figure 29 Reeving for various parts of line.

3.3.0 Range Diagram

A range diagram (*Figure 31*) is a side view of a crane with its full range of boom lengths, jibs, and boom extensions. Manufacturers include a range diagram in their load charts. It serves three important functions:

- It determines the configuration of the crane for a lift. The operator can determine the boom length, boom angle, boom point height or hook elevation, jib length, and jib offset required to suit the lift conditions at the site.
- It supports determining the main boom angle of telescopic booms for a given partial extension when only the load radius is known.
- It may identify the minimum allowable clearances between the load block and boom tip.

The operator should consult the range diagram at the beginning of the lift planning to establish the best configuration to conduct the lift. Planners must give consideration to the location of the load, where it needs to go, any site restrictions, and the dimensions and capabilities of the crane. The range diagram does not show crane lift capacities. After a configuration and basic setup has been chosen, the operator will consult the load chart to determine if the crane has the needed capacity in the proposed configuration.

> **NOTE**
> Boom angles shown in a range diagram are for booms under no-load conditions. During lift planning, the operator must account for load-radius and boom-angle changes due to boom deflection under load.

3.4.0 Calculating Crane Capacity

There can be many factors involved in accurately determining a crane's capacity in a given configuration. This section explains, step by step, how to find the net capacity for each of the four crane categories.

Crane manufacturers provide operational notes with the load charts for each crane model. (Refer again to *Figure 28* for an example of these notes.) They provide very important information about the use of the chart as well as operating and safety guidance. The operator must read and understand these notes before making capacity selections.

Work through the following four examples and their accompanying load charts to become familiar with the factors involved in determining gross and net crane capacities. A standard method for determining the net capacity of a crane follows this simple pattern:

- Establish the gross capacity based on the planned configuration.
- Identify and sum all the deductions the configuration requires.
- Subtract all deductions from the gross capacity to determine net capacity.

WIRE ROPE CAPACITY

Maximum Lifting Capacities Based On Wire Rope Strength			
Parts of Line	5/8" Type RB	5/8" Type RB	Notes
1	9,080	11,080	Capacities shown are in pounds and working loads must not exceed the ratings on the capacity charts in the Crane Rating Manual.
2	18,160	22,160	
3	27,240	33,240	
4	36,320	44,320	
5	45,400	55,400	
6	54,480	66,480	Study Operator's Manual for wire rope inspection procedures and single part of line applications.
7	63,560	77,560	
8	72,640	88,640	
9	81,720		
LBCE	**DESCRIPTION**		
TYPE RB	18 × 19 Rotation Resistant – Compact Strand, High Strength Preformed, Right Regular Lay		
TYPE ZB	36 × 7 Rotation Resistant – Extra Improved Plow Steel – Right Regular Lay		

Figure 30 Wire rope capacity chart.

Table 2 Sheave Ratio Factors for Parts of Line

Ratio A: Bronze Bushed Sheaves	Ratio B: Anti-friction Bearing Sheaves	Number Of Line Parts
.96	.98	1
1.87	1.94	2
2.75	2.88	3
3.59	3.81	4
4.39	4.71	5
5.16	5.60	6
5.90	6.47	7
6.60	7.32	8
7.27	8.16	9
7.91	8.98	10
8.52	9.79	11
9.11	10.60	12
9.68	11.40	13
10.20	12.10	14
10.70	12.90	15
11.20	13.60	16
11.70	14.30	17
12.20	15.00	18
12.60	15.70	19
13.00	16.40	20

Link-Belt
CONSTRUCTION EQUIPMENT

WORKING RANGE DIAGRAM

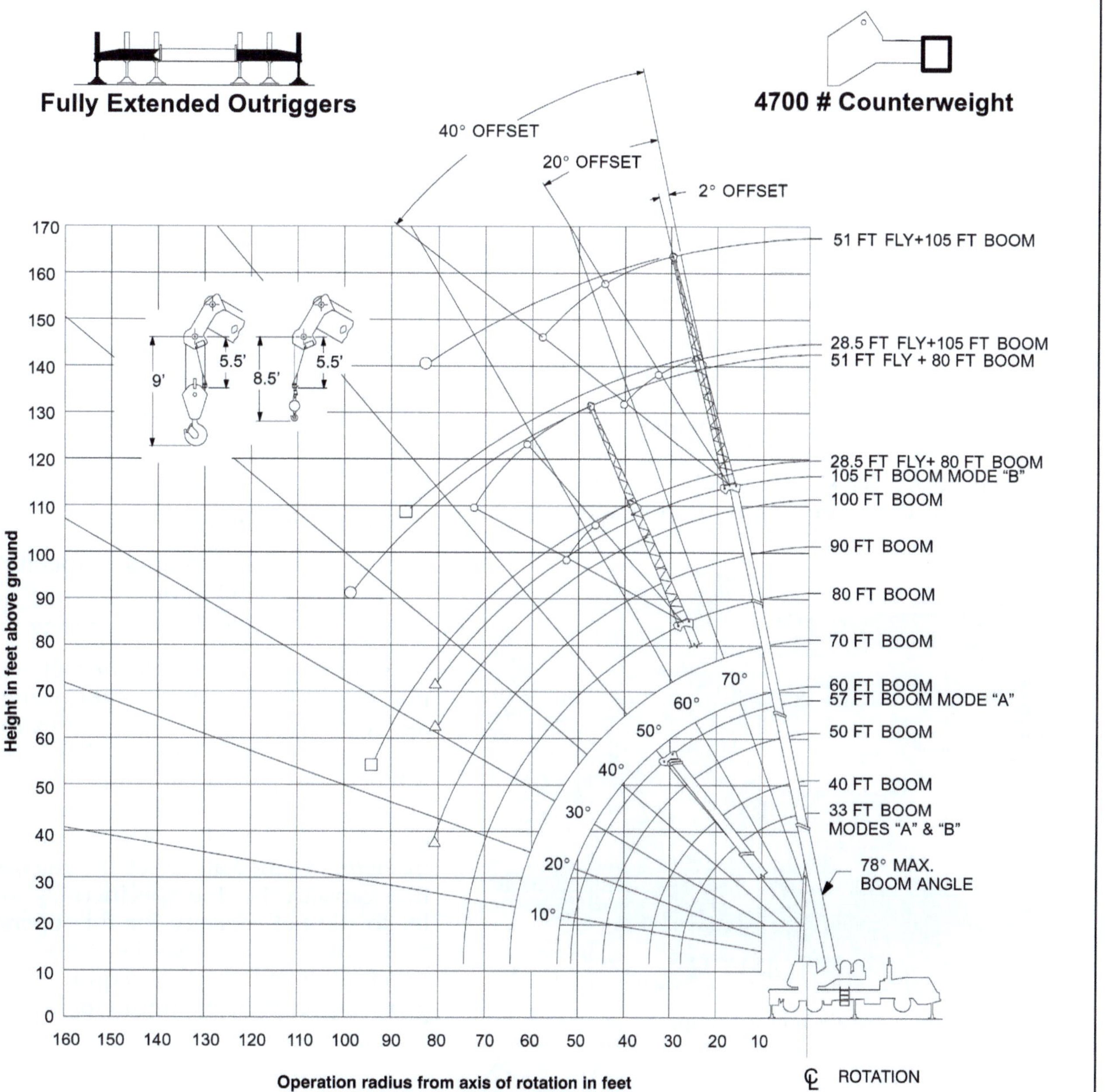

○ **Denotes Main Boom + 51 FT Offset Fly–Boom Mode "B"**

□ **Denotes Main Boom + 28.5 FT Offset Fly–Boom Mode "B"**

△ **Denotes Main Boom – Boom Mode "B"**

Note: Boom and fly geometry shown are for unloaded condition and crane standing level on firm supporting surface. Boom deflection, subsequent radius and boom angle change must be accounted for when applying load to hook.

⚠ WARNING

Do Not Lower The Boom Below The Minimum Boom Angle For No Load Stability As Shown In The Lift Charts For The Boom Lengths Given. Loss Of Stability Will Occur Causing A Tipping Condition.

Figure 31 Range diagram.

3.4.1 Example One: Industrial/All-Purpose Crane

Find the net capacity of a Grove YB4409 industrial crane (*Figure 32*) for a pick-and-carry operation, and its maximum allowable travel speed, configured as follows:

- On-rubber operation
- Over-front carry; boom centered
- 4-section main boom with fourth section retracted
- Load radius of 12 feet
- 6-foot jib stowed
- Equipped with manufacturer-specified tires at the specified pressure
- Hook and headache ball
- 40 lb of rigging

A simple graphic illustration of the configuration is shown in *Figure 33*.

The first step is to determine the crane's gross capacity in the planned configuration. Refer to the Grove YB4409 load chart in *Figure 34*. Verify that the crane's configuration agrees with the operating notes. Then, locate the column 4-Section Boom with 4th Retracted. Select the sub-column for On-Rubber, and its sub-column, F/R (Front/Rear). Read down the column to the row for the 12-foot load radius. The intersection of these two values indicates that the gross capacity for this crane is 3,400 lbs for over-front or over-rear operation.

Then, identify the required deductions from the gross capacity. Locate the deduction chart in the lower left corner of the load chart. Review the items on the chart and determine if any deductions are required. There is no load block in this configuration. The crane is equipped with a hook and ball, resulting in a deduction of 100 lb. The jib is stowed and the main boom is in use, so no deduction is required there.

Finally, determine the net capacity by subtracting any deductions from the gross capacity. You must deduct the headache ball (100 lb) and the rigging components (40 lb) listed in the configuration (*Figure 33*), for a total deduction of 140 lb (since 100 lb + 40 lb = 140 lb). The calculation is completed as follows:

3,400 lb gross capacity
– 140 lb deductions
3,260 lb net capacity

According to the deduction chart, no deduction is required for the weight of the hoist cable; its weight has been considered in the chart values. Note that a reeved load block is required to achieve the highest capacities for the crane. The operator should check the maximum line pull allowed for a single line. The rigging chart shown in *Figure 34* indicates that the maximum single-line capacity for the specified ½" rope is 8,500 lb, so the net capacity for this operation is well within this limit; a load block reeved with additional parts of line is not required for the 3,400 lb gross capacity the configuration allows.

Figure 32 Grove YB4409 industrial crane.

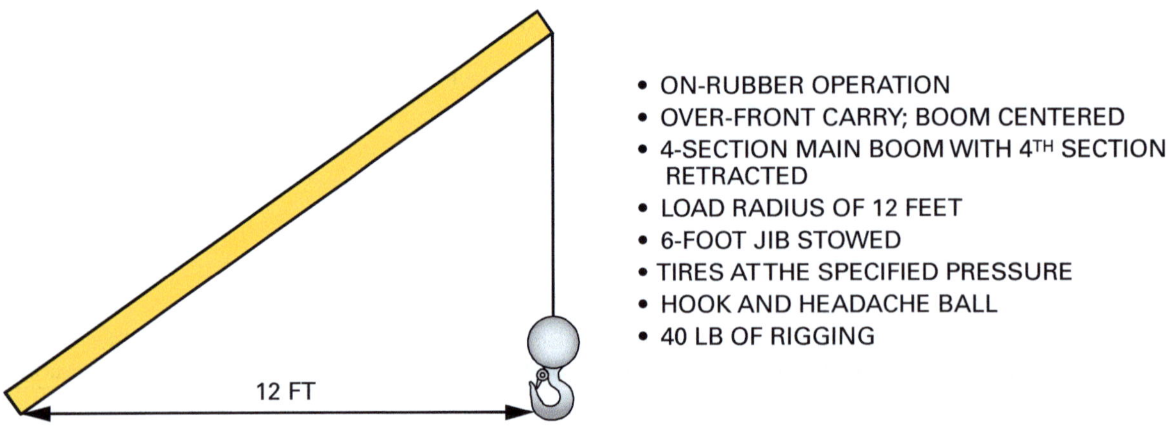

Figure 33 Graphic illustration of YB4409 configuration.

The maximum travel speed allowed during the operation is also found on the load chart. Note 5 in *Figure 34* specifies that loads on rubber may be transported at a maximum speed of 2.5 mph on a smooth surface, with the boom retracted to the shortest possible length and centered over the front.

3.4.2 Example Two: Telescopic-Boom Crane

When lifting from the main boom, find the net capacity of a Grove TM-1500 truck-mounted telescopic-boom crane (*Figure 35*) configured as follows:

- Outriggers properly set, fully extended, and level
- Over-the-rear lift
- 33'–58' extension stowed
- Boom length of 52 feet
- Load radius of 23 feet
- 150-ton, 8-sheave block
- Auxiliary boom head attached
- 10-ton headache ball reeved over auxiliary boom head
- Minimum reeving for the load weight will be used
- 195 lb of rigging and other hoisting accessories

A graphic illustration of the configuration is provided in *Figure 36*.

Again, the first step is to determine the gross capacity. Review the load chart notes shown in *Figure 37*. Per Note 2, no additional weight needs to be deducted for wire rope because the minimum reeving required for the load weight will be used. Also refer to Note 5, which indicates that if a dimension falls between the listed values, the next larger radius and/or boom length should be selected.

Study the load chart in *Figure 38*. Read down the Radius in Feet column to locate 23 ft. Since this specific radius isn't listed, go to the next longer load radius, which is 25 ft. Read across the row to the boom length column for 52 ft. This boom length isn't listed on the chart either, so go to the next longer boom length of 58 ft. The gross capacity where these two values intersect is 131,500 lb.

Now determine which deductions are appropriate for the configuration (refer to *Figure 39*). Include the 195 lb of rigging, which won't be shown on this chart. Adding the applicable deductions results in the following:

1,246	(33'–58' extension stowed)
5,254	(150-ton, 8-sheave block)
261	(auxiliary boom head)
560	(10-ton headache ball)
+ 195	(rigging)
7,516	**lb total deduction**

Finally, determine the net capacity by subtracting the deductions from the gross capacity, as shown here:

131,500	lb gross capacity
– 7,516	lb total deduction
123,984	**lb net capacity**

3.4.3 Example Three: Lattice-Boom Crane

Find the net capacity of the Manitowoc 4100W-2 lattice-boom crawler crane (*Figure 40*) in the following configuration:

- Counterweights totaling 206,400 lb installed, including (2) 30,000-lb carbody counterweights and (2) 12,000-lb side counterweights
- Crawlers extended
- #22C boom, 230' long
- 15-ton hook and ball, suspended 20 feet below the boom point on whip line
- 100-ton block reeved with four parts of line
- Lifting from the main hoist
- 125' load radius
- 345 lb of rigging

A graphic illustration of the basic configuration is shown in *Figure 41*.

Determine the gross capacity by examining the load charts. Start by reading the notes on the first page (*Figure 42A*) to ensure the configuration matches the information on the chart.

There are several points for discussion in the notes. Point 1 lists the deductions for single- and two-sheave upper boom points; they are not a part of this configuration, so no deduction is needed. Point 2 indicates that everything below the boom tip is part of the load weight, including the wire rope. Point 3 states that the crawlers must be fully extended, which matches the planned configuration. Point 4 lists the equipment and counterweight that must be installed for this chart to be valid; all the listed counterweight has been accounted for in the configuration. At Point 5, the load and whip lines are specified, along with the weight per foot; the total weight of the rope below the boom tip must be determined and included in the deductions.

The capacity chart begins with *Figure 42B*. Look for the section that applies to the #22C 230-foot boom; this is on the following page, in *Figure 42C*. Find the needed 125-foot operating radius and read across to the column for Capacity: Crawlers Extended. At 125 feet, the gross capacity is 24,100 lb. Note that the boom tip height is 202.6 feet. This information will be needed to calculate the weight of the hoist line.

RADIUS (ft)	3-SECTION BOOM OR 4-SECTION BOOM WITH 4TH RETRACTED			
	ON OUTRIGGERS (lb)		ON RUBBER (lb)	
	F/R	360°	F/R	360°
5.0	* 17,000	* 17,000	10,000	10,000
6.0	* 15,700	* 14,400	10,000	7450
8.0	* 12,400	10,800	7150	4400
10.0	10,200	8600	4700	2950
12.0	8500	7000	3400	2175
14.0	7200	5400	2650	1700
16.0	6100	4300	2150	1325
18.0	5000	3550	1775	1100
20.0	2700	2700	1525	950
22.0	2600	2600	1350	825
24.5	2200	2200	1510	675
26.0	–	–	–	–
28.0	–	–	–	–
30.0	–	–	–	–
31.0	–	–	–	–

	RADIUS (ft)	4-SECTION BOOM WITH 4TH RETRACTED			
		ON OUTRIGGERS (lb)		ON RUBBER (lb)	
		F/R	360°	F/R	360°
MAIN BOOM	5.0	6100	6100	6100	6100
	6.0	6100	6100	6100	6100
	8.0	5300	5300	5300	5250
	10.0	4700	4700	4700	3750
	12.0	4400	4400	3800	2450
	14.0	4400	4400	2900	1875
	16.0	4400	4400	2325	1525
	18.0	4400	4100	1900	1225
	20.0	4300	3150	1600	1000
	22.0	3700	2700	1375	850
	24.5	3100	2300	1200	700
JIB	26.0	2700	2125	1010	650
	28.0	2500	1900	975	575
	30.0	2300	1700	850	475
	31.0	2200	1625	800	425

MAIN BOOM ANGLE (deg)	JIB STRUCTURAL CAPACITIES (lb)	
	3-SECTION BOOM OR 4-SECTION BOOM WITH 4TH RETRACTED	4-SECTION BOOM WITH 4TH EXTENDED
60	3500	3500
55	3325	3325
50	3150	3150
45	3000	3000
40	2875	2875
35	2800	2800
30	2700	2700
25	2600	2600
20	2500	2500
15	2400	2400
10	2325	2325
5	2250	2250
0	2200	2200

* LIMIT RATINGS TO 11,800 LB WHEN 3RD SECTION IS EXTENDED ANY AMOUNT.

NOTES:
JIB CAPACITY IS LIMITED BY BOTH STRUCTURAL CAPACITY CHART AND MAIN CAPACITY CHART.

SHADED AREAS ARE FOVERNED BY STRUCTURAL STRENGTH, DO NOT RELY ON TIPPING.

OPERATION OF THIS EQUIPMENT IN EXCESS OF RATING CHARTS AND DISREGARD OF INSTRUCTIONS IS DANGEROUS AND VOIDS WARRANTY.

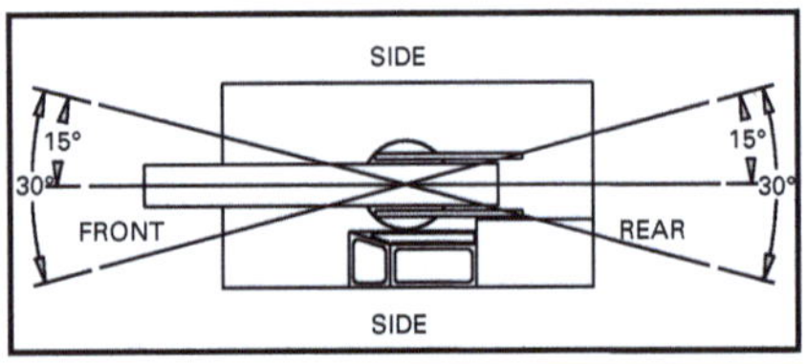

REDUCTION CHART	FROM MAIN BOOM RATINGS	FROM JIB RATINGS
MAIN BLOCK	140 lb	N/A
HOOK & BALL	100 lb	100 lb
JIB, STOWED	0 lb	N/A
JIB, DEPLOYED	150 lb	0 lb

RIGGING CHART	
1-PART 0-8500 lb	2-PART 0-17,000 lb
WIRE ROPE: 1/2 inch diameter Min. breaking strength 29,750 lb	

1) The rated loads are the maximum lift capacities as determined by operating radius, boom extension and boom angle. The operating radius is the horizontal distance from a projection of the axis of rotation to the supporting surface, before loading, to the center of vertical hoist line or tackle with load applied.

2) The rated loads shown on outriggers do not exceed 85% of actual tipping. The rated loads shown on rubber do not exceed 75% of actual tipping. These ratings are based on freely suspended loads with the crane leveled, standing on a firm, uniform supporting surface. Practical working loads depend on supporting surface, operating radius and other factors affecting stability. Hazardous surroundings, climatic conditions, experience of personnel and proper training must all be taken into account by the operator.

3) The weights of all load handling devices such as hooks, hook blocks, slings, etc., except the hoist rope, shall be considered part of the load. See reduction chart.

4) Ratings on outriggers are for either outriggers fully extended and down or fully retracted and down. Ratings for outriggers fully retracted and down will apply for any intermediate outrigger setting.

5) Ratings on rubber depend on tire capacity, condition of tires and proper inflation pressure (100 psi). Loads on rubber may be transported at a maximum seed of 2.5 mph on a smooth, hard, level surface with boom retracted to the shortest length possible and centered over front.

6) For operating radius not shown, use load rating of next larger radius.

7) The maximum combined total boom and deck load is 12,000 lb. The maximum deck load only is 14,000 lb.

8) Do not induce any external side loads to boom or jib.

1301286YB

Figure 34 Grove YB4409 load chart.

Figure 35 Grove TM-1500 crane.

Structural Failure Before Tipping

You might recall from an earlier section that modern cranes are more likely than older cranes to experience structural failure before tipping. The notes section of the load chart shown in *Figure 34* highlights this fact. The related note reads "Shaded areas are governed by structural strength; do not rely on tipping." This means that exceeding the crane's capacity in the shaded areas will likely result in structural failure before any sort of tipping action occurs.

NOTES FOR LIFTING CAPACITIES

WARNING: THIS CHART IS ONLY A GUIDE. The Notes below are for illustration only and should not be relied upon to operate the crane. The individual crane's load chart, operating instructions and other instruction plates must be read and understood prior to operating the crane.

1. All rated loads have been tested to and meet minimum requirements of SAE J1063 OCT80 - Cantilevered Boom Crane Structures - Method of Test, and do not exceed 85% of the tipping load on outriggers (75% of the tipping load on rubber) as determined by SAE J765 OCT 80 Crane Stability Test Code.
2. Capacities given do not include the weight of hookblocks, slings, auxiliary lifting equipment and load handling devices. Their weights MUST be added to the load to be lifted. When more than minimum required reeving is used, the additional rope weight shall be considered part of the load.
3. Capacities appearing above the bold line are based on structural strength and tipping should not be relied upon as a capacity limitation.
4. All capacities are for crane on firm, level surface. It may be necessary to have structural supports under the outrigger floats or tires to spread the load to a larger bearing surface.
5. When either boom length or radius or both are between values listed, the smallest load shown at either the next larger radius or boom length shall be used.
6. Tires shall be inflated to the recommended pressure before lifting on rubber.
7. For outrigger operation, ALL outriggers shall be fully extended with tires raised free of ground before raising the boom or lifting loads.
8. Unless otherwise stated, capacities are with powered boom sections equally extended.
9. With the tele boom extension in working position and main boom length greater than 133 ft. boom angle must not be less than 46.5°, since loss of stability will occur causing a tipping condition.

Figure 37 TM-1500 load chart notes.

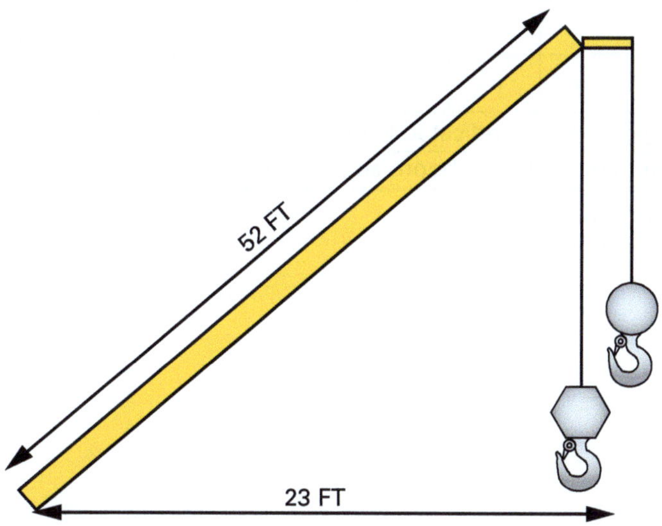

- OUTRIGGERS FULLY EXTENDED
- OVER-THE-REAR LIFT
- 33'–58' EXTENSION STOWED
- BOOM LENGTH: 52 FEET
- LOAD RADIUS: 23 FEET
- 150-TON, 8-SHEAVE BLOCK
- AUXILIARY BOOM HEAD ATTACHED
- 10-TON HEADACHE BALL REEVED OVER AUXILIARY BOOM HEAD
- MINIMUM REEVING FOR THE LOAD WILL BE USED
- 195 LB OF RIGGING

Figure 36 Graphic illustration of the TM-1500 configuration.

Rated Lifting Capacities in Pounds
46 Ft – 173 Ft Boom on Outriggers - Over Rear

Radius in Feet	#0001									#0002
	Main Boom Length in Feet (Power Pinned Fly Retracted)									Power Pin. Fly Ext. & 141 ft.
	46	58	70	82	94	106	118	130	141	173
10	300,000 (74.5)									
12	280,000 (72)	143,500 (76)	142,000 (79)							
15	235,000 (67.5)	143,500 (72.5)	141,500 (76.5)	130,000 (78.5)						
20	173,500 (60.5)	143,500 (67.5)	123,500 (72)	112,000 (75)	102,000 (77.5)	90,300 (79.5)				
25	136,500 (52)	131,500 (61.5)	110,500 (67.5)	98,650 (71)	89,250 (74)	78,550 (76.5)	73,700 (78.5)	69,300 (80)		
30	106,000 (43)	106,000 (55.5)	98,000 (63)	88,350 (67.5)	78,750 (71)	69,250 (73.5)	65,100 (76)	61,000 (77.5)	60,000 (79.5)	
35	84,700 (30.5)	84,700 (49)	84,700 (58)	80,150 (63.5)	69,000 (67.5)	60,750 (70.5)	57,150 (73)	54,000 (75.5)	52,150 (77.5)	
40		70,500 (41)	70,500 (52.5)	70,500 (59.5)	61,300 (64)	54,000 (67.5)	50,600 (70.5)	48,300 (73)	45,850 (75)	38,000 (79)
45		58,850 (32)	58,850 (47)	58,850 (55)	55,000 (60.5)	48,500 (64.5)	46,200 (68)	43,050 (71)	40,400 (73)	35,750 (77)
50		49,600 (17.5)	49,600 (40.5)	49,600 (50.5)	48,750 (57)	43,050 (61.5)	40,700 (65)	38,250 (68.5)	35,750 (71)	32,100 (75.5)
60			36,200 (22.5)	36,200 (39.5)	36,200 (48.5)	34,300 (55)	33,600 (59.5)	30,750 (63.5)	28,500 (66.5)	26,350 (72)
70				26,050 (25)	26,050 (39.5)	26,050 (47.5)	26,050 (53)	24,750 (58)	23,100 (61.5)	22,000 (68.5)
80					18,850 (27)	18,850 (39)	18,850 (46.5)	18,850 (52.5)	18,700 (56.5)	18,500 (64.5)
90						13,500 (28)	13,500 (38.5)	13,500 (46.5)	13,500 (51.5)	15,250 (60.5)
100							9,390 (29)	9,390 (38)	9,390 (45.5)	12,600 (56.5)
110							6,080 (12.5)	6,080 (30.5)	6,080 (39)	10,100 (52)
120								3,390 (17.5)	3,390 (31)	7,530 (47.5)
130									1,150 (19.5)	5,390 (42.5)
140										3,610 (36.5)
150										2,100 (30)
Minimum boom angle (deg.) for indicated length (no load)									10	19
Maximum boom length (ft.) at 0 deg. boom angle (no load)									140	167

Note: (   ) Boom angles are in degrees.
#LMI operating code. Refer to LMI manual for instructions.

Figure 38 Grove TM-1500 load chart.

WEIGHT REDUCTIONS FOR LOAD HANDLING DEVICES

33 ft. Extension	
*Stowed -	892 lbs.
*Erected -	5,704 lbs.

33 ft. - 58 ft. Extension	
*Stowed -	1,246 lbs.
*Erected (Ret.) -	8,412 lbs.
*Erected (Ext.) -	11,406 lbs.

46 ft. - 173 ft. Boom with	
*46 ft. Jib Erected -	9,613 lbs.
*60 ft. Jib Erected -	14,571 lbs.
*74 ft. Jib Erected -	20,443 lbs.
*88 ft. Jib Erected -	27,199 lbs.
*Fixed Jib Accessories -	327 lbs.

*Reduction of main boom capacities

HOOKBLOCKS:	
30 Ton, 1 Sheave	1,022 lbs.
150 Ton, 8 Sheave	5,254 lbs.
Auxiliary Boom Head	261 lbs.
10 Ton Headache Ball	560 lbs.
15 Ton Headache Ball	803 lbs.

Figure 39 Grove TM-1500 weight deduction table.

Now, gather the information needed for deductions. Refer to the fourth page of the load chart (*Figure 42D*). The 15-ton hook and ball requires a deduction of 865 lb. The 100-ton block requires a deduction of 2,065 lb.

The weight of the whip and load lines can be calculated using several pieces of information. Note that the weight per foot is the same for both lines. The whip line hook and ball is suspended 20 feet. With 20 feet of rope and a weight of 2.34 lb/ft, the deduction for the whip line is calculated as follows. (round the result up to the nearest whole number):

Figure 40 Manitowoc 4100W-2 lattice-boom crawler crane.

20.00 ft
× 2.34 lb/ft
47.00 lb

For the load line, it should be assumed that it will need to extend to the ground. The boom tip height at this operating radius is 202.6 feet. Recall that there are four parts of line reeved on the load block, which can make a significant difference in the figures. Calculate the deduction for the load line by first multiplying the weight (in pounds per foot) of the rope by the operating radius, as follows:

202.60 ft
× 2.34 lb/ft
475.00 lb

Multiply the result by the number of parts of line to find the total line deduction:

475 ft
× 4 parts of line
1,900 lb load line deduction

If you are using a calculator, notice that the multiplication results are rounded up to ensure that the end result remains conservative.

There is now enough information to calculate the total deductions from gross capacity, as shown below (be sure to add the weight of the rigging):

865.00 lb (15-ton hook and ball)
2,065.00 lb (100-ton block)
47.00 lb (whip line)
1,900.00 lb (load line)
\+ 345.00 lb (rigging)
5,222.00 lb total deduction

The net capacity can now be calculated by subtracting the deductions from the gross capacity:

24,100.00 lb gross capacity
− 5,222.00 lb total deduction
18,878.00 lb net capacity

3.4.4 Example Four: Boom Truck

Find the net capacity of the National Crane 18103 Series truck-mounted, telescopic-boom truck (*Figure 43*) in the following configuration:

- Fully extended 103-foot main boom, lifting from a 31-foot installed jib
- Auxiliary boom head
- 35-ton, 3-sheave block
- Outriggers fully extended
- Load radius of 68 feet
- 287 lb of rigging

A graphic illustration of this configuration is shown in*Figure 44*. In this case, the 103-foot fully extended main-boom length was taken from the range diagram in *Figure 45*. The total boom length increases to 134 feet when fully extended and the 31-foot jib is added. The appropriate arc is identified on the range diagram.

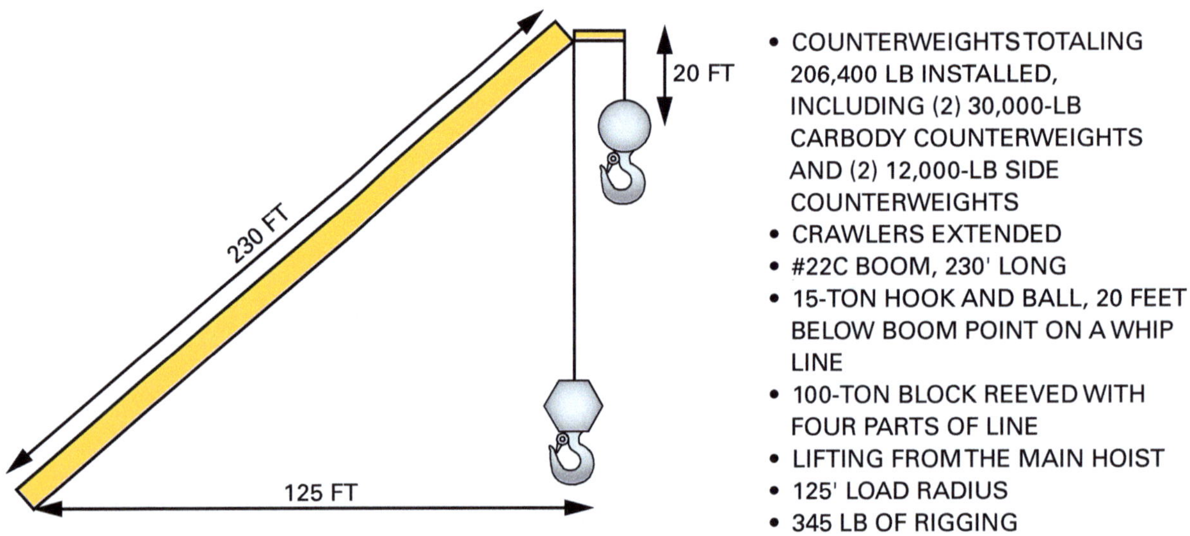

Figure 41 Graphic illustration of the Manitowoc 4100-W2 configuration.

MANITOWOC 4100 W-2

LIFTCRANE CAPACITIES

BOOM NO. 22C WITH OPEN THROAT TOP
146,400 LB. CRANE COUNTERWEIGHT
60,000 LB. CARBODY COUNTERWEIGHT
26'6" CRAWLERS EXTENDED

WARNING: This chart will apply only when two 12,000 lb. side ctwts. and two 30,000 lb. carbody ctwts. bear MEC registered Serial Numbers.

LIFTING CAPACITIES: Capacities for various boom lengths and operating radii may be based on percent of tipping, strength of structural components, operating speeds and other factors.

Capacities are for freely suspended loads and do not exceed 75% of a static tipping load. Capacities based on structural competence are shown by shaded areas.

1 Capacities are shown in pounds. Deduct 1200 pounds from capacities listed when single sheave upper boom point is attached and 1500 pounds when two sheave upper boom point is attached. To comply with B30.5 requirements,
2 upper boompoint cannot be used on the 260 ft. boom. Weight of jib, (see chart A), all load blocks, hooks, weight ball, slings, hoist lines beneath boom and jib point sheaves, etc., is considered part of the main boom load. Boom is not to be lowered beyond radii where combined weights are greater than rated capacity. Where no capacity is shown, operation is not intended or approved.

3 **OPERATING CONDITIONS:** Machine to operate in a level position on a firm surface with crawlers fully extended and gantry in working position and be rigged in accordance with and under conditions referred to in rigging drawing No. 190693 and load line specification chard No. 6592-A.

Crane operator judgement must be used to allow for dynamic load effects of swinging, hoisting or lowering, travel, as well as adverse operating conditions & physical machine depreciation.

OPERATOR RADIUS: Operating is the horizontal distance from the axis of rotation to the center of vertical hoist line or load block with the load freely suspended. Add 14" to boom point radius for radius of sheave when using single part hoist line.

Boom angle is the angle between horizontal and centerline of boom butt and inserts and is an indication of operating radius. In all cases, operating radius shall govern capacity.

BOOM POINT ELEVATION: Boom point elevation, in feet, is the vertical distance from ground level to centerline of boom point shaft.

MACHINE EQUIPMENT: Machine equipped with 26'6" extendible crawlers, 48" treads, 17' retractable gantry, 12 part boom hoist reeving, four 1 3/8" boom pendants, 1st ctwt. 41,900 lbs., 2nd ctwt. 41,500 lbs., 3rd ctwt. 39,000 lbs., two 12,000 lbs. side ctwt's. and two 30,000 lbs. carbody ctwt's. 4

LOAD AND WHIP LINE SPECIFICATIONS 5	
LOAD LINE:	1-1/8" - 6 x 31 Warrington-Seale, Extra Improved Plow Steel, Regular Lay, IWRC. Minimum Breaking Strength 65 Ton. (Approx.Weight Per Ft. in Lbs. 2.34)
WHIP LINE:	1-1/8" - Warrington-Seale, Improved Plow Steel, Regular Lay, IWRC. Minimum Breaking Strength 56.5 Ton. Maximum Load - 28,300 Lbs. Per Line. (Approx. Weight Per Ft. in Lbs. 2.34)

HOIST REEVING FOR MAIN LOAD BLOCK						
No.Parts Of Line	1	2	3	4	5	6
Max Load - Lbs.	32,500	65,000	97,500	130,000	162,500	195,000
No. Parts of Line	7	8	9	10	11	12
Max. Load - Lbs.	227,500	260,000	292,500	325,000	357,500	400,000
No. Parts of Line	13					
Max. Load - Lbs.	430,000					

MAXIMUM BOOM AND JIB LENGTH LIFTED UNASSISTED				DEDUCT FROM CAPACITIES WHEN JIB IS ATTACHED	
OVER FRONT OF BLOCKED CRAWLERS		OVER SIDE OF EXTENDED CRAWLERS			
BOOM LENGTH	JIB NO. 123	BOOM LENGTH	JIB NO. 123	JIB LENGTH	JIB NO. 123
260'	--	260'	--	30'	3,000 lbs.
250'	--	250'	--	40'	3,600 lbs.
240'	40'	240'	40'	50'	4,200 lbs.
230'	60'	230'	60'	60'	4,900 lbs.
Load block, hook and weight ball on ground to start.					

FOR JIB CAPACITIES, CONSULT JIB CHART.

BOOM LGTH FEET	OPER. RAD. FEET	BOOM ANG. DEG.	BOOM POINT ELEV.	CAPACITY: CRAWLERS EXTENDED
70	16.5	79.7	75.9	460,000
	17	79.3	75.8	400,000
	18	78.5	75.6	380,100
	19	77.6	75.4	363,000
	20	76.8	75.1	347,300
	22	75.1	74.6	319,600
	24	73.4	74.1	293,400
	26	71.7	73.5	266,100
	28	69.9	72.8	237,500
	30	68.2	72.0	214,300
	32	66.4	71.2	195,100
	34	64.6	70.2	178,900
	36	62.8	69.3	165,200
	38	60.9	68.2	153,300
	40	59.1	67.0	143,000
	45	54.1	63.7	122,100
	50	48.9	59.8	106,300
	55	43.2	54.9	93,900
	60	36.9	49.0	84,000
	65	29.4	41.3	75,800
	70	19.5	30.3	63,900

BOOM LGTH FEET	OPER. RAD. FEET	BOOM ANG. DEG.	BOOM POINT ELEV.	CAPACITY: CRAWLERS EXTENDED
80	17	80.6	85.9	392,800
	18	79.9	85.8	378,900
	19	79.2	85.6	361,800
	20	78.5	85.4	346,100
	22	77.0	84.9	318,400
	24	75.5	84.5	292,500
	26	74.0	83.9	265,600
	28	72.5	83.3	237,000
	30	71.0	82.7	213,800
	32	69.5	81.9	194,600
	34	68.0	81.2	178,500
	36	66.4	80.3	164,700
	38	64.8	79.4	152,800
	40	63.3	78.4	142,400
	45	59.2	75.7	121,500
	50	54.9	72.5	105,700
	55	50.4	68.6	93,300
	60	45.6	64.1	83,400
	65	40.3	58.8	75,200
	70	34.4	52.2	68,300
	75	27.4	43.9	52,500
	80	18.2	32.0	53,900

BOOM LGTH FEET	OPER. RAD. FEET	BOOM ANG. DEG.	BOOM POINT ELEV.	CAPACITY: CRAWLERS EXTENDED
90	18	81.1	95.9	355,400
	19	80.4	95.7	346,900
	20	79.8	95.6	336,900
	22	78.5	95.2	317,400
	24	77.2	94.7	291,700
	26	75.9	94.3	264,800
	28	74.5	93.7	236,600
	30	73.2	93.2	213,400
	32	71.9	92.5	194,200
	34	70.5	91.9	178,000
	36	69.2	91.1	164,200
	38	67.8	90.3	152,300
	40	66.4	89.5	142,000
	45	62.9	87.1	121,100
	50	59.3	84.4	105,200
	55	55.5	81.2	92,800
	60	51.5	77.5	82,900
	65	47.3	73.2	74,700
	70	42.8	68.2	67,800
	75	37.9	62.3	62,000
	80	32.4	55.2	57,000
	85	25.8	46.2	52,600
	90	17.1	33.5	45,900

BOOM LGTH FEET	OPER. RAD. FEET	BOOM ANG. DEG.	BOOM POINT ELEV.	CAPACITY: CRAWLERS EXTENDED
100	19	81.4	105.9	332,900
	20	80.8	105.7	327,100
	22	79.6	105.4	316,200
	24	78.5	105.0	290,800
	26	77.3	104.5	263,900
	28	76.1	104.1	236,200
	30	74.9	103.6	212,900
	32	73.7	103.0	193,700
	34	72.5	102.4	177,500
	36	71.3	101.7	163,700
	38	70.1	101.0	151,800
	40	68.9	100.3	141,400
	45	65.8	98.2	120,500
	50	62.6	95.8	104,700
	55	59.3	93.0	92,300
	60	55.9	89.8	82,300
	65	52.4	86.2	74,100
	70	48.7	82.1	67,200
	75	44.8	77.4	61,400
	80	40.5	72.0	56,400
	85	35.9	65.6	52,000
	90	30.7	58.0	48,200
	95	24.5	48.5	44,900
	100	16.3	35.0	39,300

CAUTION! CHECK AMOUNT OF COUNTERWEIGHT ON MACHINE BEFORE USE OF THIS CHART

Figure 42A Manitowoc 4100 W-2 load chart (1 of 4).

LIFTCRANE CAPACITIES – 4100W SERIES 2 BOOM NO 22C WITH OPEN-THROAT TOP, CONTINUED

BOOM LGTH FEET	OPER. RAD. FEET	BOOM ANG. DEG.	BOOM POINT ELEV.	CAPACITY: CRAWLERS EXTENDED
110	22	82.3	115.7	240,500
	24	81.3	115.4	232,100
	26	80.2	115.0	224,400
	28	79.1	114.6	217,300
	30	78.1	114.1	210,700
	32	77.0	113.6	193,300
	34	75.9	113.1	177,100
	36	74.8	112.5	163,300
	38	73.7	111.8	151,400
	40	72.6	111.2	141,000
	45	69.9	109.3	120,100
	50	67.0	107.2	104,200
	55	64.1	104.7	91,800
	60	61.2	101.9	81,800
	65	58.1	98.8	73,600
	70	54.9	95.3	66,800
	75	51.6	91.3	60,900
	80	48.1	86.8	55,900
	85	44.4	81.7	51,600
	90	40.4	75.9	47,800
	95	36.1	69.2	44,400
	100	31.1	61.1	41,400
	105	25.3	51.1	38,700
	110	17.6	37.2	33,900
120	22	83.0	125.9	228,700
	24	82.0	125.5	220,500
	26	81.0	125.2	213,000
	28	80.0	124.8	206,100
	30	79.1	124.4	199,700
	32	78.1	123.9	192,800
	34	77.1	123.4	176,600
	36	76.1	122.9	162,800
	38	75.1	122.3	150,900
	40	74.1	121.7	140,500
	45	71.6	120.0	119,500
	50	69.1	118.1	103,700
	55	66.5	115.9	91,300
	60	63.8	113.4	81,300
	65	61.1	110.6	73,100
	70	58.3	107.5	66,200
	75	55.4	104.0	60,400
	80	52.4	100.1	55,300
	85	49.2	95.8	51,000
	90	45.9	91.0	47,200
	95	42.4	85.5	43,800
	100	38.6	79.4	40,800
	105	34.4	72.2	38,100
	110	29.7	63.7	35,700
	115	24.1	53.2	33,500
130	24	82.6	135.7	212,700
	26	81.7	135.3	205,300
	28	80.8	135.0	198,600
	30	79.9	134.6	192,300
	32	79.0	134.2	186,500
	34	78.1	133.7	176,300
	36	77.2	133.2	162,500
	38	76.3	132.7	150,600
	40	75.4	132.1	140,200
	45	73.1	130.6	119,300
	50	70.8	128.8	103,400
	55	68.4	126.8	91,000
	60	66.0	124.6	81,000
	65	63.5	122.0	72,800
	70	61.0	119.2	65,900
	75	58.4	116.1	60,000
	80	55.7	112.7	55,000
	85	53.0	108.9	50,700
	90	50.1	104.8	46,800
	95	47.1	100.1	43,500
	100	43.9	95.0	40,500
	105	40.6	89.2	37,800
	110	36.9	82.6	35,300
	115	33.0	75.1	33,100
	120	28.5	66.1	31,100
	125	23.1	55.1	29,300
140	26	82.3	145.5	198,400
	28	81.5	145.1	191,800
	30	80.7	144.8	185,600
	32	79.8	144.4	179,900
	34	79.0	144.0	174,700
	36	78.1	143.5	162,000
	38	77.3	143.0	150,100
	40	76.5	142.5	139,700
	45	74.3	141.1	118,700
	50	72.2	139.4	102,800
	55	70.0	137.6	90,400
	60	67.8	135.5	80,400
	65	65.6	133.2	72,200
	70	63.3	130.7	65,300
	75	60.9	127.9	59,500
	80	58.5	124.8	54,400
	85	56.1	121.4	50,100
	90	53.5	117.7	46,200
	95	50.9	113.7	42,900
	100	48.1	109.2	39,900
140	105	45.3	104.2	37,200
	110	42.2	98.8	34,700
	115	39.0	92.7	32,500
	120	35.5	85.7	30,500
	125	31.7	77.8	28,700
	130	27.4	68.5	27,000
	135	22.2	57.0	25,500
150	28	82.1	155.3	185,600
	30	81.3	154.9	179,600
	32	80.5	154.6	174,000
	34	79.7	154.2	168,800
	36	79.0	153.8	161,600
	38	78.2	153.3	149,600
	40	77.4	152.8	139,200
	45	75.4	151.5	118,200
	50	73.4	150.0	102,400
	55	71.4	148.3	89,900
	60	69.4	146.4	79,900
	65	67.3	144.3	71,700
	70	65.2	141.9	64,800
	75	63.0	139.4	59,000
	80	60.9	136.6	53,900
	85	58.6	133.5	49,600
	90	56.3	130.1	45,700
	95	54.0	126.5	42,400
	100	51.5	122.5	39,400
	105	49.0	118.2	36,700
	110	46.4	113.4	34,200
	115	43.6	108.2	32,000
	120	40.7	102.4	30,000
	125	37.6	96.0	28,200
	130	34.2	88.8	26,500
	135	30.5	80.5	25,000
	140	26.4	70.8	23,500
	145	21.4	58.8	22,200
160	28	82.6	165.4	179,800
	30	81.8	165.1	173,900
	32	81.1	164.7	168,400
	34	80.4	164.4	163,300
	36	79.7	164.0	158,600
	38	78.9	163.6	149,100
	40	78.2	163.1	138,700
	45	76.3	161.9	117,700
	50	74.5	160.5	101,800
	55	72.6	158.9	89,300
	60	70.7	157.1	79,300
	65	68.8	155.1	71,100
	70	66.8	153.0	64,200
	75	64.9	150.6	58,400
	80	62.9	148.0	53,300
	85	60.8	145.2	49,000
	90	58.7	142.2	45,100
	95	56.6	138.8	41,800
	100	54.4	135.2	38,700
	105	52.1	131.4	36,000
	110	49.7	127.1	33,600
	115	47.3	122.5	31,400
	120	44.8	117.5	29,400
	125	42.1	112.0	27,600
	130	39.3	105.9	25,900
	135	36.3	99.2	24,300
	140	33.1	91.7	22,900
	145	29.5	83.0	21,600
	150	25.5	72.9	20,400
	155	20.7	60.5	19,200
170	30	82.3	175.2	169,000
	32	81.6	174.9	163,600
	34	81.0	174.5	158,700
	36	80.3	174.2	154,000
	38	79.6	173.8	148,800
	40	78.9	173.3	138,400
	45	77.2	172.2	117,400
	50	75.4	170.9	101,500
	55	73.7	169.4	89,100
	60	71.9	167.7	79,000
	65	70.1	165.9	70,800
	70	68.3	163.9	63,900
	75	66.4	161.7	58,100
	80	64.6	159.3	53,000
	85	62.7	156.7	48,700
	90	60.8	153.9	44,900
	95	58.8	150.8	41,500
	100	56.8	147.5	38,500
	105	54.7	144.0	35,800
	110	52.6	140.2	33,300
	115	50.4	136.0	31,100
	120	48.1	131.5	29,100
	125	45.8	126.7	27,300
	130	43.3	121.4	25,600
	135	40.8	115.7	24,100
	140	38.1	109.3	22,600
	145	35.2	102.3	21,300
	150	32.0	94.5	20,100
	155	28.6	85.5	18,900
	160	24.7	75.1	17,800
	165	20.0	62.2	16,400
180	32	82.1	185.0	158,700
	34	81.5	184.7	154,000
	36	80.8	184.3	149,500
	38	80.2	184.0	145,200
	40	79.5	183.6	137,900
	45	77.9	182.5	116,900
	50	76.3	181.2	101,000
	55	74.6	179.8	88,500
	60	72.9	178.3	78,500
	65	71.3	176.6	70,300
	70	69.6	174.7	63,400
	75	67.8	172.6	57,500
	80	66.1	170.4	52,500
	85	64.3	168.0	48,100
	90	62.5	165.4	44,300
	95	60.7	162.5	40,900
	100	58.9	159.5	37,900
	105	57.0	156.2	35,200
	110	55.0	152.7	32,700
	115	53.0	149.0	30,500
	120	51.0	144.9	28,500
	125	48.9	140.5	26,700
	130	46.7	135.8	25,000
	135	44.4	130.7	23,500
	140	42.0	125.2	22,000
	145	39.6	119.2	20,700
	150	36.9	112.6	19,500
	155	34.1	105.3	18,300
	160	31.1	97.2	17,200
	165	27.7	87.9	16,200
	170	23.9	77.1	15,300
	175	19.4	63.8	13,500
190	32	82.5	195.1	153,100
	34	81.9	194.8	148,800
	36	81.3	194.5	144,800
	38	80.7	194.1	140,800
	40	80.1	193.8	137,000
	45	78.5	192.7	116,400
	50	77.0	191.5	100,500
	55	75.4	190.2	88,000
	60	73.9	188.8	78,000
	65	72.3	187.1	69,700
	70	70.7	185.4	62,800
	75	69.1	183.4	57,000
	80	67.4	181.4	52,000
	85	65.8	179.1	47,600
	90	64.1	176.6	43,800
	95	62.4	174.0	40,400
	100	60.7	171.2	37,400
	105	58.9	168.2	34,700
	110	57.1	164.9	32,200
	115	55.3	161.4	30,000
	120	53.4	157.7	28,000
	125	51.5	153.7	26,200
	130	49.5	149.5	24,500
	135	47.5	144.9	22,900
	140	45.3	140.0	21,500
	145	43.1	134.6	20,200
	150	40.8	128.9	18,900
	155	38.4	122.6	17,800
	160	35.9	115.8	16,700
	165	33.2	108.3	15,700
	170	30.2	99.8	14,800
	175	27.0	90.3	13,900
	180	23.3	79.1	13,100
	185	18.9	65.4	11,000
200	34	82.3	204.9	143,100
	36	81.7	204.6	139,500
	38	81.2	204.3	136,000
	40	80.6	203.9	132,500
	45	79.1	203.0	115,900
	50	77.7	201.8	100,000
	55	76.2	200.6	87,500
	60	74.7	199.2	77,400
	65	73.2	197.7	69,200
	70	71.7	196.0	62,300
	75	70.2	194.2	56,400
	80	68.6	192.2	51,400
	85	67.1	190.1	47,000
	90	65.5	187.8	43,200
	95	63.9	185.3	39,800
	100	62.3	182.7	36,800
	105	60.6	179.8	34,100
	110	59.0	176.8	31,600
	115	57.3	173.6	29,400
	120	55.5	170.2	27,400
	125	53.0	166.5	25,600
	130	51.9	162.6	23,900
	135	50.1	158.4	22,300
	140	48.1	153.9	20,900
	145	46.2	149.1	19,600
	150	44.1	144.0	18,300
	155	42.0	138.4	17,200
	160	39.7	132.5	16,100
	165	37.4	126.0	15,100
	170	34.9	118.9	14,200
200	175	32.3	111.1	13,300
	180	29.4	102.4	12,400
	185	26.2	92.5	11,700
	190	22.6	81.0	10,900
	195	18.4	67.0	8,500
210	36	82.1	214.7	134,200
	38	81.6	214.4	131,000
	40	81.0	214.1	127,900
	45	79.6	213.2	115,500
	50	78.3	212.1	99,600
	55	76.9	210.9	87,100
	60	75.4	209.6	77,100
	65	74.0	208.1	68,800
	70	72.6	206.6	61,900
	75	71.2	204.8	56,100
	80	69.7	203.0	51,000
	85	68.2	201.0	46,700
	90	66.7	198.8	42,800
	95	65.2	196.5	39,400
	100	63.7	194.0	36,400
	105	62.2	191.3	33,700
	110	60.6	188.5	31,300
	115	59.0	185.5	29,100
	120	57.4	182.3	27,100
	125	55.7	178.9	25,200
	130	54.1	175.2	23,500
	135	52.3	171.4	22,000
	140	50.6	167.3	20,600
	145	48.8	162.9	19,200
	150	46.9	158.2	18,000
	155	45.0	153.2	16,800
	160	43.0	147.9	15,800
	165	40.9	142.1	14,800
	170	38.7	135.9	13,800
	175	36.5	129.2	12,900
	180	34.0	121.9	12,100
	185	31.5	113.9	11,300
	190	28.7	104.9	10,600
	195	25.6	94.7	9,900
	200	22.1	82.9	8,800
	205	17.9	68.5	6,500
220	38	82.0	224.6	125,700
	40	81.4	224.2	122,800
	45	80.1	223.3	115,000
	50	78.8	222.3	99,100
	55	77.5	221.2	86,600
	60	76.1	220.0	76,600
	65	74.8	218.6	68,300
	70	73.4	217.1	61,400
	75	72.0	215.4	55,500
	80	70.7	213.7	50,500
	85	69.3	211.8	46,100
	90	67.9	209.7	42,300
	95	66.4	207.5	38,900
	100	65.0	205.2	35,800
	105	63.6	202.7	33,100
	110	62.1	200.0	30,700
	115	60.6	197.2	28,500
	120	59.1	194.2	26,500
	125	57.5	191.0	24,600
	130	55.9	187.6	23,000
	135	54.3	184.0	21,400
	140	52.7	180.2	20,000
	145	51.0	176.1	18,600
	150	49.3	171.8	17,400
	155	47.6	167.3	16,300
	160	45.7	162.4	15,200
	165	43.9	157.2	14,200
	170	41.9	151.7	13,200
	175	39.9	145.7	12,300
	180	37.8	139.3	11,500
	185	35.6	132.4	10,700
	190	33.2	124.8	10,000
	195	30.7	116.6	9,300
	200	28.0	107.3	8,600
	205	25.0	96.9	8,000
	210	21.5	84.8	6,500

CAPACITIES CONTINUED ON NEXT PAGE

These load charts are intended for instructional purposes only. They were derived from manufacturer sales information which may not be complete or machine specific. Not responsible for typographical errors.

Figure 42B Manitowoc 4100 W-2 load chart (2 of 4).

LIFTCRANE CAPACITIES – 4100W SERIES 2 BOOM NO 22C
WITH OPEN-THROAT TOP, CONTINUED

SEE CONDITIONS ON FRONT PAGE

BOOM LGTH FEET	OPER. RAD. FEET	BOOM ANG. DEG.	BOOM POINT ELEV.	CAPACITY: CRAWLERS EXTENDED
230	38	81.5	234.5	132,900
	40	81.0	234.2	125,200
	45	79.7	233.3	114,500
	50	78.5	232.4	98,600
	55	77.2	231.3	86,100
	60	75.9	230.1	76,100
	65	74.6	228.8	67,800
	70	73.3	227.3	60,900
	75	72.0	225.8	55,000
	80	70.7	224.1	50,000
	85	69.4	222.3	45,600
	90	68.0	220.3	41,700
	95	66.7	218.2	38,400
	100	65.3	216.0	35,300
	105	64.0	213.6	32,600
	110	62.6	211.1	30,200
	115	61.1	208.4	28,000
	120	59.7	205.6	26,000
	125	58.3	202.6	24,100
	130	56.8	199.4	22,400
	135	55.3	196.0	20,900
	140	53.8	192.5	19,400
	145	52.2	188.7	18,100
	150	50.6	184.7	16,900
	155	49.0	180.5	15,700
	160	47.3	176.0	14,700
	165	45.6	171.3	13,600
	170	43.8	166.2	12,700
	175	42.0	160.8	11,800
	180	40.1	155.1	11,000
	185	38.1	148.9	10,200
	190	36.0	142.3	9,400
	195	33.9	135.1	8,700
	200	31.6	127.3	8,100
	205	29.1	118.8	7,400
	210	26.4	109.3	6,800
240	40	81.4	244.3	123,400
	45	80.2	243.5	112,600
	50	78.9	242.6	98,100
	55	77.7	241.5	85,600
	60	76.5	240.4	75,600
	65	75.3	239.1	67,300
	70	74.0	237.7	60,400
	75	72.8	236.3	54,500
	80	71.5	234.6	49,500
	85	70.3	232.9	45,100
	90	69.0	231.1	41,200
	95	67.7	229.1	37,800
	100	66.4	227.0	34,800
	105	65.1	224.7	32,100
	110	63.8	222.3	29,700
	115	62.5	219.8	27,400
	120	61.1	217.1	25,400
	125	59.7	214.3	23,600
	130	58.3	211.3	21,900
	135	56.9	208.1	20,300
	140	55.5	204.7	18,900
	145	54.0	201.2	17,600
	150	52.5	197.5	16,300
	155	51.0	193.5	15,200
	160	49.5	189.4	14,100
	165	47.9	185.0	13,100
	170	46.2	180.3	12,200
	175	44.6	175.4	11,300
	180	42.8	170.2	10,400
	185	41.0	164.6	9,600
	190	39.2	158.7	8,900
	195	37.3	152.3	8,200
	200	35.2	145.5	7,500
	205	33.1	138.1	6,700
	210	30.9	130.1	6,000

BOOM LGTH FEET	OPER. RAD. FEET	BOOM ANG. DEG.	BOOM POINT ELEV.	CAPACITY: CRAWLERS EXTENDED
250	45	80.6	253.6	106,600
	50	79.4	252.7	97,800
	55	78.2	251.7	85,200
	60	77.1	250.6	75,200
	65	75.9	249.4	66,900
	70	74.7	248.1	60,000
	75	73.5	246.7	54,100
	80	72.3	245.2	49,100
	85	71.1	243.5	44,700
	90	69.9	241.7	40,800
	95	68.7	239.8	37,400
	100	67.4	237.8	34,400
	105	66.2	235.7	31,700
	110	64.9	233.4	29,300
	115	63.6	231.0	27,100
	120	62.4	228.5	25,000
	125	61.1	225.8	23,200
	130	59.7	222.9	21,500
	135	58.4	219.9	20,000
	140	57.0	216.8	18,500
	145	55.7	213.4	17,200
	150	54.3	209.9	16,000
	155	52.8	206.2	14,800
	160	51.4	202.4	13,700
	165	49.9	198.3	12,700
	170	48.4	193.9	11,800
	175	46.8	189.4	10,900
	180	45.3	184.5	10,100
	185	43.6	179.4	9,300
	190	41.9	174.0	8,500
	195	40.2	168.3	7,800
	200	38.4	162.2	7,000
	205	36.5	155.7	6,200
	210	34.5	148.6	5,500
260	45	80.9	263.7	104,800
	50	79.8	262.9	95,700
	55	78.7	261.9	84,700
	60	77.6	260.9	74,700
	65	76.4	259.7	66,400
	70	75.3	258.5	59,500
	75	74.2	257.1	53,600
	80	73.0	255.6	48,500
	85	71.8	254.1	44,100
	90	70.7	252.4	40,300
	95	69.5	250.6	36,900
	100	68.3	248.6	33,900
	105	67.1	246.6	31,100
	110	65.9	244.4	28,700
	115	64.7	242.1	26,500
	120	63.5	239.7	24,500
	125	62.3	237.1	22,600
	130	61.0	234.4	20,900
	135	59.7	231.6	19,400
	140	58.5	228.6	17,900
	145	57.2	225.4	16,600
	150	55.8	222.1	15,400
	155	54.5	218.7	14,200
	160	53.1	215.0	13,100
	165	51.7	211.2	12,100
	170	50.3	207.1	11,200
	175	48.9	202.9	10,300
	180	47.4	198.4	9,500
	185	45.9	193.7	8,700
	190	44.3	188.7	7,900
	195	42.7	183.4	7,000
	200	41.1	177.8	6,200
	205	39.4	171.9	5,400
	210	37.6	165.6	4,700

Figure 42C Manitowoc 4100 W-2 load chart (3 of 4).

DESCRIPTION	APPROX. WEIGHT (IN LBS.)
JIB NO. 123	
Jib Top - 15' (w/Jib point)	695
Jib Butt - 15'	690
Jib Insert - 10'	340
Basic Pendant - 33' 3-3/4" (2 Req'd)	115 each
Pendant - 10' (2 Per insert)	65 each
Jib Backstay Pendant	155 each
Jib Strut - 12'- 6"	365
COMPONENTS	
Hook Rollers (6) - w/shafts	1,020
Light Plant - 6.5KW - w/mounting platform	1,390
Catwalk - left and right side w/rails	1,320
Lagging - 27-5/8" dia. plain	1,410
Boom Hoist Rope - 12 Part - 760' of 7/8" - 6 x 26	1,080
Wire Rope Guide Assembly - Lower	325
Wire Rope Guide Assembly - Upper	510
Rope Guide Roller Assembly	55 each
2-Part Gantry w/Telescopic Backhitch	7,805
Equalizer	2,000
Hoist Line - 1-1/8" - 6 x 31	2.34 lbs./ft.
Whip Line - 1-1/8" - 6 x 31	2.34 lbs./ft.
15-Ton Hook and Weight Ball	865
100-Ton Hook Block Assembly	2,065
200-Ton Hook Block Assembly	4,900
230-Ton Hook Block Assembly	5,375
Boom Stop - Telescopic Air Cushioned	675
Dragline Fairlead - Revolving	1,910
Dragline Fairlead - Hinged	9,330

NOTE: The above weights may fluctuate up or down 5% due to manufacturing tolerances.

Figure 42D Manitowoc 4100 W-2 load chart (4 of 4).

Figure 43 NC 1800 Series boom truck.

Always begin by determining the gross capacity. The best practice is to review the load chart notes first. The notes in *Figure 45* indicate that any accessories attached to the boom or load line must be deducted from the load chart capacities. The notes also indicate that when operating with the jib, the operator must refer to the jib-rated load sections of the chart. That section of the load chart has been highlighted.

The chart doesn't have an entry for a 68-foot load radius, so select the next higher radius of 70 feet. The gross capacity shown for a 70-foot load radius is 4,100 lb.

Next, identify all the applicable deductions. The deduction table shown in *Figure 46* indicates a deduction of 100 lb for the auxiliary boom head, and 870 lb for the 35-ton load block on the crane.

The required deductions are added together, as shown here (again, include the weight of the rigging):

100.00 lb (auxiliary boom head)
870.00 lb (3-sheave block)
\+ 287.00 lb (rigging)
1,257.00 lb total deduction

The final step of determining the net capacity can now be completed:

4,100.00 lb gross capacity
− 1,257.00 lb total deduction
2,843.00 lb net capacity

These have been basic examples of different crane types and configurations. Operators will certainly encounter more complex scenarios. While it is not unusual to find load charts a little confusing at first, remember that most crane operators aren't switching between drastically different cranes on a daily basis. In many cases, an operator works with the same crane, or a similar one, for days and perhaps weeks and months.

As an operator, you will become familiar with your crane's documentation and will soon be able to navigate the charts quickly and accurately. You may even begin to memorize load chart information and gross capacities. However, it is never a good idea to rely on your memory when it comes to load charts. Always take the time to review the load chart carefully, since errors can be catastrophic.

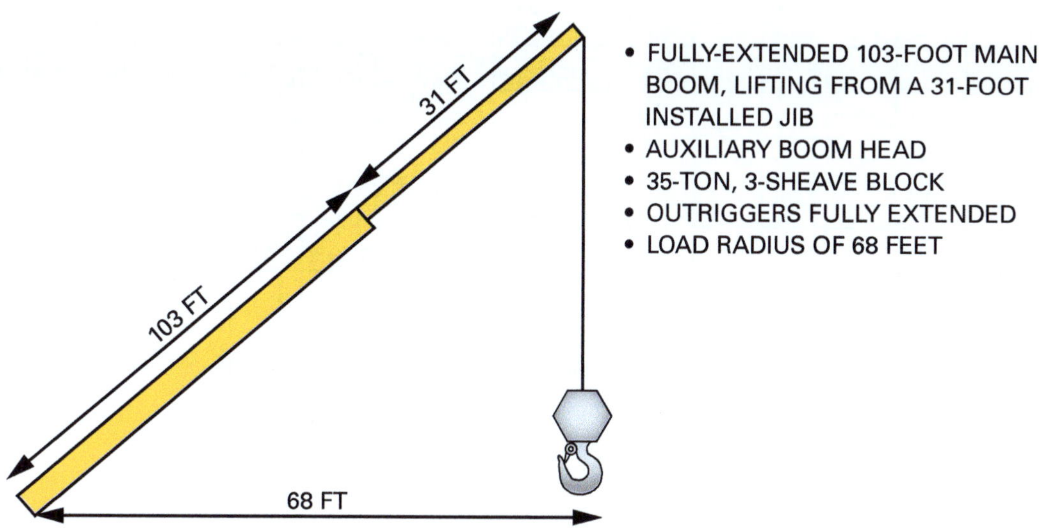

Figure 44 Graphic illustration of the National 18103 boom truck configuration.

Series 18103: 31,39 m boom with 9,45 m-16,76 m (31 ft - 55 ft) jib/full span outrigger 7,6 m (24.7 ft)

National Crane will send you a chart on request – or you may secure needed load rating information through your nearest National Crane dealer.

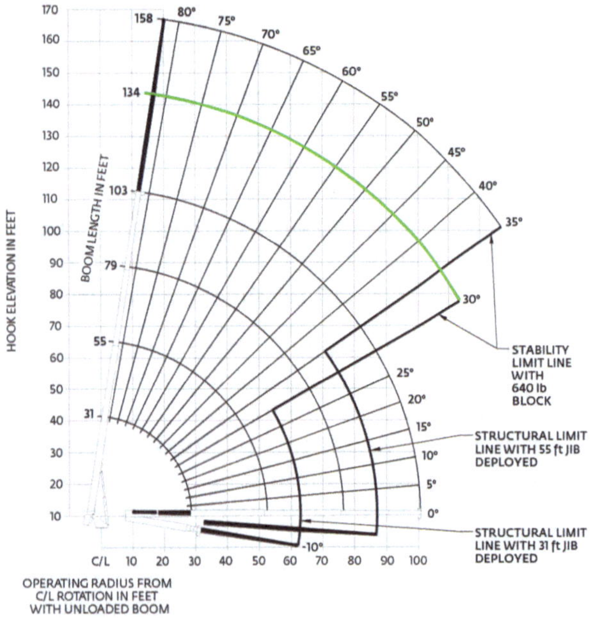

CAUTION:

- **Do not operate crane booms, jib extensions, any accessories or loads within 3 m (10 ft) of live power lines or other conductors of electricity.**
- Jib and boom capacities shown are maximum for each section.
- Do not exceed capacities at reduced radii.
- Load ratings shown on the load rating charts are maximum allowable loads with the outriggers properly extended on a firm, level surface and the crane leveled and mounted on a factory recommended truck.
- Always level the crane with the level indicator located on the crane.
- The operator must reduce load to allow for factors such as wind, ground conditions, operating speeds and their effects on freely suspended loads.
- Overloading this crane may cause structural collapse or instability.
- Weights on any accessories attached to the boom or loadline must be deducted from the load chart capacities.
- Do not exceed jib capabilities at any reduced boom lengths.
- Do not deadhead lineblock against boom tip when extending boom or winching up.
- Keep at least three wraps of loadline on drum at all times.
- Use only specified cable with this machine.

Load chart

31 ft - 103 ft BOOM RATED LOADS WITHOUT JIB

LOAD RADIUS (ft)	LOADED BOOM ANGLE	31 FT BOOM (lb)	LOADED BOOM ANGLE	55 FT BOOM (lb)	LOADED BOOM ANGLE	79 FT BOOM (lb)	LOADED BOOM ANGLE	103 FT BOOM (lb)
7	73.9	80,000						
8	71.9	74,000						
10	67.7	65,000	78.9	50,000				
12	63.5	57,000	76.6	45,000				
15	56.7	44,000	73.3	38,000	79.6	30,000		
20	44.1	30,800	67.7	31,500	75.9	26,000	79.5	17,000
25	27.4	23,200	61.7	23,800	72.1	22,000	76.7	15,200
30			55.3	18,800	68.1	18,500	73.8	13,500
35			48.3	15,200	64	15,500	70.9	12,000
40			40.5	12,500	59.6	12,800	67.8	10,500
45			31.2	10,500	55.1	10,700	65	9300
50			19.3	9000	50.7	9000	61.8	8300
55					45.5	7600	58.5	7400
60					39.9	6600	55.1	6500
65					33.4	5600	51.4	5600
70					25.5	4800	47.5	4800
75					13.4	4050	43.4	4100
80							38.9	3500
85							33.8	2950
90							28	2450
95							20.7	2050
100							7.9	1650
	0	19,700	0	8200	0	3800	0	1600

31 ft JIB RATED LOADS

RADIUS FULLY EXTENDED	LOADED BOOM ANGLE	RATED LOADS ALL BOOM LENGTHS
25	80	8800
38	75	8000
49	70	6500
60	65	5100
70	60	4100
79	55	3300
88	50	2600
96	45	1900
103	40	1350
110	35	950
115	30	650

55 ft JIB RATED LOADS

RADIUS FULLY EXTENDED	LOADED BOOM ANGLE	RATED LOADS ALL BOOM LENGTHS
29	80	4000
45	75	3700
59	70	3300
73	65	3000
85	60	2600
96	55	2100
106	50	1700
115	45	1300
123	40	950
130	35	650

NOTE:

1. Operate with jib by radius when main boom is fully extended. If necessary increase boom angle to maintain loaded radius.
2. Operate with jib by boom angle when main boom is not fully extended. Do not exceed rated jib capacities at any reduced boom lengths.
3. Capacities do not exceed 85% stability.
4. Shaded areas are structurally limited capacities.

NOTE:

1. All capacities are in pounds, angles in degrees, radius in feet.
2. Loaded boom angles are given as reference only.
3. Shaded areas are structurally limited capacities.

RATED LOAD REDUCTIONS WITH JIB

BOOM LENGTH	31 ft-55 ft JIB STOWED	31 ft-55 ft JIB ERECTED AT 31 ft LENGTH
31 ft	Reduce load 800 lb	Reduce load 2300 lb
55 ft	Reduce load 450 lb	Reduce load 2000 lb
79 ft	Reduce load 350 lb	Reduce load 1900 lb
103 ft	Reduce load 250 lb	Reduce load 1800 lb

Figure 45 NC 1800 Series load chart.

	Loadline deduct	
	Aux boom head	45 kg (100 lb)
5 USt	Downhaul weight	82 kg (180 lb)
15 USt	1-sheave block	170 kg (375 lb)
25 USt	2-sheave block	290 kg (640 lb)
35 USt	3-sheave block	395 kg (870 lb)
40 USt	4-sheave block	440 kg (970 lb)

Figure 46 NC 1800 Series boom truck deductions.

Additional Resources

ASME Standard B30.5, Mobile and Locomotive Cranes. Current edition. New York, NY: American Society of Mechanical Engineers.

Cranes: Design, Practice, and Maintenance, Ing J. Verschoof. 2002. Hoboken, NJ: John Wiley and Sons, Inc.

29 *CFR* 1926, Subpart CC, *Cranes and Derricks in Construction.* **www.ecfr.gov**

3.0.0 Section Review

1. Load chart capacity deductions are usually required for all of the following *except* _____.
   a. a jib extension when hoisting on the main boom
   b. the main boom as built by the manufacturer
   c. multi-sheave blocks and headache balls
   d. an auxiliary boom head

2. Rotation-resistant wire rope is made by _____.
   a. using fiber-core wire rope
   b. using synthetic rope for the intermediate strands
   c. applying lubricants after each strand is applied
   d. applying strand layers in opposing directions

3. At what point in a lift operation is the best time to consult the crane's range diagram?
   a. Just prior to commencing the lift
   b. At the start of lift planning before any capacity determinations are done
   c. Only if a problem arises during the lift that requires adjusting the crane configuration
   d. Not required if the crane has already been erected

4. Refer to the YB4409 load chart shown in *Figure 34*. Which of the following is a *true* statement?
   a. When the jib is extended, the maximum load radius is 26 feet, regardless of load weight.
   b. At a 16-foot load radius on rubber, the crane has more gross capacity when the fourth section of the main boom is extended.
   c. Deploying the jib results in a gross capacity deduction of 250 lb.
   d. Exchanging the hook and ball for a load block results in an additional 140 lb deduction from gross capacity.

SUMMARY

Crane operators must understand how to read and interpret load charts for every crane they operate. The range diagram section of load charts illustrates the allowable operating radius, boom-tip heights, and boom angles possible with the crane in any allowed configuration. The load charts are the first thing operators should reference when planning a lift. The notes and specifications in the charts provide a wealth of information related to crane capacity and operation. The capacity tables of load charts provide the gross capacity for a given crane configuration.

Manufacturers establish the gross capacities listed on the charts under ideal lifting conditions and well within structural and tipping limits. The purpose of the charts is to provide the operator with reliable and authoritative guidance to operate the crane safely. Although operators will become thoroughly familiar with the charts of a crane after a period of use, they should never rely on memory with so much at stake. A crane operator must study and use the load charts for every new situation to ensure that every lift is within the capabilities of the equipment.

Review Questions

1. The point where you can assume an object's weight to be located is the _____.
   a. moment
   b. center of gravity
   c. tipping axis
   d. gross weight
2. If a crane's load chart becomes lost or illegible, _____.
   a. only the manufacturer can replace it
   b. the company engineer can recreate it
   c. an OSHA-qualified person may reproduce it
   d. the company must sell the associated crane for scrap
3. Which of the following items is no longer required to be included in crane load charts?
   a. Range chart showing allowed boom angles and load radius
   b. Work area chart
   c. Load limitations based on strength of materials, hydraulics, stability, etc.
   d. Cautions, warnings, and notes relating to load limitations
4. The quantity obtained from crane gross capacity minus weight deductions is _____.
   a. crane leverage
   b. crane net capacity
   c. load leverage
   d. net load
5. The boom of industrial/all-purpose cranes can usually rotate _____.
   a. 360 degrees
   b. 180 degrees
   c. 90 degrees
   d. 12 degrees (6 degrees to each side of center)
6. When operating a crane by the load charts (rather than an LMI), always _____.
   a. round the results of calculations down
   b. interpolate the chart data as needed
   c. enter the chart at the next highest load radius or next lowest boom angle
   d. operate the crane at less than 75 percent of the net capacity
7. An operational quadrant is defined _____.
   a. by the boom angle above the horizontal
   b. by the tipping axis for a given configuration and the direction of the crane boom
   c. by the front, side, and rear faces of the crane when on rubber or crawlers
   d. only when outriggers are in use
8. Load chart capacities for on-outrigger lifting do not exceed _____.
   a. 65 percent of the tipping loads
   b. 75 percent of the tipping loads
   c. 85 percent of the tipping loads
   d. 95 percent of the tipping loads
9. The offset of a jib and boom refers to _____.
   a. the angle formed by an erected jib in relation to the center line of the main boom
   b. how far off-center the jib is mounted on the end of the main boom
   c. the length of a boom extension between the main boom and the jib-boom itself
   d. the distance between the jib-boom tip and the end of the main boom
10. What is the industry standard for the ratio of a crane's hoist-line minimum breaking strength to its maximum single-line pull?
    a. 1:5
    b. 1:1
    c. 2:1
    d. 5:1
11. What information regarding crane configuration would you *not* find on a range diagram?
    a. Permitted crane boom configurations and angles
    b. Crane lift capacities for various boom angles and configurations
    c. Load radiuses for allowed boom configurations
    d. Telescopic boom-tip heights and boom angles at partial extensions for a given load radius

12. Referring to the Terex RT555-1 load charts in *Appendix A*, find the net capacity of the crane configured as follows:

    - On outriggers, fully extended
    - Lift over side
    - 50-foot main boom length
    - 22-foot load radius
    - 4-sheave block on main boom (690 lb)
    - Counterweights installed (included)
    - Hook and ball (239 lb)
    - Rigging and accessories (65 lb)
    - Lifting from the main boom

    a. 42,626 lb
    b. 41,876 lb
    c. 40,856 lb
    d. 40,106 lb

13. Referring to the load charts for the Manitowoc 4100W-2 in *Appendix B*, find the *gross capacity* for the following configuration:

    - All required counterweights installed, totaling 206,400 lb
    - #22C open-throat top boom
    - 180-foot main boom
    - Boom angle is 60 degrees
    - Crawlers fully extended

    a. 37,900 lb
    b. 40,900 lb
    c. 45,300 lb
    d. 52,100 lb

14. Referring to the load charts for the Manitowoc 4100W-2 in *Appendix B*, find the *net capacity* for the following configuration:

    - All required counterweights installed, totaling 206,400 lb
    - Crawlers fully extended
    - #22C open-throat top boom; 190-ft main boom
    - Lifting from the main boom hoist
    - 130-ft load radius
    - 2-sheave boom point attached
    - 15-ton hook block (on whip line), 20 ft below boom point
    - 100-ton block reeved with 4 parts of line
    - 345 lb of rigging

    a. 17,219 lb
    b. 18,209 lb
    c. 18,278 lb
    d. 19,029 lb

15. Referring to the load charts for the Grove TM-1500 truck-mounted telescopic-boom crane in *Appendix C*, find the net capacity for the following configuration:

    - Outriggers properly set, fully extended, and level
    - Over-the-side lift
    - 33-foot fixed extension erected at 15-degree offset
    - Boom length of 173 feet
    - Load radius of 68 feet
    - 30-ton, 1-sheave block
    - Auxiliary boom head attached
    - 10-ton headache ball reeved over auxiliary boom head
    - 167 lb of rigging and other hoisting accessories

    a. 10,460 lb
    b. 10,660 lb
    c. 10,840 lb
    d. 10,940 lb

Trade Terms Introduced in This Module

Boom angle: In load charts, the angle between the boom relative to the horizontal plane. On telescopic-boom cranes, boom angle instruments indicate the angle between the bottom of the base section and the horizontal with the boom loaded. On lattice-boom cranes, it is the angle between the center line of the boom and a horizontal line.

Boom length: Measured from the boom foot pins to the center of the boom-point sheaves. It does not include the length of an attached jib.

Boom-point elevation: Distance from the ground to the center of the boom-point sheaves.

Configuration: A term that addresses all aspects of a crane's condition for a given lift that can include boom assembly and position, hoisting line and accessories, base arrangement, counterweights, and other factors addressed in a load chart.

Effective weight: With reference to jib/boom extensions, the weight calculated by the manufacturer which, when applied at the boom tip, has the same effect on the crane capacity as the jib and/or boom extension itself.

Gross capacity: A crane capacity value as displayed in the crane's load chart for a specified configuration.

Gross load: The total weight of all the items lifted by the crane. The gross capacity of the crane minus any deductions for the crane configuration results in the gross load that can be lifted. Gross load is directly related to net capacity.

Interpolate: To estimate or calculate an unknown intermediate value that falls between two known values.

Offset: The angle formed by an erected boom, measured from an imaginary line extending through the center line of the main boom. The crane's load chart provides capacities for allowable jib or boom-extension offset angles.

Line pull: The maximum pulling force a crane hoist drum can apply to a hoist cable, accounted for by the torque that the motor can apply to the drum and the working radius of the rope as it leaves the drum.

Load operating radius: The horizontal distance from a vertical line through the axis of crane rotation to the center of the vertical hoist line or tackle, with a load applied. Also called load radius, lifting radius, or operating radius.

Net capacity: The crane's hoisting force available after taking into account the crane's and/or load's configurations relating to a lift. Equal to load chart gross capacity minus weight deductions.

Ringer crane: A large, high-capacity crane erected on a circular track (the ring, usually installed directly on the ground) that supports the entire crane base.

Tipping axis: In crane operations, an imaginary horizontal line around which a crane's center of gravity would rotate as it tips.

Terex RT555-1 Load Chart

Lifting Capacities – Pounds (33' – 110' boom)

MODEL RT555-1

COUNTERWEIGHT:
W/AUX. WINCH 13,100 LBS.
W/O AUX. WINCH 14,200 LBS.
BOOM LENGTH 33-110 FT.
OUTRIGGER SPREAD 22 FT.

STABILITY PCT.
ON OUTRIGGERS 85%
ON TIRES 75%
PCSA CLASS 10-210

CAUTION: Do not use this specification sheet as a load rating chart. The format of data is not consistent with the machine chart and may be subject to change.

ON OUTRIGGERS - FULLY EXTENDED

LOAD RADIUS (FT)	BOOM LENGTH 35 FT			BOOM LENGTH 50 FT			BOOM LENGTH 65 FT			LOAD RADIUS (FT)
	BOOM ANGLE (DEG)	LOADED OVER FRONT (LB)	360° (LB)	BOOM ANGLE (DEG)	LOADED OVER FRONT (LB)	360° (LB)	BOOM ANGLE (DEG)	LOADED OVER FRONT (LB)	360° (LB)	
10	66.7	110,000*	110,000*	73.9	60,100*	60,100*				10
12	63.1	96,700*	93,700*	71.5	60,100*	60,100*				12
15	57.5	75,200*	73,100*	67.9	60,100*	60,100*	73.2	58,800*	58,800*	15
20	47.1	53,600*	52,300*	61.5	54,900*	53,600*	68.5	52,200*	52,200*	20
25	34.5	40,700*	39,700*	54.8	42,000*	41,100*	63.7	42,700*	41,700*	25
30	14.8	31,900*	31,200*	47.4	33,400*	32,700*	58.6	34,100*	33,400*	30
35	**			39.0	27,300*	26,700*	53.3	28,000*	27,400*	35
40				28.8	22,000	21,000	47.6	22,700	21,700	40
45				12.4	17,400	16,500	41.3	18,300	17,400	45
50				**			34.1	14,900	14,200	50
55							25.2	12,300	11,700	55
60							10.9	10,100	9,600	60
65							**			65
70										70
75										75
80										80
85										85
90										90
95										95
100										100
105										105
110										110

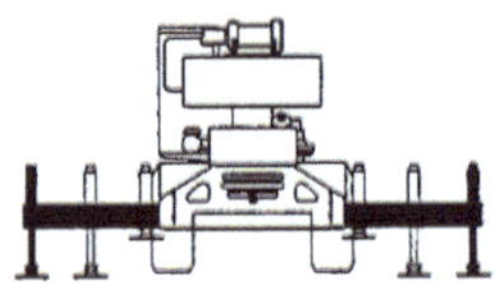

USE THESE CHARTS ONLY WHEN ALL OUTRIGGERS ARE FULLY EXTENDED

ON OUTRIGGERS - FULLY EXTENDED

LOAD RADIUS (FT)	BOOM LENGTH 80 FT			BOOM LENGTH 95 FT			BOOM LENGTH 110 FT			LOAD RADIUS (FT)
	LOADED BOOM ANGLE (DEG)	OVER REAR (LB)	360° (LB)	LOADED BOOM ANGLE (DEG)	OVER REAR (LB)	360° (LB)	LOADED BOOM ANGLE (DEG)	OVER REAR (LB)	360° (LB)	
10										10
12										12
15										15
20	72.7	38,700*	38,700*							20
25	68.9	33,600*	33,600*	72.3	29,300*	29,300*				25
30	65.0	29,600*	29,600*	69.1	25,900*	25,900*	72.1	22,900*	22,900*	30
35	61.0	26,500*	26,500*	65.9	23,000*	23,000*	69.3	20,500*	20,500*	35
40	56.8	23,000	22,000	62.5	20,800*	20,800*	66.5	18,400*	18,400*	40
45	52.4	18,600	17,700	59.1	18,800	17,900	63.6	16,500*	16,500*	45
50	47.7	15,300	14,600	55.5	15,500	14,800	60.7	14,900*	14,900	50
55	42.7	12,700	12,100	51.7	12,900	12,300	57.7	13,000	12,400	55
60	37.1	10,700	10,100	47.8	10,900	10,400	54.5	11,000	10,500	60
65	30.6	9,000	8,500	43.6	9,200	8,800	51.3	9,400	8,900	65
70	22 6	7,500	7,100	39.0	7,900	7,400	47.8	8,000	7,600	70
75	9.8	6,300	5,900	33.9	6,700	6,300	44.2	6,800	6,500	75
80	**			28.1	5,700	5,300	40.4	5,900	5,500	80
85				20.8	4,800	4,400	36.1	5,000	4,700	85
90				9.0	3,900	3,600	31.5	4,200	3,900	90
95				**			26.5	3,500	3,200	95
100							19.3	2,900	2,400	100
105							8.4	2,300	2,100	105
110							**			110

**** MAXIMUM CAPACITY AT 0 DEGREE BOOM ANGLE**

BOOM LENGTH 35 FT			BOOM LENGTH 50 FT			BOOM LENGTH 65 FT			BOOM LENGTH 80 FT			BOOM LENGTH 95 FT			BOOM LENGTH 110 FT		
LOAD RADIUS (FT)	OVER FRONT (LB)	360° (LB)	LOAD RADIUS (FT)	OVER FRONT (LB)	360° (LB)	LOAD RADIUS (FT)	OVER FRONT (LB)	360° (LB)	LOAD RADIUS (FT)	OVER FRONT (LB)	360° (LB)	LOAD RADIUS (FT)	OVER FRONT (LB)	360° (LB)	LOAD RADIUS (FT)	OVER FRONT (LB)	360° (LB)
31.2	20,900*	20,800*	46.2	12,600*	12,700*	61.2	8,200*	8,200*	76.2	5,400*	5,400*	91.2	3,500*	3,300	106.17	2,100	1,800

GENERAL NOTES

GENERAL

1. Rated loads as shown on Lift Charts pertain to this machine as originally manufactured and equipped. Modifications to the machine or use of optional equipment other than that specified can result in a reduction of capacity.
2. Construction equipment can be hazardous if improperly operated or maintained. Operation and maintenance of this machine shall be in compliance with the information in the Operator's, Parts and Safety Manuals supplied with this machine. If these manuals are missing, order replacements from the manufacturer through your distributor.
3. These warnings do not constitute all of the operating conditions for the crane. The operator and job site supervision must read the OPERATORS MANUAL, CIMA SAFETY MANUAL, APPLICABLE OSHA REGULATIONS, AND SOCIETY OF MECHANICAL ENGINEERS (ASME) SAFETY STANDARDS FOR CRANES.
4. This crane and its load ratings are in accordance with POWER CRANE & SHOVEL ASSOCIATION, STANDARD NO. 4, SAE CRANE LOAD STABILITY TEST CODE J765A, SAE METHOD OF TEST FOR CRANE STRUCTURE J1063 AND APPLICABLE SAFETY CODE FOR CRANES, DERRICKS AND HOISTS, ASME/ANSI B30.5.

DEFINITIONS

1. LOAD RADIUS – The horizontal distance from the axis of rotation before loading to the center of the vertical hoist line or tackle with a load applied.
2. LOADED BOOM ANGLE – It is the angle between the boom base section and the horizontal, after lifting the rated load at the rated radius. The boom angle before loading should be greater to account for deflections. The loaded boom angle combined with boom length give only an approximation of the operating radius.
3. WORKING AREA – Areas measured in a circular arc about the centerline of rotation as shown in the diagram.
4. FREELY SUSPENDED LOAD – Load hanging free with no direct external force applied except by the hoist rope.
5. SIDE LOAD – Horizontal force applied to the lifted load either on the ground or in the air.
6. NO LOAD STABILITY LIMIT – The stability limit radius shown on the range diagrams is the radius beyond which it is not permitted to position the boom, when the boom angle is less than the minimum shown on the applicable load chart, because the machine can overturn without any load.
7. BOOM SIDE OF CRANE – The side of the crane over which the boom is positioned when in an OVER SIDE working position.

SET–UP

1. Crane load ratings are based on the crane being leveled and standing on a firm, uniform supporting surface.
2. Crane load ratings on outriggers are based on all outrigger beams being fully extended or in the case of partial extension ratings mechanically pinned in the appropriate position, and the tires free of the supporting surface.
3. Crane load ratings on tires depend on appropriate inflation pressure and the tire conditions. Caution must be exercised when increasing air pressures in tires. Consult Operator's Manual for precautions.
4. Use of jibs, lattice–type boom extensions, or fourth section pullouts extended is not permitted for pick and carry operations.
5. Consult appropriate section of the Operator's and Service Manual for more exact description of hoist line reeving.
6. The use of more parts of line than required by the load may result in having insufficient rope to allow the hook block to reach the ground.
7. Properly maintained wire rope is essential for safe crane operation. Consult Operator's Manual for proper maintenance and inspection requirements.
8. When spin-resistant wire rope is used, the allowable rope loading shall be the breaking strength divided by five (5), unless otherwise specified by the wire rope manufacturer.
9. Do not elevate the boom above 60° unless the boom is positioned in-line with the crane's chassis or the outriggers are extended. Failure to observe this warning may result in loss of stability.

OPERATION

1. CRANE LOAD RATINGS MUST NOT BE EXCEEDED. DO NOT ATTEMPT TO TIP THE CRANE TO DETERMINE ALLOWABLE LOADS.
2. When either radius or boom length, or both, are between listed values, the smaller of the two listed load ratings shall be used.
3. Do not operate at longer radii than those listed on the applicable load rating chart (cross hatched areas shown on range diagrams).
4. The boom angles shown on the Capacity Chart give an approximation of the operating radius for a specified boom length. The boom angle, before loading, should be greater to account for boom deflection. It may be necessary to retract the boom if maximum boom angle is insufficient to maintain rated radius.
5. Power telescoping boom sections must be extended equally.
6. Rated loads include the weight of hook block, slings, and auxiliary lifting devices. Their weights shall be subtracted from the listed rated load to obtain the net load that can be lifted.

   When lifting over the jib the weight of any hook block, slings, and auxiliary lifting devices at the boom head must be added to the load.

   When jibs are erected but unused add two (2) times the weight of any hook block, slings, and auxiliary lifting devices at the jib head to the load.
7. Rated loads do not exceed 85% on outriggers or 75% on tires, of the tipping load as determined by SAE Crane Stability Test Code J765a. Structural strength ratings in chart are indicated with an asterisk (*).
8. Rated loads are based on freely suspended loads. No attempt shall be made to drag a load horizontally on the ground in any direction.
9. The user shall operate at reduced ratings to allow for adverse job conditions, such as: Soft or uneven ground, out of level conditions, high winds, side loads, pendulum action, jerking or sudden stopping of loads, hazardous conditions, experience of personnel, two machine lifts, traveling with loads, electric wires, etc., (side pull on boom or jib is hazardous). Derating of the cranes lifting capacity is required when wind speed exceeds 20 MPH, the center of the lifted load must never be allowed to move more than 3* feet off the center line of the base boom section due to the effects of wind, inertia, or any combination of the two.

   *"Use 2 feet off the center line of the base boom for a two section boom, 3 feet for a three section boom, or 4 feet for a four section boom."
10. The maximum load which can be telescoped is not definable, because of variations in loadings and crane maintenance, but it is permissible to attempt retraction and extension if load ratings are not exceeded.
11. Load ratings are dependent upon the crane being maintained according to manufacturer's specifications.
12. It is recommended that load handling devices, including hooks, and hook blocks, be kept away from boom head at all times.
13. FOR TRUCK CRANES ONLY: 360° capacities apply only to machines equipped with a front outrigger jack and all five (5) outrigger jacks properly set. If the front (5th) outrigger jack is not properly set, the work area is restricted to the over side and over rear areas as shown on the Crane Working Positions diagram. Use the 360° load ratings in the overside work areas.
14. Do not lift with outrigger beams positioned between the fully extended and intermediate (pinned) positions.
15. Truck Cranes not equipped with equalizing (bogie) beams between the rear axles may not be used for lifting "on tires". Truck Cranes equipped with equalizing beams and rear air suspension should "dump" the air before lifting "on tires".

CLAMSHELL, MAGNET, AND CONCRETE BUCKET SERVICE

1. Maximum boom length for clamshell and magnet service is 50 feet.
2. Weight of clamshell or magnet, plus contents are not to exceed 6,000 pounds or 90% of rated lifting capacities, whichever is less. For concrete bucket operation, weight of bucket and load must not exceed 90% of rated lifting capacity.

WE RESERVE THE RIGHT TO AMEND THESE SPECIFICATIONS AT ANY TIME WITHOUT NOTICE. THE ONLY WARRANTY APPLICABLE IS OUR STANDARD WRITTEN WARRANTY APPLICABLE TO THE PARTICULAR PRODUCT AND SALE. WE MAKE NO OTHER WARRANTY, EXPRESSED OR IMPLIED.

TEREX

TEREX Cranes
106 12th Street S.E.
Waverly, IA 50677-9466 USA
TEL: +1 (319) 352-3920
FAX: +1 (319) 352-5727
EMAIL: inquire@terexwaverly.com
WEB: terex.com

Manitowoc 4100W-2 Load Chart

WEIGHTS	4100W-S1 4100W-S2
DESCRIPTION	APPROX. WEIGHT (IN LBS.)
LIFTCRANE- w/70' No. 22C boom, machine counterweight, universal gantry w/telescopic backhitch, 26'-6" crawlers w/48" treads, full width tandem drums, ind. swing, ind. boom hoist, Cummins NTA-855-C360 engine, 27-5/8" dia. lagging for rear drum, boom hoist rope, equalizer, telescopic air cushioned boom stop, 200 ton load block, 15 ton hook and weight ball, single sheave upper boom point, and upper wire rope guide	S1 363,795* S2 446,295*
UPPERWORKS- w/Cummins NTA-855-C360 engine, ind. boom hoist, ind. swing, and 27-5/8" dia. lagging for rear drum; LESS boom, gantry and backhitch, equalizer, load block, weight ball, counter weights, telescopic air cushioned boom stop, upper wire rope guide, and catwalk	80,500*
UPPERWORKS as above - w/gantry and backhitch, equalizer, boom hoist rope, and carbody; LESS crawlers	138,685*
CARBODY-w/roller path, ring gear, and king pin; LESS crawlers	49,700
CRAWLERS, 26'-6" w/48" treads	37,965 each
COUNTERWEIGHT	
Inner (Self-Removing)	41,900
Middle (Self-Removing)	41,500
Outer (Self-Removing)	39,000
Side (2 Req'd)	12,000 each
Carbody (2 Req'd)	30,000 each
BOOM No. 22C	
Boom Butt - 30'	6,150
Boom Top - 40' (w/Lower boom point assembly)	8,445
Upper Boom Point (Removable - single sheave)	1,260
double sheave	1,505
Jib Adapter (Removable)	545
Boom Insert - 10' (w/rope guide roller assembly)	1,350
Boom Insert - 20' (w/rope guide roller assembly)	2,435
Boom Insert - 40' (w/rope guide roller assembly)	4,460
Boom Insert - 40' (w/jib backstay, and rope guide roller assembly)	4,560
Basic Pendant - 40' 9-3/4" (4 Req'd)	255 each
Pendant - 10' (4 per insert)	115 each
Pendant - 20' (4 per insert)	155 each
Pendant - 40' (4 per insert)	215 each
Pendant Spreader Bar	320

*Weights do not include hoist line, whip line, or fuel. For CAT. D-343TA add 1,170 lbs., for CAT. 3406 PCTA add 100 lbs. and for CM 12 V-71N engines add 600 lbs.

DESCRIPTION	APPROX. WEIGHT (IN LBS.)
JIB NO. 123	
Jib Top - 15' (w/Jib point)	695
Jib Butt - 15'	690
Jib Insert - 10'	340
Basic Pendant - 33' 3-3/4" (2 Req'd)	115 each
Pendant - 10' (2 Per insert)	65 each
Jib Backstay Pendant	155 each
Jib Strut - 12'-6"	365
COMPONENTS	
Hook Rollers (6) - w/shafts	1,020
Light Plant - 6.5KW - w/mounting platform	1,390
Catwalk - left and right side w/rails	1,320
Lagging - 27-5/8" dia. plain	1,410
Boom Hoist Rope - 12 Part - 760' of 7/8" - 6 x 26	1,080
Wire Rope Guide Assembly - Lower	325
Wire Rope Guide Assembly - Upper	510
Rope Guide Roller Assembly	55 each
2-Part Gantry w/Telescopic Backhitch	7,805
Equalizer	2,000
Hoist Line - 1-1/8" - 6 x 31	2.34 lbs./ft.
Whip Line - 1-1/8" - 6 x 31	2.34 lbs./ft.
15-Ton Hook and Weight Ball	865
100-Ton Hook Block Assembly	2,065
200-Ton Hook Block Assembly	4,900
230-Ton Hook Block Assembly	5,375
Boom Stop - Telescopic Air Cushioned	675
Dragline Fairlead - Revolving	1,910
Dragline Fairlead - Hinged	9,330

NOTE: The above weights may fluctuate up or down 5% due to manufacturing tolerances.

MANITOWOC 4100 W-2

LIFTCRANE CAPACITIES

BOOM NO. 22C WITH OPEN THROAT TOP
146,400 LB. CRANE COUNTERWEIGHT
60,000 LB. CARBODY COUNTERWEIGHT
26'6" CRAWLERS EXTENDED

WARNING: This chart will apply only when two 12,000 lb. side ctwts. and two 30,000 lb. carbody ctwts. bear MEC registered Serial Numbers.

LIFTING CAPACITIES: Capacities for various boom lengths and operating radii may be based on percent of tipping, strength of structural components, operating speeds and other factors.

Capacities are for freely suspended loads and do not exceed 75% of a static tipping load. Capacities based on structural competence are shown by shaded areas.

Capacities are shown in pounds. Deduct 1200 pounds from capacities listed when single sheave upper boom point is attached and 1500 pounds when two sheave upper boom point is attached. To comply with B30.5 requirements, upper boompoint cannot be used on the 260 ft. boom. Weight of jib, (see chart A), all load blocks, hooks, weight ball, slings, hoist lines beneath boom and jib point sheaves, etc., is considered part of the main boom load. Boom is not to be lowered beyond radii where combined weights are greater than rated capacity. Where no capacity is shown, operation is not intended or approved.

OPERATING CONDITIONS: Machine to operate in a level position on a firm surface with crawlers fully extended and gantry in working position and be rigged in accordance with and under conditions referred to in rigging drawing No. 190693 and load line specification chard No. 6592-A.

HOIST REEVING FOR MAIN LOAD BLOCK						
No.Parts Of Line	1	2	3	4	5	6
Max Load - Lbs.	32,500	65,000	97,500	130,000	162,500	195,000
No. Parts of Line	7	8	9	10	11	12
Max. Load - Lbs.	227,500	260,000	292,500	325,000	357,500	400,000
No. Parts of Line	13					
Max. Load - Lbs.	430,000					

Crane operator judgement must be used to allow for dynamic load effects of swinging, hoisting or lowering, travel, as well as adverse operating conditions & physical machine depreciation.

OPERATOR RADIUS: Operating is the horizontal distance from the axis of rotation to the center of vertical hoist line or load block with the load freely suspended. Add 14" to boom point radius for radius of sheave when using single part hoist line.

Boom angle is the angle between horizontal and centerline of boom butt and inserts and is an indication of operating radius. In all cases, operating radius shall govern capacity.

BOOM POINT ELEVATION: Boom point elevation, in feet, is the vertical distance from ground level to centerline of boom point shaft.

MACHINE EQUIPMENT: Machine equipped with 26'6" extendible crawlers, 48" treads, 17' retractable gantry, 12 part boom hoist reeving, four 1 3/8" boom pendants, 1st ctwt. 41,900 lbs., 2nd ctwt. 41,500 lbs., 3rd ctwt. 39,000 lbs., two 12,000 lbs. side ctwt's. and two 30,000 lbs. carbody ctwt's.

LOAD AND WHIP LINE SPECIFICATIONS	
LOAD LINE:	1-1/8" - 6 x 31 Warrington-Seale, Extra Improved Plow Steel, Regular Lay, IWRC. Minimum Breaking Strength 65 Ton. (Approx.Weight Per Ft. in Lbs. 2.34)
WHIP LINE:	1-1/8" - Warrington-Seale, Improved Plow Steel, Regular Lay, IWRC. Minimum Breaking Strength 56.5 Ton. Maximum Load - 28,300 Lbs. Per Line. (Approx. Weight Per Ft. in Lbs. 2.34)

MAXIMUM BOOM AND JIB LENGTH LIFTED UNASSISTED				DEDUCT FROM CAPACITIES WHEN JIB IS ATTACHED	
OVER FRONT OF BLOCKED CRAWLERS		OVER SIDE OF EXTENDED CRAWLERS			
BOOM LENGTH	JIB NO. 123	BOOM LENGTH	JIB NO. 123	JIB LENGTH	JIB NO. 123
260'	--	260'	--	30'	3,000 lbs.
250'	--	250'	--	40'	3,600 lbs.
240'	40'	240'	40'	50'	4,200 lbs.
230'	60'	230'	60'	60'	4,900 lbs.
Load block, hook and weight ball on ground to start.					

FOR JIB CAPACITIES, CONSULT JIB CHART.

BOOM LGTH FEET	OPER. RAD. FEET	BOOM ANG. DEG.	BOOM POINT ELEV.	CAPACITY: CRAWLERS EXTENDED
70	16.5	79.7	75.9	460,000
	17	79.3	75.8	400,000
	18	78.5	75.6	380,100
	19	77.6	75.4	363,000
	20	76.8	75.1	347,300
	22	75.1	74.6	319,600
	24	73.4	74.1	293,400
	26	71.7	73.5	266,100
	28	69.9	72.8	237,500
	30	68.2	72.0	214,300
	32	66.4	71.2	195,100
	34	64.6	70.2	178,900
	36	62.8	69.3	165,200
	38	60.9	68.2	153,300
	40	59.1	67.0	143,000
	45	54.1	63.7	122,100
	50	48.9	59.8	106,300
	55	43.2	54.9	93,900
	60	36.9	49.0	84,000
	65	29.4	41.3	75,800
	70	19.5	30.3	63,900

BOOM LGTH FEET	OPER. RAD. FEET	BOOM ANG. DEG.	BOOM POINT ELEV.	CAPACITY: CRAWLERS EXTENDED
80	17	80.6	85.9	392,800
	18	79.9	85.8	378,900
	19	79.2	85.6	361,800
	20	78.5	85.4	346,100
	22	77.0	84.9	318,400
	24	75.5	84.5	292,500
	26	74.0	83.9	265,600
	28	72.5	83.3	237,000
	30	71.0	82.7	213,800
	32	69.5	81.9	194,600
	34	68.0	81.2	178,500
	36	66.4	80.3	164,700
	38	64.8	79.4	152,800
	40	63.3	78.4	142,400
	45	59.2	75.7	121,500
	50	54.9	72.5	105,700
	55	50.4	68.6	93,300
	60	45.6	64.1	83,400
	65	40.3	58.8	75,200
	70	34.4	52.2	68,300
	75	27.4	43.9	52,500
	80	18.2	32.0	53,900

BOOM LGTH FEET	OPER. RAD. FEET	BOOM ANG. DEG.	BOOM POINT ELEV.	CAPACITY: CRAWLERS EXTENDED
90	18	81.1	95.9	355,400
	19	80.4	95.7	346,900
	20	79.8	95.6	336,900
	22	78.5	95.2	317,400
	24	77.2	94.7	291,700
	26	75.9	94.3	264,800
	28	74.5	93.7	236,600
	30	73.2	93.2	213,400
	32	71.9	92.5	194,200
	34	70.5	91.9	178,000
	36	69.2	91.1	164,200
	38	67.8	90.3	152,300
	40	66.4	89.5	142,000
	45	62.9	87.1	121,100
	50	59.3	84.4	105,200
	55	55.5	81.2	92,800
	60	51.5	77.5	82,900
	65	47.3	73.2	74,700
	70	42.8	68.2	67,800
	75	37.9	62.3	62,000
	80	32.4	55.2	57,000
	85	25.8	46.2	52,600
	90	17.1	33.5	45,900

BOOM LGTH FEET	OPER. RAD. FEET	BOOM ANG. DEG.	BOOM POINT ELEV.	CAPACITY: CRAWLERS EXTENDED
100	19	81.4	105.9	332,900
	20	80.8	105.7	327,100
	22	79.6	105.4	316,200
	24	78.5	105.0	290,800
	26	77.3	104.5	263,900
	28	76.1	104.1	236,200
	30	74.9	103.6	212,900
	32	73.7	103.0	193,700
	34	72.5	102.4	177,500
	36	71.3	101.7	163,700
	38	70.1	101.0	151,800
	40	68.9	100.3	141,400
	45	65.8	98.2	120,500
	50	62.6	95.8	104,700
	55	59.3	93.0	92,300
	60	55.9	89.8	82,300
	65	52.4	86.2	74,100
	70	48.7	82.1	67,200
	75	44.8	77.4	61,400
	80	40.5	72.0	56,400
	85	35.9	65.6	52,000
	90	30.7	58.0	48,200
	95	24.5	48.5	44,900
	100	16.3	35.0	39,300

CAUTION! CHECK AMOUNT OF COUNTERWEIGHT ON MACHINE BEFORE USE OF THIS CHART

BOOM LGTH FEET	OPER. RAD. FEET	BOOM ANG. DEG.	BOOM POINT ELEV.	CAPACITY: CRAWLERS EXTENDED
110	22	82.3	115.7	240,500
	24	81.3	115.4	232,100
	26	80.2	115.0	224,400
	28	79.1	114.6	217,300
	30	78.1	114.1	210,700
	32	77.0	113.6	193,300
	34	75.9	113.1	177,100
	36	74.8	112.5	163,300
	38	73.7	111.8	151,400
	40	72.6	111.2	141,000
	45	69.9	109.3	120,100
	50	67.0	107.2	104,200
	55	64.1	104.7	91,800
	60	61.2	101.9	81,800
	65	58.1	98.8	73,600
	70	54.9	95.3	66,800
	75	51.6	91.3	60,900
	80	48.1	86.8	55,900
	85	44.4	81.7	51,600
	90	40.4	75.9	47,800
	95	36.1	69.2	44,400
	100	31.1	61.1	41,400
	105	25.3	51.1	38,700
	110	17.6	37.2	33,900
120	22	83.0	125.9	228,700
	24	82.0	125.5	220,500
	26	81.0	125.2	213,000
	28	80.0	124.8	206,100
	30	79.1	124.4	199,700
	32	78.1	123.9	192,800
	34	77.1	123.4	176,600
	36	76.1	122.9	162,800
	38	75.1	122.3	150,900
	40	74.1	121.7	140,500
	45	71.6	120.0	119,500
	50	69.1	118.1	103,700
	55	66.5	115.9	91,300
	60	63.8	113.4	81,300
	65	61.1	110.6	73,100
	70	58.3	107.5	66,200
	75	55.4	104.0	60,400
	80	52.4	100.1	55,300
	85	49.2	95.8	51,000
	90	45.9	91.0	47,200
	95	42.4	85.5	43,800
	100	38.6	79.4	40,800
	105	34.4	72.2	38,100
	110	29.7	63.7	35,700
	115	24.1	53.2	33,500
130	24	82.6	135.7	212,700
	26	81.7	135.3	205,300
	28	80.8	135.0	198,600
	30	79.9	134.6	192,300
	32	79.0	134.2	186,500
	34	78.1	133.7	176,300
	36	77.2	133.2	162,500
	38	76.3	132.7	150,600
	40	75.4	132.1	140,200
	45	73.1	130.6	119,300
	50	70.8	128.8	103,400
	55	68.4	126.8	91,000
	60	66.0	124.6	81,000
	65	63.5	122.0	72,800
	70	61.0	119.2	65,900
	75	58.4	116.1	60,000
	80	55.7	112.7	55,000
	85	53.0	108.9	50,700
	90	50.1	104.8	46,800
	95	47.1	100.1	43,500
	100	43.9	95.0	40,500
	105	40.6	89.2	37,800
	110	36.9	82.6	35,300
	115	33.0	75.1	33,100
	120	28.5	66.1	31,100
	125	23.1	55.1	29,300
140	26	82.3	145.5	198,400
	28	81.5	145.1	191,800
	30	80.7	144.8	185,600
	32	79.8	144.4	179,900
	34	79.0	144.0	174,700
	36	78.1	143.5	162,000
	38	77.3	143.0	150,100
	40	76.5	142.5	139,700
	45	74.3	141.1	118,700
	50	72.2	139.4	102,800
	55	70.0	137.6	90,400
	60	67.8	135.5	80,400
	65	65.6	133.2	72,200
	70	63.3	130.7	65,300
	75	60.9	127.9	59,500
	80	58.5	124.8	54,400
	85	56.1	121.4	50,100
	90	53.5	117.7	46,200
	95	50.9	113.7	42,900
	100	48.1	109.2	39,900

BOOM LGTH FEET	OPER. RAD. FEET	BOOM ANG. DEG.	BOOM POINT ELEV.	CAPACITY: CRAWLERS EXTENDED
140	105	45.3	104.2	37,200
	110	42.2	98.8	34,700
	115	39.0	92.7	32,500
	120	35.5	85.7	30,500
	125	31.7	77.8	28,700
	130	27.4	68.5	27,000
	135	22.2	57.0	25,500
150	28	82.1	155.3	185,600
	30	81.3	154.9	179,600
	32	80.5	154.6	174,000
	34	79.7	154.2	168,800
	36	79.0	153.8	161,600
	38	78.2	153.3	149,600
	40	77.4	152.8	139,200
	45	75.4	151.5	118,200
	50	73.4	150.0	102,400
	55	71.4	148.3	89,900
	60	69.4	146.4	79,900
	65	67.3	144.3	71,700
	70	65.2	141.9	64,800
	75	63.0	139.4	59,000
	80	60.9	136.6	53,900
	85	58.6	133.5	49,600
	90	56.3	130.1	45,700
	95	54.0	126.5	42,400
	100	51.5	122.5	39,400
	105	49.0	118.2	36,700
	110	46.4	113.4	34,200
	115	43.6	108.2	32,000
	120	40.7	102.4	30,000
	125	37.6	96.0	28,200
	130	34.2	88.8	26,500
	135	30.5	80.5	25,000
	140	26.4	70.8	23,500
	145	21.4	58.8	22,200
160	28	82.6	165.4	179,800
	30	81.8	165.1	173,900
	32	81.1	164.7	168,400
	34	80.4	164.4	163,300
	36	79.7	164.0	158,600
	38	78.9	163.6	149,100
	40	78.2	163.1	138,700
	45	76.3	161.9	117,700
	50	74.5	160.5	101,800
	55	72.6	158.9	89,300
	60	70.7	157.1	79,300
	65	68.8	155.1	71,100
	70	66.8	153.0	64,200
	75	64.9	150.6	58,400
	80	62.9	148.0	53,300
	85	60.8	145.2	49,000
	90	58.7	142.2	45,100
	95	56.6	138.8	41,800
	100	54.4	135.2	38,700
	105	52.1	131.4	36,000
	110	49.7	127.1	33,600
	115	47.3	122.5	31,400
	120	44.8	117.5	29,400
	125	42.1	112.0	27,600
	130	39.3	105.9	25,900
	135	36.3	99.2	24,300
	140	33.1	91.7	22,900
	145	29.5	83.0	21,600
	150	25.5	72.9	20,400
	155	20.7	60.5	19,200
170	30	82.3	175.2	169,000
	32	81.6	174.9	163,600
	34	81.0	174.5	158,700
	36	80.3	174.2	154,000
	38	79.6	173.8	148,800
	40	78.9	173.3	138,400
	45	77.2	172.2	117,400
	50	75.4	170.9	101,500
	55	73.7	169.4	89,100
	60	71.9	167.7	79,000
	65	70.1	165.9	70,800
	70	68.3	163.9	63,900
	75	66.4	161.7	58,100
	80	64.6	159.3	53,000
	85	62.7	156.7	48,700
	90	60.8	153.9	44,900
	95	58.8	150.8	41,500
	100	56.8	147.5	38,500
	105	54.7	144.0	35,800
	110	52.6	140.2	33,300
	115	50.4	136.0	31,100
	120	48.1	131.5	29,100
	125	45.8	126.7	27,300
	130	43.3	121.4	25,600
	135	40.8	115.7	24,100
	140	38.1	109.3	22,600
	145	35.2	102.3	21,300
	150	32.0	94.5	20,100
	155	28.6	85.5	18,900
	160	24.7	75.1	17,800
	165	20.0	62.2	16,400

BOOM LGTH FEET	OPER. RAD. FEET	BOOM ANG. DEG.	BOOM POINT ELEV.	CAPACITY: CRAWLERS EXTENDED
180	32	82.1	185.0	158,700
	34	81.5	184.7	154,000
	36	80.8	184.3	149,500
	38	80.2	184.0	145,200
	40	79.5	183.6	137,900
	45	77.9	182.5	116,900
	50	76.3	181.2	101,000
	55	74.6	179.8	88,500
	60	72.9	178.3	78,500
	65	71.3	176.6	70,300
	70	69.6	174.7	63,400
	75	67.8	172.6	57,500
	80	66.1	170.4	52,500
	85	64.3	168.0	48,100
	90	62.5	165.4	44,300
	95	60.7	162.5	40,900
	100	58.9	159.5	37,900
	105	57.0	156.2	35,200
	110	55.0	152.7	32,700
	115	53.0	149.0	30,500
	120	51.0	144.9	28,500
	125	48.9	140.5	26,700
	130	46.7	135.8	25,000
	135	44.4	130.7	23,500
	140	42.0	125.2	22,000
	145	39.6	119.2	20,700
	150	36.9	112.6	19,500
	155	34.1	105.3	18,300
	160	31.1	97.2	17,200
	165	27.7	87.9	16,200
	170	23.9	77.1	15,300
	175	19.4	63.8	13,500
190	32	82.5	195.1	153,100
	34	81.9	194.8	148,800
	36	81.3	194.5	144,800
	38	80.7	194.1	140,800
	40	80.1	193.8	137,000
	45	78.5	192.7	116,400
	50	77.0	191.5	100,500
	55	75.4	190.2	88,000
	60	73.9	188.8	78,000
	65	72.3	187.1	69,700
	70	70.7	185.4	62,800
	75	69.1	183.4	57,000
	80	67.4	181.4	52,000
	85	65.8	179.1	47,600
	90	64.1	176.6	43,800
	95	62.4	174.0	40,400
	100	60.7	171.2	37,400
	105	58.9	168.2	34,700
	110	57.1	164.9	32,200
	115	55.3	161.4	30,000
	120	53.4	157.7	28,000
	125	51.5	153.7	26,200
	130	49.5	149.5	24,500
	135	47.5	144.9	22,900
	140	45.3	140.0	21,500
	145	43.1	134.6	20,200
	150	40.8	128.9	18,900
	155	38.4	122.6	17,800
	160	35.9	115.8	16,700
	165	33.2	108.3	15,700
	170	30.2	99.8	14,800
	175	27.0	90.3	13,900
	180	23.3	79.1	13,100
	185	18.9	65.4	11,000
200	34	82.3	204.9	143,100
	36	81.7	204.6	139,500
	38	81.2	204.3	136,000
	40	80.6	203.9	132,500
	45	79.1	203.0	115,900
	50	77.7	201.8	100,000
	55	76.2	200.6	87,500
	60	74.7	199.2	77,400
	65	73.2	197.7	69,200
	70	71.7	196.0	62,300
	75	70.2	194.2	56,400
	80	68.6	192.2	51,400
	85	67.1	190.1	47,000
	90	65.5	187.8	43,200
	95	63.9	185.3	39,800
	100	62.3	182.7	36,800
	105	60.6	179.8	34,100
	110	59.0	176.8	31,600
	115	57.3	173.6	29,400
	120	55.5	170.2	27,400
	125	53.0	166.5	25,600
	130	51.9	162.6	23,900
	135	50.1	158.4	22,300
	140	48.1	153.9	20,900
	145	46.2	149.1	19,600
	150	44.1	144.0	18,300
	155	42.0	138.4	17,200
	160	39.7	132.5	16,100
	165	37.4	126.0	15,100
	170	34.9	118.9	14,200

BOOM LGTH FEET	OPER. RAD. FEET	BOOM ANG. DEG.	BOOM POINT ELEV.	CAPACITY: CRAWLERS EXTENDED
200	175	32.3	111.1	13,300
	180	29.4	102.4	12,400
	185	26.2	92.5	11,700
	190	22.6	81.0	10,900
	195	18.4	67.0	8,500
210	36	82.1	214.7	134,200
	38	81.6	214.4	131,000
	40	81.0	214.1	127,900
	45	79.6	213.2	115,500
	50	78.3	212.1	99,600
	55	76.9	210.9	87,100
	60	75.4	209.6	77,100
	65	74.0	208.1	68,800
	70	72.6	206.6	61,900
	75	71.2	204.8	56,100
	80	69.7	203.0	51,000
	85	68.2	201.0	46,700
	90	66.7	198.8	42,800
	95	65.2	196.5	39,400
	100	63.7	194.0	36,400
	105	62.2	191.3	33,700
	110	60.6	188.5	31,300
	115	59.0	185.5	29,100
	120	57.4	182.3	27,100
	125	55.7	178.9	25,200
	130	54.1	175.2	23,500
	135	52.3	171.4	22,000
	140	50.6	167.3	20,600
	145	48.8	162.9	19,200
	150	46.9	158.2	18,000
	155	45.0	153.2	16,800
	160	43.0	147.9	15,800
	165	40.9	142.1	14,800
	170	38.7	135.9	13,800
	175	36.5	129.2	12,900
	180	34.0	121.9	12,100
	185	31.5	113.9	11,300
	190	28.7	104.9	10,600
	195	25.6	94.7	9,900
	200	22.1	82.9	8,800
	205	17.9	68.5	6,500
220	38	82.0	224.6	125,700
	40	81.4	224.2	122,800
	45	80.1	223.3	115,000
	50	78.8	222.3	99,100
	55	77.5	221.2	86,600
	60	76.1	220.0	76,600
	65	74.8	218.6	68,300
	70	73.4	217.1	61,400
	75	72.0	215.4	55,500
	80	70.7	213.7	50,500
	85	69.3	211.8	46,100
	90	67.9	209.7	42,300
	95	66.4	207.5	38,900
	100	65.0	205.2	35,800
	105	63.6	202.7	33,100
	110	62.1	200.0	30,700
	115	60.6	197.2	28,500
	120	59.1	194.2	26,500
	125	57.5	191.0	24,600
	130	55.9	187.6	23,000
	135	54.3	184.0	21,400
	140	52.7	180.2	20,000
	145	51.0	176.1	18,600
	150	49.3	171.8	17,400
	155	47.6	167.3	16,300
	160	45.7	162.4	15,200
	165	43.9	157.2	14,200
	170	41.9	151.7	13,200
	175	39.9	145.7	12,300
	180	37.8	139.3	11,500
	185	35.6	132.4	10,700
	190	33.2	124.8	10,000
	195	30.7	116.6	9,300
	200	28.0	107.3	8,600
	205	25.0	96.9	8,000
	210	21.5	84.8	6,500

CAPACITIES CONTINUED ON NEXT PAGE

These load charts are intended for instructional purposes only. They were derived from manufacturer sales information which may not be complete or machine specific. Not responsible for typographical errors.

Grove TM1500 Load Chart

RATED LIFTING CAPACITIES IN POUNDS
46 FT. - 173 FT. BOOM ON OUTRIGGERS - OVER REAR

Radius in Feet	#0001 Main Boom Length in Feet (Power Pinned Fly Retracted)									#0002 Power Pin Fly Ext. & 141 ft.
	46	58	70	82	94	106	118	130	141	173
10	300,000 (74.5)									
12	280,000 (72)	143,500 (76)	142,000 (79)							
15	235,000 (67.5)	143,500 (72.5)	141,500 (76.5)	130,000 (78.5)						
20	173,500 (60.5)	143,500 (67.5)	123,500 (72)	112,000 (75)	102,000 (77.5)	90,300 (79.5)				
25	136,500 (52)	131,500 (61.5)	110,500 (67.5)	98,650 (71)	89,250 (74)	78,550 (76.5)	73,700 (78.5)	69,300 (80)		
30	106,000 (43)	106,000 (55.5)	98,000 (63)	88,350 (67.5)	78,750 (71)	69,250 (73.5)	65,100 (75)	61,000 (77.5)	60,000 (79.5)	
35	84,700 (30.5)	84,700 (49)	84,700 (58)	80,150 (63.5)	69,000 (67.5)	60,750 (70.5)	57,150 (73)	54,000 (75.5)	52,150 (77.5)	
40		70,500 (41)	70,500 (52.5)	70,500 (59.5)	51,300 (64)	54,000 (67.5)	50,600 (70.5)	48,300 (73)	45,850 (75)	38,000 (79)
45		58,850 (32)	58,850 (47)	58,850 (55)	55,000 (60.5)	48,500 (64.5)	46,200 (68)	43,050 (71)	40,400 (73)	35,750 (77)
50		49,600 (17.5)	49,600 (40.5)	49,600 (50.5)	48,750 (57)	43,050 (61.5)	40,700 (65)	38,250 (68.5)	35,750 (71)	32,100 (75.5)
60			36,200 (22.5)	36,200 (38.5)	36,200 (48.5)	34,300 (55)	33,600 (58.5)	30,750 (63.5)	28,500 (66.5)	26,350 (72)
70				26,050 (25)	26,050 (39.5)	26,050 (47.5)	26,050 (53)	24,750 (58)	23,100 (61.5)	22,000 (68.5)
80					18,850 (27)	18,850 (39)	18,850 (46.5)	18,850 (52.5)	18,700 (56.5)	18,500 (64.5)
90						13,500 (28)	13,500 (38.5)	13,500 (46.5)	13,500 (51.5)	15,250 (60.5)
100							9,390 (29)	9,390 (39)	9,390 (45.5)	12,600 (56.5)
110							6,080 (12.5)	6,080 (30.5)	6,080 (39)	10,100 (52)
120								3,390 (17.5)	3,390 (31)	7,530 (47.5)
130									1,150 (19.5)	5,390 (42.5)
140										3,610 (36.5)
150										2,100 (30)
Minimum boom angle (deg.) for indicated length (no load)									10	19
Maximum boom length (ft) at 0 deg boom angle (no load)									140	167

NOTE: ( ) Boom angles are in degrees.
#LMI operating code. Refer to LMI manual for instructions.

A6-829-007136 & -007143

WEIGHT REDUCTIONS FOR LOAD HANDLING DEVICES

33 ft. Extension	
* Stowed -	892 lbs.
* Erected -	5,704 lbs.

33 ft. - 58 ft. Extension	
* Stowed -	1,246 lbs.
* Erected (Ret.) -	8,412 lbs.
* Erected (Ext.) -	11,406 lbs.

46 ft. - 173 ft. Boom with	
* 46 ft. Jib Erected -	9,613 lbs.
* 60 ft. Jib Erected -	14,571 lbs.
* 74 ft. Jib Erected -	20,443 lbs.
* 88 ft. Jib Erected -	27,199 lbs.
* Fixed Jib Accessories -	327 lbs.

* Reduction of main boom capacities

HOOKBLOCKS:	
30 Ton, 1 Sheave	1,022 lbs.
150 Ton, 8 Sheave	5,254 lbs.
Auxiliary Boom Head	261 lbs.
10 Ton Headache Ball	560 lbs.
15 Ton Headache Ball	803 lbs.

RATED LIFTING CAPACITIES IN POUNDS
46 FT. - 173 FT. BOOM ON OUTRIGGERS - 360°

Radius in Feet	#0001									#0002
	Main Boom Length in Feet (Power Pinned Fly Retracted)									Power Pin Fly Ext. & 141 ft.
	46	58	70	82	94	106	118	130	141	173
10	300,000 (74.5)									
12	280,000 (72)	143,500 (76)	142,000 (79)							
15	235,000 (67.5)	143,500 (72.5)	141,500 (76.5)	130,000 (78.5)						
20	173,500 (60.5)	143,500 (67.5)	123,500 (72)	112,000 (75)	102,000 (77.5)	90,300 (79.5)				
25	135,500 (52)	131,500 (61.5)	110,500 (67.5)	98,650 (71)	89,250 (74)	78,550 (76.5)	73,700 (78.5)	69,300 (80)		
30	106,000 (43)	106,000 (55.5)	98,000 (63)	88,350 (67.5)	78,750 (71)	69,250 (73.5)	65,100 (76)	61,000 (77.5)	60,000 (79.5)	
35	84,700 (30.5)	84,700 (49)	84,700 (58)	80,150 (63.5)	69,000 (67.5)	60,750 (70.5)	57,150 (73)	54,000 (75.5)	52,150 (77.5)	
40		70,500 (41)	70,500 (52.5)	70,500 (59.5)	61,300 (64)	54,000 (67.5)	50,600 (70.5)	48,300 (73)	45,850 (75)	38,000 (79)
45		57,250 (32)	57,250 (47)	57,250 (55)	55,000 (60.5)	48,500 (64.5)	45,200 (68)	43,050 (71)	40,400 (73)	35,750 (77)
50		46,550 (17.5)	46,550 (40.5)	46,550 (50.5)	46,550 (57)	43,050 (61.5)	40,700 (65)	38,250 (68.5)	35,750 (71)	32,100 (75.5)
60			32,000 (22.5)	32,000 (39.5)	32,000 (48.5)	32,000 (55)	32,600 (58.5)	30,750 (63.5)	28,500 (66.5)	26,350 (72)
70				22,550 (25)	22,550 (39.5)	22,550 (47.5)	22,550 (53)	22,550 (58)	22,550 (61.5)	22,000 (68.5)
80					15,900 (27)	15,900 (39)	15,900 (46.5)	15,900 (52.5)	15,900 (56.5)	18,500 (64.5)
90						11,000 (28)	11,000 (38.5)	11,000 (46.5)	11,000 (51.5)	15,250 (60.5)
100							7,280 (29)	7,280 (39)	7,260 (45.5)	11,750 (56.5)
110							4,260 (12.5)	4,260 (30.5)	4,260 (39)	8,520 (52)
120								1,830 (17.5)	1,830 (31)	5,850 (47.5)
130										3,610 (42.5)
140										1,710 (36.5)
Minimum boom angle (deg.) for indicated length (no load)									20	30
Maximum boom length (ft) at 0 dog boom angle (no load)									133	166

NOTE: ( ) Boom angles are in degrees.
#LMI operating code. Refer to LMI manual for instructions.

A6-829-007136 & -007143

33 FT. FIXED LENGTH EXTENSION ON OUTRIGGERS - 360°

Main Boom Angle (Deg.)	#0051		#0052		#0053	
	2° OFFSET		15° OFFSET		30° OFFSET	
	Rad. Ref. (ft.)	Cap. lbs.	Rad. Ref. (ft.)	Cap. lbs.	Rad. Ref. (ft.)	Cap. lbs.
80	39.1	22,850	45.9	20,200	52.0	14,750
75	56.6	17,400	63.0	16,100	68.7	12,750
70	73.7	14,000	79.6	12,950	84.8	10,700
65	90.2	11,850	95.5	10,750	100.3	9,210
60	106.0	10,750	110.8	9,140	115.0	8,050
55	121.1	7,240	125.2	6,160	128.8	5,320
50	135.2	4,330	138.6	3,700	141.8	3,150

A6-829-008157

DIMENSIONS ARE FOR LARGEST GROVE FURNISHED HOOK BLOCK AND HEADACHE BALL, WITH ANTI-TWO BLOCK ACTIVATED.

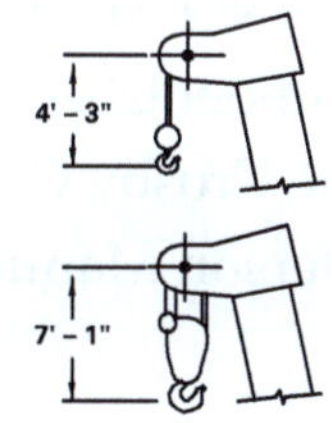

33 FT. - 58 FT. TELE EXTENSION ON OUTRIGGERS - 360°

Main Boom Angle (Deg.)	33 ft. EXTENSION						48 ft. EXTENSION						58 ft. EXTENSION					
	#0021		#0022		#0023		#0031		#0032		#0033		#0041		#0042		#0043	
	2° OFFSET		15° OFFSET		30° OFFSET		2° OFFSET		15° OFFSET		30° OFFSET		2° OFFSET		15° OFFSET		30° OFFSET	
	Rad. Ref. (ft.)	Cap. lbs.	Rad. Ref. (ft.)	Cap. lbs.	Rad. Ref. (ft.)	Cap. lbs.	Rad. Ref. (ft.)	Cap. lbs.	Rad. Ref. (ft.)	Cap. lbs.	Rad. Ref. (ft.)	Cap. lbs.	Rad. Ref. (ft.)	Cap. lbs.	Rad. Ref. (ft.)	Cap. lbs.	Rad. Ref. (ft.)	Cap. lbs.
80	39.1	22,300	45.9	19,650	52.0	14,200	43.9	14,800	55.8	12,700	65.0	9,040	45.1	9,600	59.1	8,300	69.6	7,000
75	56.6	16,850	63.0	15,550	68.7	12,200	62.5	13,700	73.4	10,900	81.8	8,320	65.0	9,150	77.9	7,950	87.8	6,440
70	73.7	13,450	78.6	12,400	84.8	10,150	80.7	11,600	90.3	8,840	97.9	7,110	84.4	8,550	96.2	7,200	105.5	5,760
65	90.2	11,300	95.5	10,200	100.3	8,870	98.2	9,230	106.7	7,330	113.3	6,120	103.2	7,630	113.8	6,000	122.3	4,970
60	106.0	10,200	110.8	8,600	115.0	7,450	115.1	7,550	122.1	8,210	127.9	5,340	121.2	6,260	130.5	5,100	138.2	4,380
55	121.1	6,400	125.2	5,370	128.8	4,540	131.1	5,200	136.7	3,860	141.5	2,950	138.2	4,650	146.2	3,265	153.1	2,380
50	135.2	3,520	138.6	2,810	141.8	2,360	146.1	2,580	150.2	1,790			154.3	2,235				

#LMI operating code. Refer to LMI manual for instructions.

A6-829-007158

Additional Resources

This module is intended as a thorough resource for task training. The following reference materials are recommended for further study.

ASME Standard B30.5, Mobile and Locomotive Cranes. Current edition. New York, NY: American Society of Mechanical Engineers.

Cranes: Design, Practice, and Maintenance, Ing J. Verschoof. 2002. Hoboken, NJ: John Wiley and Sons, Inc.

29 *CFR* 1926, Subpart CC,*Cranes and Derricks in Construction*. **www.ecfr.gov**

Figure Credits

Link-Belt Construction Equipment Company, Module opener, Figures 3, 4, 6A, 8, 11– 13, 16, 25, 27, 30, 31

Terex Cranes, used with permission of the owner, Figures 6B, 14, Appendix A

The Manitowoc Company, Inc., Figures 6C, 7, 9, 10, 18– 22, 28, 32, 34, 35, 37–40, 42, 43, 45, 46, Appendix B, Appendix C

TADANO America Corporation, Figure 15

Mammoet USA South Inc., Figures 23, 24, 26

© The Crosby Group LLC, Table 2

Broderson Manufacturing Corp., Exam art figure 1

Section Review Answer Key

Answer	Section Reference	Objective
Section One		
1. a	1.1.2	1a
2. d	1.2.0	1b
Section Two		
1. a	2.1.1	2a
2. b	2.2.0	2b
3. d	2.3.0	2c
4. a	2.4.0	2d
Section Three		
1. b	3.0.0	3a
2. d	3.2.1	3b
3. b	3.3.0	3c
4. b	3.4.1	3d

NCCER CURRICULA — USER UPDATE

NCCER makes every effort to keep its textbooks up-to-date and free of technical errors. We appreciate your help in this process. If you find an error, a typographical mistake, or an inaccuracy in NCCER's curricula, please fill out this form (or a photocopy), or complete the online form at **www.nccer.org/olf**. Be sure to include the exact module ID number, page number, a detailed description, and your recommended correction. Your input will be brought to the attention of the Authoring Team. Thank you for your assistance.

Instructors – If you have an idea for improving this textbook, or have found that additional materials were necessary to teach this module effectively, please let us know so that we may present your suggestions to the Authoring Team.

NCCER Product Development and Revision

13614 Progress Blvd., Alachua, FL 32615

Email: curriculum@nccer.org
Online: www.nccer.org/olf

❑ Trainee Guide ❑ Lesson Plans ❑ Exam ❑ PowerPoints Other ____________________

Craft / Level: Copyright Date:

Module ID Number / Title:

Section Number(s):

Description:

Recommended Correction:

Your Name:

Address:

Email: Phone:

Lift Planning

Overview

Industrial and safety standards require some form of lift planning before a lift takes place. The level of detail required for a lift plan depends on the complexity of the lift, and the real and potential hazards involved. The more hazardous or sensitive the lift, the more detailed and formal the plan must be. A lift plan contains information relating to the crane(s) used, the load and its rigging, and all the site coordination issues. This module explains what lift planning involves and describes how to prepare a lift plan.

Module 21304

Trainees with successful module completions may be eligible for credentialing through the NCCER Registry. To learn more, go to **www.nccer.org** or contact us at 1.888.622.3720. Our website has information on the latest product releases and training, as well as online versions of our *Cornerstone* magazine and Pearson's product catalog.

Your feedback is welcome. You may email your comments to **curriculum@nccer.org**, send general comments and inquiries to **info@nccer.org**, or fill in the User Update form at the back of this module.

This information is general in nature and intended for training purposes only. Actual performance of activities described in this manual requires compliance with all applicable operating, service, maintenance, and safety procedures under the direction of qualified personnel. References in this manual to patented or proprietary devices do not constitute a recommendation of their use.

21304 V3

21304

Lift Planning

Objective

When you have completed this module, you will be able to do the following:

1. Describe lift planning, development, and implementation.
   a. Discuss the considerations that go into lift planning.
   b. Describe the development and contents of standard and critical lift plans.
   c. Describe how to implement a lift plan.

Performance Tasks

Under the supervision of your instructor, you should be able to do the following:

1. Perform a net-load capacity calculation for the crane specified in a scenario.
2. Fill in and complete a lift data sheet for a given lift scenario.

Trade Terms

Ground bearing pressure (GBP)
Lift data sheet (LDS)

Industry Recognized Credentials

If you are training through an NCCER-accredited sponsor, you may be eligible for credentials from NCCER's Registry. The ID number for this module is 21304. Note that this module may have been used in other NCCER curricula and may apply to other level completions. Contact NCCER's Registry at 888.622.3720 or go to **www.nccer.org** for more information.

Contents

1.0.0 Lift Plans 1
1.1.0 Lift Planning Considerations 1
1.1.1 Purposes and Standards 1
1.1.2 Lift Personnel and Responsibilities 1
1.1.3 Potential Hazards to Personnel 2
1.1.4 Hazards Near Work Area 2
1.1.5 Complexity of the Activity 2
1.1.6 Commercial Impacts 2
1.1.7 Crane and Rigging Capacities 4
1.1.8 Environmental Conditions 4
1.1.9 Repetitive Lifting Operations 4
1.1.10 Standard and Critical Lifts 4
1.2.0 Developing a Lift Plan 5
1.2.1 The Lift Data Sheet 6
1.2.2 The Crane Configuration 6
1.2.3 The Load and Rigging 10
1.2.4 Crane Capacity Calculations 10
1.2.5 Load and Crane Considerations 12
1.2.6 Reference Documents and Attachments 13
1.2.7 Notes, Review, and Approvals 15
1.2.8 Critical Lift Plans 15
1.3.0 Implementing a Lift Plan 16
1.3.1 Pre-Lift Review/Meeting 16
1.3.2 Executing the Lift 16
1.3.3 Post-Lift Review 18

Figures

Figure 1 Lifting personnel by crane requires special precautions. 2
Figure 2 Two or more cranes involved in a lift increase the operation's complexity. 3
Figure 3 Damage to this nuclear reactor vessel would be a serious setback. 3
Figure 4 Filling concrete forms may require repetitive lifts 4
Figure 5 Demolition lifting operations may be designated critical lifts 5
Figure 6 Plan view showing the position of the crane for placing the HVAC condenser 7
Figure 7 Elevation view of building on the crane's range diagram. 8
Figure 8 Evaluating the crane's available boom configurations for the lift. 9
Figure 9 Crane capacities at the available boom lengths. 9
Figure 10 Rigging diagram for the HVAC unit. 10
Figure 11 Completing the Crane Capacities section of the LDS 11
Figure 12 Positions of the adjustable suspension lugs on an equalizer lifting beam. 12
Figure 13 When positioning a load becomes complicated, two or more cranes may be required. 12
Figure 14 Determining building clearance from the crane range diagram 14
Figure 15 Inadequate ground bearing pressure (GBP) information can lead to crane tipping incidents 15
Figure 16 A formal pre-lift meeting is necessary for complicated or critical lifts 16
Figure 17 At the onset of dangerous weather, stop the lifting operation 17

SECTION ONE

1.0.0 LIFT PLANS

Objective

Describe lift planning, development, and implementation.

a. Discuss the considerations that go into lift planning.
b. Describe the development and contents of standard and critical lift plans.
c. Describe how to implement a lift plan.

Performance Tasks

1. Perform a net-load capacity calculation for the crane specified in a scenario.
2. Fill in and complete a lift data sheet for a given lift scenario.

Trade Terms

Ground bearing pressure (GBP): The maximum force per unit area (pressure), measured in pounds per square foot, that can be exerted on the ground's surface without causing deformation of the surface. Can only be determined by testing.

Lift data sheet (LDS): The essential page or pages of a written lift plan that organize and summarize the important information required to develop a lift plan.

Crane operations are essential to most construction projects, but these operations carry great potential for accidents or errors. Mistakes can be costly, both monetarily and in terms of human lives or injuries. Lift planning, when properly done, identifies and addresses all the factors that can work against a safe and successful lift. Before a lift begins, relevant standards require that all workers involved minimize potential hazards as much as feasible.

1.1.0 Lift Planning Considerations

Every lift operation is unique in one way or another. However, crane lifts generally involve raising, lowering, or moving a load between two points. OSHA regulations and industry standards address lift planning and provide lift planning guidelines to enhance safety. This section identifies the key factors that lift planning should address.

1.1.1 Purposes and Standards

The primary purpose of lift planning standards and regulations is to protect lives and prevent injuries. The American Society of Mechanical Engineers (ASME) developed the most relevant and thorough lift-planning standard in *ASME P30.1, Planning for Load Handling Activities*. The sections that follow are summaries of the relevant content found in this standard. *Nonmandatory Appendix B—Industry References* of *ASME P30.1* provides a list of additional engineering standards that pertain to lift planning.

The Occupational Safety and Health Administration (OSHA) has also issued regulations that specifically address the safety aspects of lift planning. You can find this information in the relevant sections of 29 *CFR* Part 1926, Subpart CC, *Cranes and Derricks in Construction*.

Many employers and crane owners develop internal procedures that exceed OSHA requirements and *ASME P30.1* guidelines. It is essential to follow these procedures in addition to those outlined by OSHA and ASME.

> **NOTE**
> While planning for every lift should take into consideration the topics of the following sections, *ASME P30.1* states that documentation of the evaluation process is not required.

1.1.2 Lift Personnel and Responsibilities

ASME P30.1, Chapter 3—Personnel and Responsibilities lists the titles and responsibilities of personnel involved in crane operations. The employer must determine that these individuals are qualified and competent in order to ensure a safe and successful lifting operation. Not all roles listed here are needed for every operation, and a single individual may fulfill the responsibilities of one or more of the following positions:

- *Crane Assembly/Disassembly Director* – Directs the assembly, disassembly, or erection of the crane.
- *Crane Operator* – Directly controls the crane's functions.
- *Crane Owner* – The individual or entity that has custodial control of the crane.
- *Crane User* – Arranges for the crane's presence and use at the site.
- *Engineer* – Provides engineering support and documentation for the operation.
- *General Contractor/Construction Manager* – Meets the contractual requirements for establishing and implementing acceptable safety and work performance results.

- *Lift Director* – Establishes the lift category, and reviews and implements the lift plan.
- *Lift Planner* – Develops the lift plan.
- *Rigger* – Performs rigging tasks associated with the lift.
- *Signal Person* – Directs the movements of the crane.
- *Site Safety Officer* – Enforces site safety policies.
- *Site Supervisor* – Oversees the work on the site.
- *Spotter* – Observes and reports the movement of the crane and load.
- *Transport Operator* – Operates transport equipment (dump trucks, flatbeds, barges, etc.) related to the lifting operation.

1.1.3 Potential Hazards to Personnel

Lift planning should include an evaluation of whether the lift has the potential to injure workers, bystanders, or the public. Potential hazards to consider include (but aren't limited to) the following:

- Pinch and crush points
- Personnel lifts (*Figure 1*)
- Moving or suspending loads over public areas
- Moving hazardous materials

1.1.4 Hazards Near Work Area

Lift planners should identify hazards near the work area, regardless of whether those hazards are relevant to the assigned task. Some hazards that fall into this category include the following:

- Power lines (29 *CFR* 1926.1408 defines the prohibited zones for crane operations)
- Sources of radio-frequency interference that could disrupt communications
- Electromagnetic energy sources that could induce electrical discharges
- Nearby pipes, storage tanks, and other equipment
- Structures that obscure the crane operator's view of the load and/or signal persons(s)
- Active roads and pedestrian pathways

Figure 1 Lifting personnel by crane requires special precautions.

1.1.5 Complexity of the Activity

Some lifts are not simple or straightforward, and contingency plans must be made. Planners should evaluate the following factors that can increase the complexity of the lift:

- Load instability due to configuration, shifting of contents, or location of lift points
- Special rigging requirements
- New or rarely used work practices
- Handling loads close to obstructions
- Using multiple cranes (*Figure 2*)
- Having to turn or tilt the load during the lift
- Traveling with the load

1.1.6 Commercial Impacts

Every lift has the potential to damage the load as well as nearby structures and equipment. *ASME P30.1* identifies the following factors that lift planners should consider:

- Will failure to complete the lift cause a project delay, or adversely affect access or use of the area by the public?
- Is the load extremely valuable or irreplaceable (*e.g.*, a rare or unique historical artifact, or a work of art)?
- Will replacing the load, if lost or damaged, take a significant amount of time (*Figure 3*)?
- Will damage caused by the lift to adjacent structures create an adverse commercial impact?

Figure 2 Two or more cranes involved in a lift increase the operation's complexity.

Figure 3 Damage to this nuclear reactor vessel would be a serious setback.

1.1.7 Crane and Rigging Capacities

Ensuring that the total load does not exceed the crane's load capacity at any point during the lift is a key element in lift planning. Module 21301 of this curriculum explains how to determine the gross and net capacities of a crane in a given configuration. Factors change during a lift, however, and unanticipated events can affect crane stability. Planners evaluate things that can influence crane capacity and stability, such as the following:

- Changes of the load control during operations involving demolition, suction, or friction
- Abrupt changes in the load's or crane's motion
- Changes in available line pull (*e.g.*, hoist drum effective diameter)
- Capacities of the crane's load-handling mechanisms (hydraulics, brakes, clutches, etc.)
- Accuracy of the load's weight information
- Shifting of the load center of gravity (CG) during the lift
- Shift of weight distribution between multiple cranes lifting a load
- Buoyancy effects for submerged lifts
- Out-of-vertical lifts

Less obvious but important factors to consider are the crane's work history and material condition. A manufacturer's load chart assumes the crane is in as-built condition. However, leaking hydraulic actuators or line fittings, deficient hydraulic pumps, slipping clutches and brakes, and damaged wire rope can limit a crane's capacity.

1.1.8 Environmental Conditions

Environmental conditions refer to the surroundings that can influence the lift. Environmental factors that are outside of a lift planner's control include the following:

- Wind
- Precipitation
- Lightning
- Temperature
- Conditions that can affect visibility, such as fog, wind-born dust, and the sun's position

Planning a lift for a specific time of day can minimize the effect of some natural factors, such as the sun's glare or the temperature.

1.1.9 Repetitive Lifting Operations

Examples of repetitive lifts may include tasks such as positioning construction materials (*Figure 4*), excavating fill, or performing dredging operations. Lift planners must evaluate the effects of repetitive operations on both the equipment and the operator. *ASME P30.1* notes that planners should consider the potential for operator and construction worker fatigue and complacency. Planners should also review the crane manufacturer's recommendations for equipment duty cycles and any additional maintenance required.

Figure 4 Filling concrete forms may require repetitive lifts.

1.1.10 Standard and Critical Lifts

After reviewing the results of all these evaluations, the lift director shall identify the category of the lift. There are two lift categories—standard lifts and critical lifts. If the lift director concludes that the lift does not present any undue hazards and can be completed using standard procedures, then it is classified as a standard lift. If the lift director's evaluation shows that the lift requires additional preparation, planning, or methods beyond those for a standard lift, then it is classified as a critical lift. Critical lift plans must always be written.

ASME B30.5—Mobile and Locomotive Cranes, Nonmandatory Appendix A, identifies other factors that could cause an operation to be designated a critical lift, listed as follows:

- The total load exceeds some value stipulated by company policy.
- The total load exceeds a predetermined percentage of the crane's load chart capacity (other than 75 percent).
- Hoisting personnel in a basket or platform.
- Blind lifts (any lift where the load can't be directly observed by the crane operator and/or the primary signal person).
- Any operation with two or more cranes lifting a common load. Use of tailing cranes may not require designation as a critical lift.
- Unique, costly, or irreplaceable loads.
- Demolition operations where the actual weight of the load may be in doubt (*Figure 5*).
- Especially hazardous lifts involving proximity to high voltages, lifts over people or roadways, etc.
- Other unspecified crane activities that are technically challenging.

Organizations may enforce policies that identify critical lift conditions. For example, shipyards and facilities servicing nuclear warships require that all lifts involving radioactive materials, nuclear power plant components, or weapons require a critical lift plan.

ASME P30.1 distinguishes between standard and critical lifts when describing what it requires a planner to do. For standard lift plans, it qualifies all planning actions with the term *should*, indicating that the item may be needed or recommended, depending on the circumstances, but isn't mandatory. The standard depends on the knowledge and experience of the lift director to understand the detail needed for each item.

When developing a lift plan, planners should review all the lift planning items listed in Chapters 4 (*Standard Lift Plan*) and 5 (*Critical Lift Plan*) of *ASME P30.1*, to the extent they are applicable. Chapter 5 covers the same items as Chapter 4, but describes each item in much greater detail. The standard makes it clear that in a critical lift plan, some actions are mandatory; it uses the term *shall* to identify those requirements. *ASME P30.1* also states that the critical lift planning process should address "any additional considerations identified during the planning process."

1.2.0 Developing a Lift Plan

Every lift requires a lift plan appropriate to the operation. Based on the principles found in *ASME P30.1*, this is true regardless of whether the crane is offloading bundles of 2 × 4 studs from a flatbed

Figure 5 Demolition lifting operations may be designated critical lifts.

truck or placing a 278-metric ton reactor vessel at a power plant. The overall approach is the same for either operation; only the scope and level of detail of the planning differs. Standard lift plans may be verbal or written. Critical lift plans shall be in written form.

The following sections allow you to assume the role of lift planner, documenting the information necessary to produce a lift plan based on a provided scenario. Then, in the role of the lift director, you will categorize the lift as a standard or critical lift. The scenario for this exercise is described as follows:

This site construction plan calls for the installation of an air-conditioning unit on the third floor roof of a building. *Appendix A* provides an architectural elevation and roof plan of the building, showing the unit's installed location and orientation. *Appendix B* supplies technical details of the unit related to the lift, including shipping weight, center of gravity, and minimum rigging requirements. *Appendix C* is the load chart for the Grove TM1500 crane selected for the lift. Assume that the site can provide any necessary materials, workers, and other equipment for the lift.

1.2.1 The Lift Data Sheet

Planning even a simple lift can be a challenging task. Many factors described in *ASME P30.1*, Chapter 2 are interdependent. Breaking the process into distinct steps in a logical order can help to provide clarity. Completing a **lift data sheet (LDS)** is the preferred method for ensuring that all applicable lift planning items have been considered.

An LDS is a summary of the essential technical details relating to the lift. *ASME P30.1, Nonmandatory Appendix A—Example Lift Plan*, includes a generic LDS form that can be the basis for users to develop their own customized forms, which the standard encourages. A brief internet search for the term *crane lift planning form* demonstrates the variety such forms can take. The form can contain the following informational sections, as well as others that companies, lift directors, or lift planners may consider necessary:

1. Lift description and other identifying information
2. Identification of the crane and its configuration
3. Identification and description of the load and its rigging requirements
4. Computation of crane capacities
5. Reference attachments and supporting documentation
6. Notes
7. Review and approvals

Appendix D is a blank LDS form based on this format. Refer to this form as the development of the lift plan proceeds through the following sections.

1.2.2 The Crane Configuration

After filling in the project information at the head of the LDS, the first section relates to the crane and its configuration. Enter the crane's identifying information, referring to the crane's load chart and operator's manuals. Before you can determine a crane configuration, you need to know the load's weight and the geometry of the lift in relation to the available setup areas.

The building's roof plan (*Appendix A*) shows that the condensing unit is located toward the south end of the building. This means that the crane should be positioned at that end of the building, and on a line perpendicular to the wall through the load's landing site to minimize the load radius (*Figure 6*). Minimizing the load radius maximizes capacity and crane stability. Before proceeding, check with the project supervisor or schedule to determine if placing a crane at that location on the scheduled day is acceptable.

Consult the crane's range diagram (*Appendix C*, page 1 of 6). Note that the roof edge is 34'-6" (10.5 m) above the ground. The center of the condensing unit will be 47'-8" (14.5 m) from the roof edge. It helps to sketch this information onto a sheet of graph paper or use tracing paper taped over the crane's range diagram. *Figure 7* shows the building elevation view superimposed on the range diagram.

Assuming that the crane will conduct the lift with the main boom over the rear, as shown in the range diagram, determine the position of the crane that results in the shortest load operating radius and also provides sufficient clearance to the building for the elevated boom. This will give you the boom angle and boom length for the lift. Note that these are tentative values; the crane's capacity in this configuration must also be determined.

The following working assumptions can be made from analyzing the situation and the range diagram with the building's outline and load location placed on it to scale:

- The crane will be on fully-extended outriggers, with its center of rotation about 25 feet (7.6 m) from the building wall. The crane body will be on a line perpendicular to the wall passing through the condenser's location. This arrangement allows the planner to rely on the range diagram directly, without adjustment. Oblique (angular) positions to the load in relation to the building require calculations using dimensions from the architectural plan and elevation views, or they must be physically measured on site.
- The range diagram shows that the minimum boom angle permitted in this location is about 40 degrees. A lower angle with the crane in the described position would place the boom too close to the edge of the roof. The load radius is 73 ft (25 ft + 48 ft = 73 ft), so a minimum boom length of 106 ft (32.3 m) at a 40-degree boom angle can be used to place the unit. The range diagram also shows that the load can be set with the main boom fully extended to 141 ft (43 m) at a boom angle of about 55 degrees.
- The power-pinned boom fly is retracted.

Consult the crane's load chart to determine whether the trial combination of load radius and boom length is within the capacity of the crane. If there is plenty of excess capacity, then you, as the lift planner, may confidently proceed with filling in the crane data on the LDS. If the load's weight appears to be a significant percentage of the crane's capacity (before weight deductions) then you must evaluate other possible configurations. Options may include lifting from a jib/boom extension, or simply trying the longer boom length and boom angle noted above.

When planning reveals the operation will be at the limit of the crane's capacity, finding an improved configuration might require several

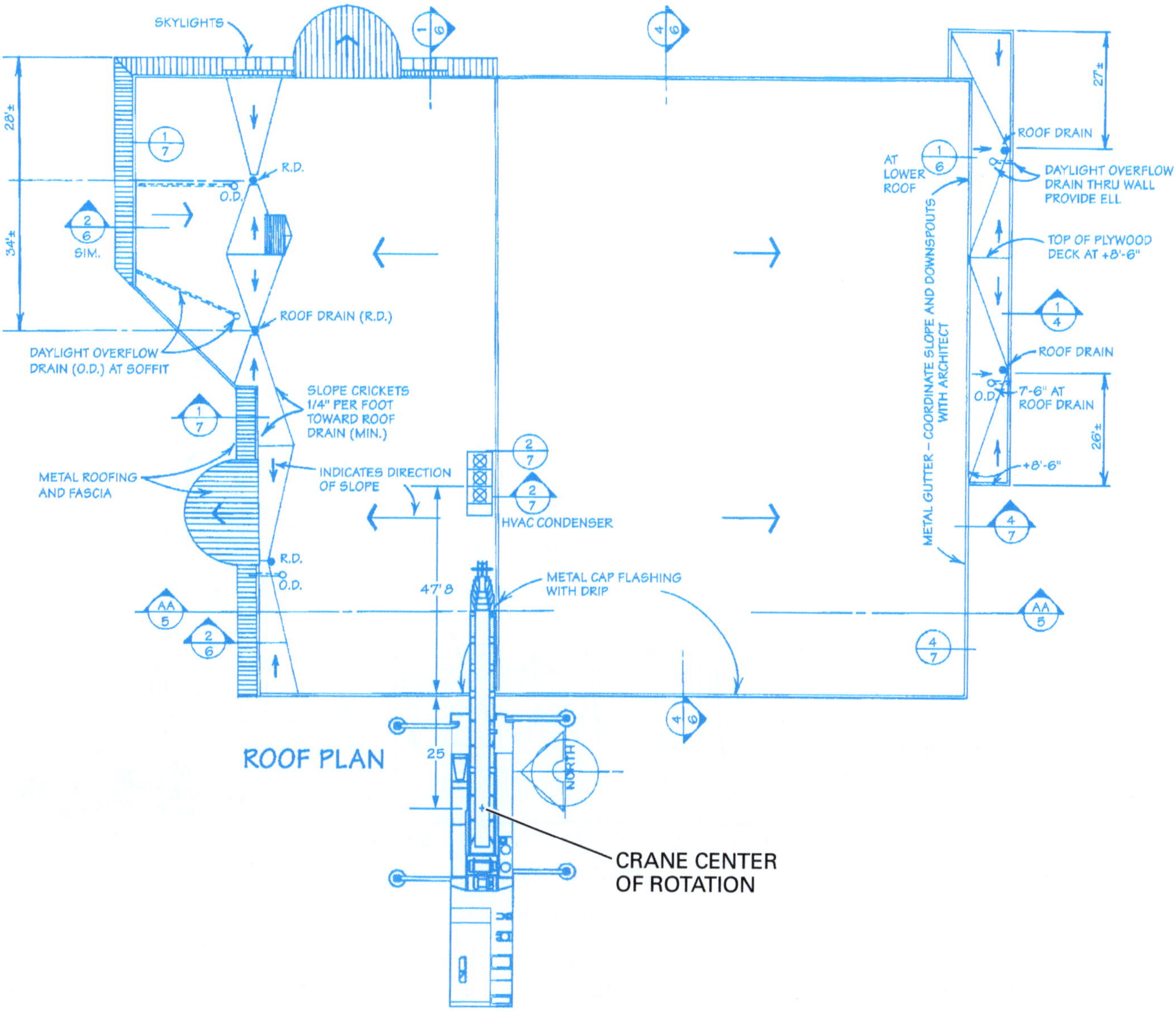

Figure 6 Plan view showing the position of the crane for placing the HVAC condenser.

Lift Planning Software

It shouldn't be surprising that there are apps available that can assist with lift planning. These apps run on desktop and laptop systems, tablets, and smartphones. Some companies offer free basic planning programs, such as 3D Lift Plan™, that run in a browser environment along with their full-featured programs. These applications allow planners to visualize the site structures, crane, and load in 3D. Users can see the geometry of the lift and the path the load will travel.

The main limitation of these apps is that they don't provide access to a specific crane's load chart to evaluate capacity variations during the lift. Some crane manufacturers, such as Manitowoc, have resolved this limitation by offering planning programs such as CraniMAX® that include an update-able crane database containing specific crane load charts.

A brief search of the internet using the search term *lift planning software* will reveal numerous low- to high-end programs offered by software companies, crane manufacturers, and crane dealers. However, as with all computer programs, the output is only as good as the input. Lift planners are still required to verify the results.

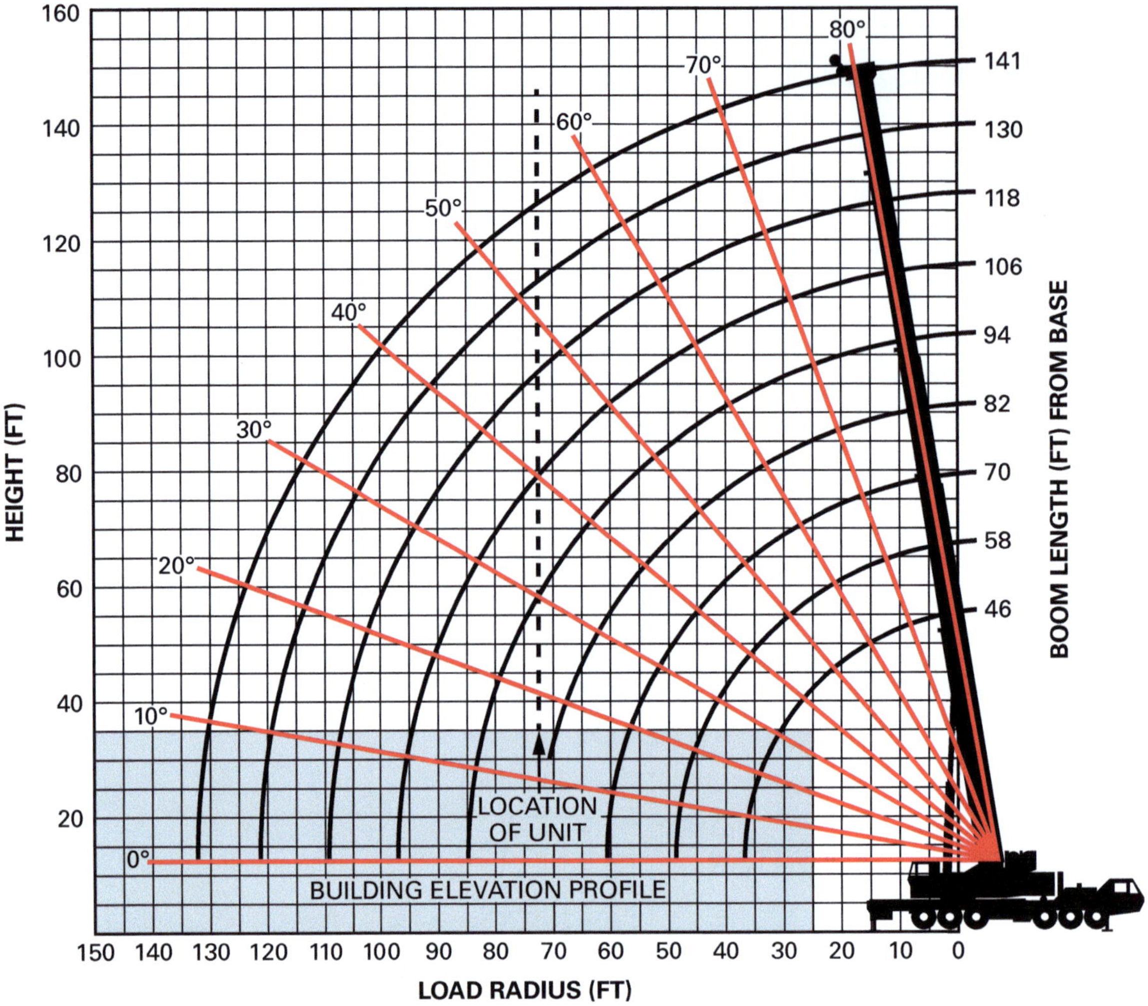

Figure 7 Elevation view of building on the crane's range diagram.

attempts. You may have to modify the crane's configuration more than once, also taking into account the suspended weights of hoist components, rigging, load, and other deductions. You may have to relocate the crane, or even choose another crane altogether. (Refer to Module 21301, which describes in detail how to perform these calculations.)

On the load chart, the load radius of 73 ft (22.3 m) falls between the listed 70 ft load radius and 80 ft load radius (*Figure 8*). Select the 80 ft load radius, because it will yield lower, more conservative crane capacities. Read across the row looking for a capacity with the boom at an angle roughly between 40 degrees and 55 degrees.

You will see that there are several options. Referring to the load chart (*Figure 9*), with the boom at 106 feet (32.3 m) long, the boom angle is only 39 degrees, which places it too close to the building's roof. At boom lengths of 118 ft (36 m) and 130 ft (39.6 m), higher boom angles are permissible. The crane's gross capacity at both boom lengths is 18,850 lb (8,550 kg). The capacity is 18,700 lb (8,482 kg) at the maximum main boom length of 141 ft (43 m), but the boom angle must be greater than 55 degrees due to structural limitations. The crane can't reach the condenser location at this angle. The 18,850 lb capacity is more than six times greater than the anticipated unit weight of 2,797 lb (1,269 kg) shown in *Appendix B*. For this lift, the load chart shows that the 130 ft boom length will provide the best stability and the greatest clearance to the building when swinging the boom.

CAUTION

Operators must not interpolate crane load chart capacities. Whenever the load radius, boom length, or boom angle falls between listed values on a load chart, the operator must select the adjacent value that yields a lower capacity. This practice provides an additional margin of safety, ensuring the load does not exceed the crane's capacity.

RATED LIFTING CAPACITIES IN POUNDS
46 FT – 173 FT BOOM ON OUTRIGGERS – OVER REAR

Radius in Feet	#0001 Main Boom Length in Feet (Power Pinned Fly Retracted)									#0002 Power Pin F'v Ext. & 141 ft
	46	58	70	82	94	106	118	130	141	172
10	300,000 (74.5)									
12	280,000 (72)	143,500 (76)	142,000 (79)							
15	235,000 (67.5)	143,500 (76)	141,500 (72)	130,000 [illegible]						
				[illegible]	(57)	[illegible] (61.5)	40,700 (65)	38,250 (68.5)	35,750 (71)	[illegible] (75.5)
60			36,299 (22.5)	36,200 (38.5)	36,200 (48.5)	34,300 (55)	33,600 (58.5)	30,750 (63.5)	28,500 (66.5)	26,350 (72)
70 73				26,050 (25)	26,050 (39.5)	26,050 (47.5)	26,050 (53)	24,750 (56)	23,100 (61.5)	22,000 (68.5)
80					18,850 (27)	18,850 (39)	18,850 (46.5)	18,850 (52.5)	18,700 (56.6)	18,500 (64.5)
90						13,500 (28)	13,500 (38.5)	13,500 (46.5)	13,500 (51.5)	15,250 (60.5)
100							9,390 (29)	9,390 (39)	9,390 (45.5)	17,500 (56.5)
110							6,080 (12.5)	6,030 (30.5)	6,080 (39)	10,100 (52)
120								3,390 (17.5)	3,390 (31)	7,530 (47.5)
130									1,150 (19.5)	5,390 (47.5)
140										3,610 (36.5)
150										2,100 (30)

Figure 8 Evaluating the crane's available boom configurations for the lift.

RATED LIFTING CAPACITIES IN POUNDS
46 FT – 173 FT BOOM ON OUTRIGGERS – OVER REAR

Radius in Feet	#0001 Main Boom Length in Feet (Power Pinned Fly Retracted)									#0002 Power Pin F'v Ext. & 141 ft
	46	58	70	82	94	106	118	130	141	172
10	300,000 (74.5)									
12	280,000 (72)	143,500 (76)	142,000 (79)							
15	235,000 (67.5)	143,500 (76)	141,500 (72)	130,000 [illegible]						
				[illegible]	(57)	[illegible] (61.5)	40,700 (65)	38,250 (68.5)	35,750 (71)	[illegible] (75.5)
60			36,299 (22.5)	36,200 (38.5)	36,200 (48.5)	34,300 (55)	33,600 (58.5)	30,750 (63.5)	28,500 (66.5)	26,350 (72)
70				26,050 (25)	26,050 (39.5)	26,050 (47.5)	26,050 (53)	24,750 (56)	23,100 (61.5)	22,000 (68.5)
80					18,850 (27)	18,850 (39)	18,850 (46.5)	18,850 (52.5)	18,700 (56.6)	18,500 (64.5)
90						13,500 (28)	13,500 (38.5)	13,500 (46.5)	13,500 (51.5)	15,250 (60.5)
100							9,390 (29)	9,390 (39)	9,390 (45.5)	17,500 (56.5)
110							6,080 (12.5)	6,030 (30.5)	6,080 (39)	10,100 (52)
120								3,390 (17.5)	3,390 (31)	7,530 (47.5)
130									1,150 (19.5)	5,390 (47.5)
140										3,610 (36.5)
150										2,100 (30)

Figure 9 Crane capacities at the available boom lengths.

Fill in the applicable blanks in the crane section of the LDS with the information that you know or can assume at this point. More data will become available later in the planning that may alter some of this information.

1.2.3 The Load and Rigging

The next section of the LDS pertains to the load and its characteristics, the required rigging, and the hoist components needed. The HVAC unit data sheet (*Appendix B*) indicates the shipping weight is 2,797 lb. You should use this value as its hoist weight, even if some shipping or packing materials are removed before the lift.

The data sheet also identifies the manufacturer's minimum lifting and rigging requirements (*Figure 10*). The condenser is factory mounted on a structural steel skid which includes four lifting lugs. The manufacturer recommends suspending the load from four slings shackled to the skid and using a spreader system to stabilize the slings. The manufacturer also suggests attaching all four slings directly to the lifting hook.

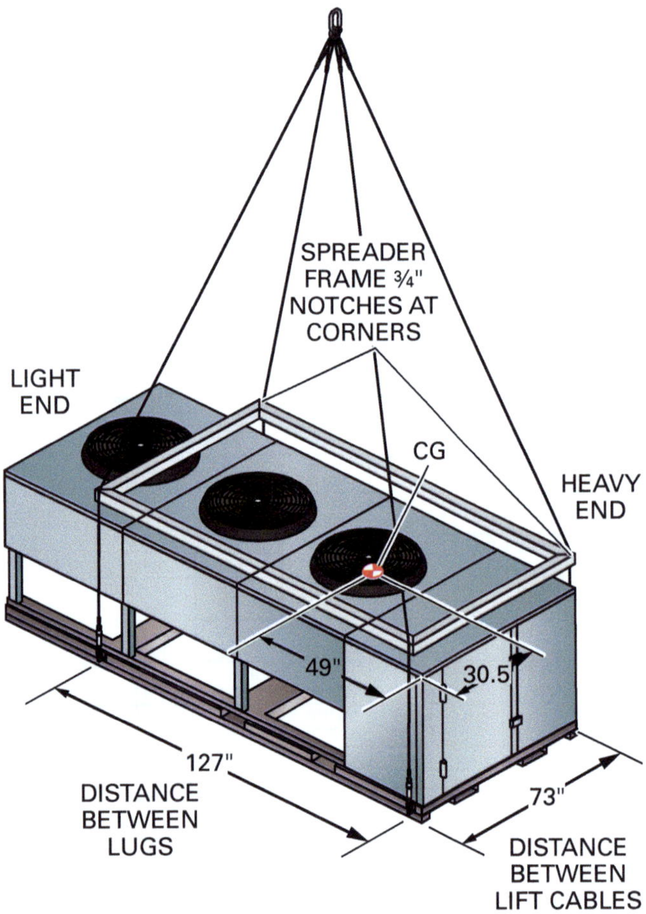

Figure 10 Rigging diagram for the HVAC unit.

Riggers have determined that the condenser will require the following rigging components:

- Four (4) 2,500 lb shackles (0.2 lb each)
- One (1) hoist shackle (2.3 lb)
- Four (4) 12-ft wire rope slings (3.9 lb each)
- One (1) 2 × 4 wooden spreader frame, 74" × 121" (100 lb)

Use this information to fill in the LDS load and rigging weight section.

Part of the planning for this lift includes determining what other hoisting hardware the crane will have to carry on the day of the lift, and the need for any alteration of the crane's configuration. Crane owners don't change the hoisting hardware configuration unless needed, so you may be able to obtain the information ahead of time for planning purposes. For this lift, assume the crane is equipped with the following hoisting components identified in the load chart:

- 150-ton, 8-sheave block
- Auxiliary boom head installed
- 10-ton headache ball (suspending the load)

Enter the weights of all applicable hoist items, then find the sum of the weights in the form to obtain the total weight acting on the crane boom:

2,797	(HVAC condenser weight)
119	(rigging)
5,254	(8-sheave block)
261	(auxiliary boom head)
+ 560	(10-ton headache ball)
8,991	**total lifted weight**

1.2.4 Crane Capacity Calculations

Accurately completing the crane capacity section of the LDS is essential to a successful lift plan. The lift director will use the capacity information as a factor to determine the lift category—standard or critical. The capacity section provides three columns, permitting calculations at different points in the lift. The planner may need these for variations in the load radius that occurs during the lift.

Before developing these estimates, a planner needs to know where the load will be staged for lifting. If the project manager doesn't know, suggest a spot that will make planning easier. One of the columns should reflect the capacity at the staging area load radius. Another column should show the capacity at the landing site. For each column of the LDS, follow these steps to determine the percent capacities:

Step 1 Identify at the column head the limiting condition for which you made calculations.

Step 2 On the next row, enter the total lifted weight calculated in the previous section.

Step 3 Enter the actual load radius for the conditions described in the column.

Step 4 Enter the load chart load radius used to determine gross capacity. This assures the lift director that you applied the proper lower capacity for a given load radius.

Step 5 Enter the load chart capacity for the configuration.

Step 6 Divide the total weight lifted by the load chart capacity in that column. Enter this value as a percentage (%).

For the condenser destination on the roof:

$$\text{\% of chart capacity} = \frac{\text{total lifted weight}}{\text{load chart capacity}} \times 100$$

$$\text{\% of chart capacity} = \frac{8{,}991 \text{ lb}}{18{,}850 \text{ lb}} \times 100$$

$$\text{\% of chart capacity} = 0.477 \times 100 = \mathbf{47.70\%}$$

Enter the largest percentage of load chart capacity you calculated in this section in the next line; the text formatting of this line draws reviewers' attention to it.

The following line evaluates the capacity of the reeved hoist line in relation to the total weight. Note that the lift will use the single-part auxiliary hoist with the 10-ton headache ball hook. Consult the crane specifications to determine the minimum breaking strength (MBS) of the hoist cable. For this crane, the manufacturer recommends $\frac{3}{4}$" IWRC rope for the auxiliary hoist, which has a MBS of 51,800 lb, per the vendors specifications.

For the total load, you need to add up only those weights lifted by the hoist cable:

2,797 (HVAC condenser weight)
119 (rigging)
+ 560 (10-ton headache ball)
3,476 total suspended weight

Enter this result in the Total Suspended Load blank. Enter the hoist line MBS in the Reeved Capacity blank. Divide the suspended weight by the line MBS and multiply by 100%:

$$\text{max \% reeved capacity} = \frac{\text{total suspended weight}}{\text{hoist-line MBS}} \times 100 =$$

$$\frac{3{,}476 \text{ lb}}{51{,}800 \text{ lb}} \times 100 =$$

$$0.067 \times 100 = \mathbf{6.70\%}$$

Enter the result in the Maximum % Reeved Capacity blank on the LDS (*Figure 11*). Industry standards specify that the ratio of maximum load to ultimate strength of a lifting system should be less than 1:5, or 20 percent.

1.2.5 Load and Crane Considerations

For the HVAC unit lift, the crane capacity and rigging configuration appear to be well within permitted limits. However, *Appendix B* shows that the condenser's CG is not exactly in the center of the lift attachment pattern. It is located closer to one end of the unit. There is some potential for the out-of-balance load to shift or become unstable during the lift. When a load's CG is off-center, the rigging supervisor may require lifting the load with an adjustable equalizer beam. Depending on the design of the beam, riggers can position the beam's lifting lug(s) so that the load's CG is directly below the load block.

Total Additional Weight:			lb
GROSS LOAD (sum of all items above):			8,991 lb
Crane Capacities Condition:	Placing Condenser		
Total Load to Crane:	8,991 lb	lb	lb
Planned Radius:	73 ft	ft	ft
Chart Radius Used:	80 ft	ft	ft
Chart Capacity:	18,850 lb	lb	lb
% of Chart Capacity:	47.7 %	%	%
MAXIMUM PERCENTAGE OF CHART CAPACITY TO BE USED IS			47.7 %
Total Suspended Load:	3,476 lb	Reeved Capacity:	51,800 lb
MAXIMUM PERCENTAGE OF REEVED CAPACITY PLANNED TO BE USED IS			6.7 %
Document Attachments			
Crane Layout	Lift Plan Category	CRP Source/Calculations	

Figure 11 Completing the Crane Capacities section of the LDS.

Note in the rigging figure of *Appendix B* that the lifting lugs at the heavy end of the condenser are 49" lengthwise from the condenser's CG. The distance between the condenser lifting lugs along the side of the unit is 127". Therefore, the distance between the unit's CG and the light-end lugs is:

127.00" – 49.00" = **78.00"**

If using an equalizer beam, the riggers position the heavy-end suspension lugs at 49" from the lifting eye, and 78" from the lifting eye for the light-end lugs (*Figure 12*). Note that the front-to-back position of the condenser's CG is also off center, but the distance is acceptable. For this lift, the condenser manufacturer has determined that a single-point lift will provide sufficient stability to place the condenser, so an equalizer beam is not required.

A single crane can place the HVAC condenser on the roof of the building as described in this scenario. However, there are situations when, either due to the excessive weight or the awkward location of the landing site, multiple cranes must handle the load (*Figure 13*). This curriculum provides information on multi-crane lifts and the relevant load calculations in Module 21303.

The lift planner may consider lifting a load with two cranes attached to each end of an equalizer beam. For loads having CGs near their geometric centers, the load lifted by each crane is nearly the same. Determining crane capacities for such lifts is fairly simple.

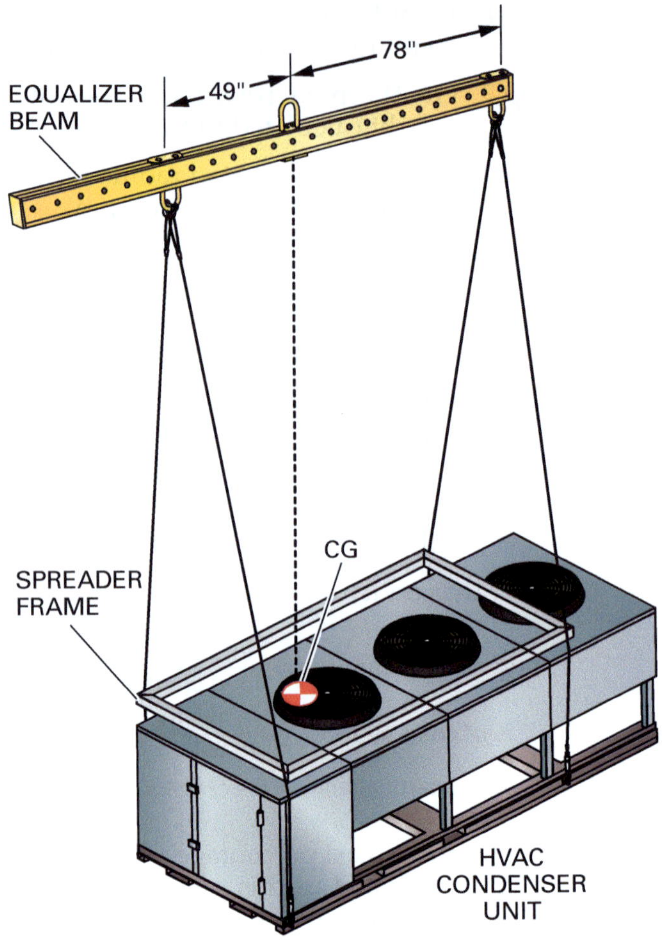

Figure 12 Positions of the adjustable suspension lugs on an equalizer lifting beam.

Figure 13 When positioning a load becomes complicated, two or more cranes may be required.

For a load with the CG well off-center, the calculations are more complicated. The crane hoisting the heavy end of the load will use a higher percentage of its gross capacity than an identical crane hoisting the lighter end, everything else being equal. If the two cranes have different gross capacities (which is common), the calculations become even more difficult. The lift planner must carefully consider the movement of the load and the capacities of the cranes at every moment and position during the lift. Planning and coordinating dual-crane lifts can be very complex. For this reason, multi-crane lifts are always in the critical lift category.

1.2.6 Reference Documents and Attachments

The Documents and Attachments section of the LDS is a checklist of typical resources that a lift planner may need to consult or include in the lift plan for supervisory review. The HVAC unit lift plan should probably include, at a minimum, the following information or documents:

- Crane layout and configuration
- Crane load chart extract
- Load/crane clearances
- Drawing of load
- Load weight and CG information
- Rigging arrangement
- Crane foundation evaluation
- **Ground bearing pressure (GBP)** data/calculations
- Job hazard analysis (JHA)
- Weather forecast

All crane-related information can be appended to the plan by including the crane's load chart or a certified copy of the relevant sections. The attachment should provide, if possible, a sketch of the crane setup for the lift, and/or include a description of the crane and lifting accessories. It should also include the crane's range diagram and tables of deducted weights. The lift director or site policies may require that each photocopied page bear a signature certifying that it is a true copy of the load chart.

The minimum building clearance for the boom can be determined from the range diagram and building elevation plan superimposed on it to scale, as shown in *Figure 14*. Policies may require a more formal estimate performed by the project engineer, who will produce a scale drawing illustrating the minimum lift clearances. The planner should add a note to the LDS stating that, before swinging the load over the roof, the operator must ensure the boom angle equals or exceeds the load chart angle specified in the planning.

Planners should reference the HVAC unit manufacturer's specifications, which will illustrate the unit in three views, as in *Appendix B*. The plans will provide the unit's dimensions, lift points, center of gravity, and the recommended rigging. If policies require a paper copy of this information, include only the pages of the manual relevant to the lift.

The rigging arrangement should be at least an accurate sketch of the intended rigging accessories, slings, and attachments. For complex lifts, the project engineer or the load's manufacturer may provide a scale drawing of the rigging and an itemized list with rigging weights. Note that the specified weights of equipment are typically dry weights. The addition of any fluids to the equipment prior to the lift must also be considered. For example, a power transformer is quite heavy alone, but a significant amount of weight is added if it is lifted into position after it is filled with oil.

Include attachments for the crane foundation, matting, and GBP if the project engineer determines these items are needed (*Figure 15*). For this lift, the ground that will support the crane is undisturbed, firm, and level. The project engineer has previously evaluated the GBP for earlier lifts at this site and has determined that ground support is adequate. The planner should include a memo to this effect in the lift plan package.

Before finalizing the plan, consult with the site safety officer for a job hazard analysis (JHA). Discuss the crane communications that will be used, the number of signal persons required, access and traffic controls needed, PPE, and any other site-specific policies that apply. Summarize this information in an attachment and include any operational notes on the LDS. Site or employer policies may require the safety officer to sign the document. *ASME P30.1, Nonmandatory Appendix A—Example Lift Plan,* includes an example of a pre-lift safety checklist used to summarize this information. The lift director can also use the form to document the safety portion of a pre-lift briefing.

An efficient lift planner always checks the weather forecast for the lift date to identify any potential problems. If the forecast is for windy conditions, thunderstorms, or heavy rains, this is clearly noted in the lift plan. Postponements for weather are relatively common. Planning should include go/no-go criteria for any contingencies that may arise.

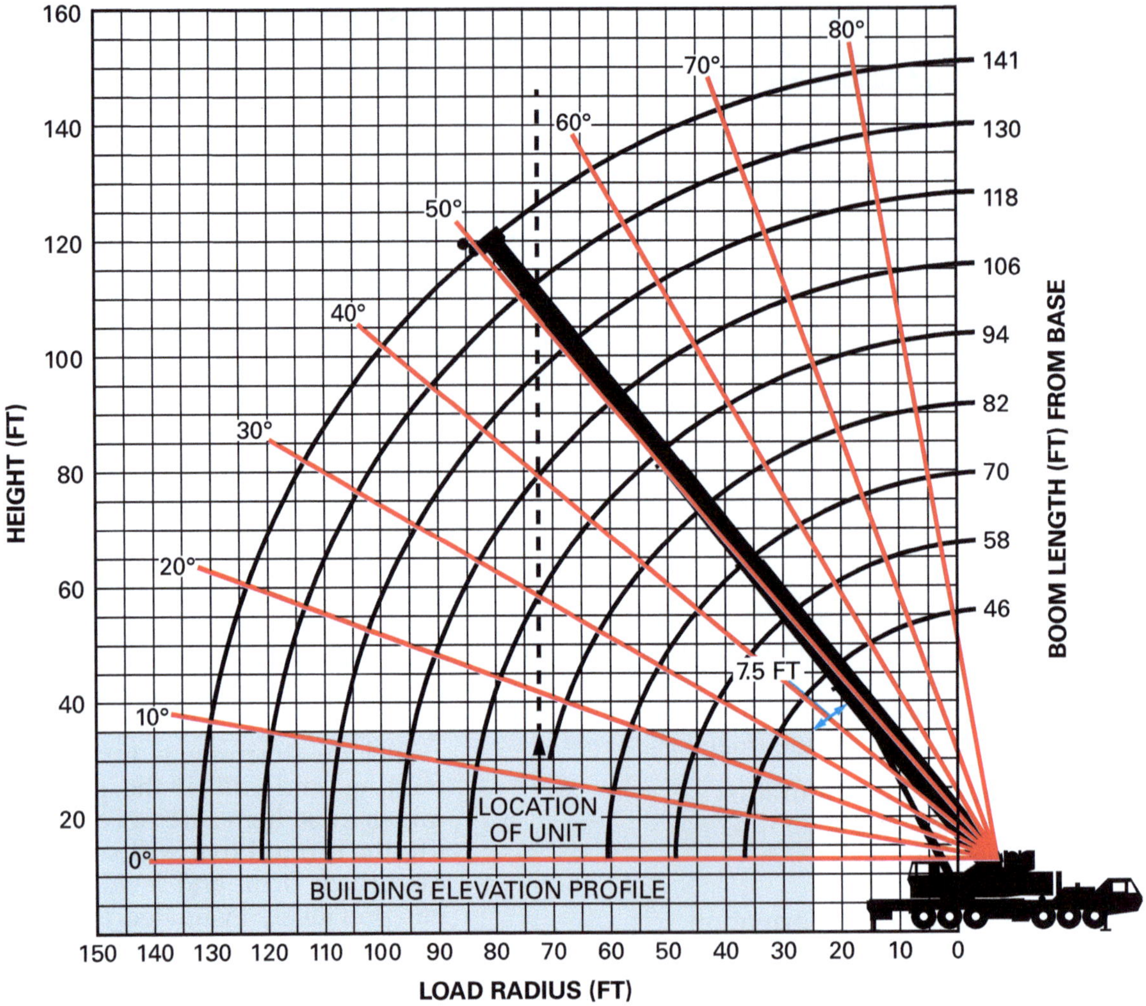

Figure 14 Determining building clearance from the crane range diagram.

Figure 15 Inadequate ground bearing pressure (GBP) information can lead to crane tipping incidents.

1.2.7 Notes, Review, and Approvals

Identify any information that reviewers should be aware of in the Notes section of the LDS. In addition to those already discussed, considerations that might be listed in this section include the following:

- Anything that might potentially limit crane capacity or re-categorize the lift
- People and equipment (*e.g.*, taglines) required for controlling the load during the lift
- Crane communication details and the signal persons/spotters required
- Establishing access and traffic controls during the lift
- Referencing emergency action plans and contact information in the event of accidents

When the planner completes the lift plan package, the next person to receive it may be the project engineer, site safety officer, or the lift director. The lift director reviews the lift plan and identifies the category of the lift.

Assume that you are the lift director for the HVAC unit lift. When categorizing the lift of the condenser, consider and answer the following questions:

- Can a single crane conduct the lift?—**Yes**
- Is the total weight lifted by the crane less than 75% of its determined capacity?—**Yes**

> **NOTE**
>
> While 29*CFR* 1926.751 classifies any lift greater than 75 percent of load chart capacity for steel erection as a critical lift, this technically applies only to steel erection. Many organizations, however, apply the same percentage of capacity to all lifts when determining if the lift is a critical lift. Organizations may establish both higher or lower percentages of capacity as the threshold for a critical lift when it is not related to steel erection, but for steel erection, the threshold can be lower than 75 percent, but not higher.

- Is the unit or its contents inherently hazardous, contain materials immediately dangerous to life or health, or extremely valuable, hard to replace, or irreplaceable?—**No, but any refrigerant it may contain is controlled by US Environmental Protection Agency (EPA) Clean Air Act, and must not be vented to atmosphere. An unintentional release, depending on the volume, may require reporting to the EPA.**
- Is the load's CG fixed and is the load considered stable?—**Yes**
- Does the lift present any unusual hazards to the public, utilities, or infrastructure?—**No**
- Are there any unique hazards to the load or crane?—**Only the building itself; accounted for in the plan**
- Are there any site or other hazards that could reclassify the lift?—**None identified**

Based on this evaluation, classifying it as a standard lift is justified. Lift directors understand that this is a tentative classification that can change at any time if an unanticipated factor is encountered.

1.2.8 Critical Lift Plans

ASME P30.1 requires that lift plans for critical lifts be in a written form. Due to the complexities or potential hazards involved, critical lift plans can be large, detailed packages, addressing all applicable sections discussed in Chapter 5 of *ASME P30.1*. Since there is no set format for lift procedures, critical lift plans differ in scope and content with each lift, by company policy, and from one work site to another.

Development of a critical lift plan follows a process similar to that for a standard lift. However, the necessary details usually require the input of multiple parties. *Appendix E* provides an example of a critical lift planning form.

In addition to sections devoted to definitions, training, and applicable regulations, these plans often include a detailed lift procedure. The procedure guides workers step-by-step through preparation for the lift, staging the load, and installing the load's rigging. For high-risk lifts involving dangerous materials or high-value loads, the procedure may require signatures for completion of each step. A critical lift may also require a test lift using a dummy weight. If the lift involves precision assembly or securement of the load as it is placed, the procedure may specify the need for other crafts and quality assurance inspections.

1.3.0 Implementing a Lift Plan

If the planning is thorough, putting a lift plan into action is straightforward, if not routine. Project managers work with their team to set the plan in motion. Efficient implementation of a lift plan requires frequent communication among the participants. A post-lift review is often done to identify issues to consider in future lifts.

1.3.1 Pre-Lift Review/Meeting

All participants need to review the lift plan. This review may be for final approval, or part of a progression towards a final plan. The review may identify conditions or requirements that are different than those originally considered. Reviewers inform the lift planner in such cases, especially if the information could result in reclassifying a lift as critical. Even if project supervision has reviewed and approved a lift plan prior to the scheduled activity, key participants should review the plan again before conducting a pre-lift meeting. *ASME P30.1* notes that the lift director should determine how often to review the lift plan during a series of lifts. It isn't usually necessary to review the lift plan between each lift.

The complexity and category of a lift should dictate the scope and formality of a pre-lift meeting. For simple jobs or repetitive lifts, a simple face-to-face meeting at the work site involving the crane operator, signal person, rigger, and the supervisor responsible for the load may be sufficient. For the initial meeting of a complicated and/or critical lift, gathering in an office may be necessary (*Figure 16*). Audiovisual services can make such a meeting more effective. Participants should include the site supervisor or representative, site safety officer, site engineer, lift director, rigging supervisor, craft supervisor responsible for the load (if applicable), the crane operator, all signal persons, and any other individuals that will provide site services or traffic control.

The lift director should conduct the pre-lift meeting. Items to review may include, but are not limited to, the following:

- Main sections of the LDS, as applicable
- Rigging requirements
- Site conditions and weather
- Coordination of services
- Primary and backup crane communications
- Site policies pertaining to the lift
- Contingency plans
- Job hazard analysis, if appropriate

Critical lift meetings will usually involve significantly more detail. Policies may require completion of an JHA checklist and review at the meeting, which all participants must sign as part of a formal safety briefing.

The lift director must resolve any concerns raised at the meeting before lifting begins. It is possible that a concern could result in re-categorizing the lift, which would require additional planning or preparations. Only the designated lift director has the authority to proceed with executing the lift plan.

Figure 16 A formal pre-lift meeting is necessary for complicated or critical lifts.

1.3.2 Executing the Lift

Before a lift may commence, all preparations pertaining to the lift must be complete, including the following:

- The load must be in its designated staging location.
- The rigging must be tested, inspected, and installed per the lift plan.
- The crane must be ready in the location and configuration specified by the lift plan, including any certifications, testing, and inspections.
- All designated personnel are on location with required PPE and equipment.
- The crane operator and signal person(s) have established and tested primary and backup crane communications.
- Vehicular and personnel traffic controls are in place.
- Weather and site conditions support the lifting operations.

The lift director must determine that no condition exists that would prevent the lift from proceeding as planned. If the team encounters something unexpected (*Figure 17*), the lift director must stop the lift and evaluate the situation. The load must be placed in a safe condition during the wait, which usually means lowering it to the ground.

Any participant can issue a Stop command during the lift. *Section 5-4.3* of *ASME P30.1* lists several actions the lift director should consider if the operation deviates from the plan. If a change to the plan is necessary, the lift director should communicate the change to all participants before continuing. Only the lift director can restart a lift after it has been stopped.

1.3.3 Post-Lift Review

Conducting a post-lift review is always a good management practice. *ASME P30.1* suggests that lift directors hold reviews after both standard and critical lifts. These activities provide the opportunity to learn something new from the lifting operation. Many organizations refer to this as "lessons learned".

To be truly effective, lift directors should schedule a review as soon as possible after the operation. Participants in the review should identify any problems in the development, planning, and execution of the lift. They should also suggest improvements that could apply to future lifts or site operations in general. The lift director will then communicate any action items to the responsible parties for consideration and correction, as applicable.

Figure 17 At the onset of dangerous weather, stop the lifting operation.

Additional Resources

Planning for Load Handling Activities, ASME P30.1. New York: The American Society of Mechanical Engineers.

"Load Charts for Use With Written Examinations—Grove TM1500." Manitowoc Crane Group and NCCER. **www.nccer.org/docs/default-source/Load-Charts/grovetm1500.pdf?sfvrsn=853e0e4f_029**

CFR 1926, Subpart CC, *Cranes and Derricks in Construction*. **www.ecfr.gov**

1.0.0 Section Review

1. The most important purpose of lift-planning standards is to _____.
   a. protect the careers of all lift participants
   b. protect lives and prevent injuries
   c. prosecute violations of crane safety
   d. provide guidance to lawmakers to create and enforce crane-related safety laws

2. The first step in determining the crane configuration for a particular lifting operation is _____.
   a. determining the minimum single-line pull
   b. calculating the number of parts of hoist line
   c. finding the crane's gross capacity from the load chart
   d. consulting the crane's range diagram

3. If a lift plan reviewer identifies a condition or requirement that differs from what the planner considered, the reviewer should _____.
   a. make a pen-and-ink change to the plan, then approve it
   b. request a second opinion from the supervisor
   c. inform the lift planner of the situation
   d. assume that the next reviewer will correct the problem

Summary

A lift plan is essential to the success of a lifting operation. It is the framework that brings together and organizes all aspects of mobile crane operations presented in this curriculum. Even if safety were not the most important factor, executing a lift without proper planning would inevitably result in false starts, work delays, damage to the load or other structures, and wasted time, effort, and money.

Only qualified individuals with extensive experience in crane and construction site operations should conduct and lead lift planning. Lift planners must be able to intelligently consult with specialists in the various crane, engineering, and construction professions to put together a safe, effective plan.

All participants must know their role in implementing the plan and be qualified to do so. Lift directors must be able to communicate the plan requirements to all team members, continually monitor the operation, and be open to lessons learned after its completion.

Review Questions

1. The person responsible for developing the lift plan is the _____.
   a. crane operator
   b. site supervisor
   c. lift director
   d. lift planner

2. Planners would consider any of the following to be a complex crane lift *except* _____.
   a. a single crane lifting materials from the ground to the top of a structure
   b. a lift involving two or more cranes
   c. having to turn a or tilt the load during the lift
   d. lifting an unstable load

3. Planning a lift for a specific time of day could minimize the adverse effects of _____.
   a. ground bearing pressure
   b. load buoyancy
   c. the sun's glare
   d. rain or lightning

4. The term that distinguishes many of the requirements of critical lift planning from standard lift planning is _____.
   a. *may*
   b. *shall*
   c. *can*
   d. *might*

5. During lift planning, positioning the crane where the load radius is minimized _____.
   a. places unnecessary stress on the outriggers
   b. creates the least stable situation
   c. improves stability but sharply reduces capacity
   d. maximizes capacity and crane stability

6. The best choice for stabilizing an off-balance load before lifting it is to _____.
   a. add weights to the light end of the load to balance it
   b. hook a tag line to each end so the riggers can manipulate it
   c. use an adjustable equalizer beam
   d. use a second crane

7. The individual responsible for categorizing the lift operation is the _____.
   a. site superintendent
   b. lift director
   c. lift planner
   d. crane operator

8. A concern raised during a pre-lift meeting could result in _____.
   a. re-categorizing the lift
   b. changing the lift planner
   c. revising the crane's load chart
   d. cancellation of the lift

9. Which factor, by itself, would *not* normally delay a scheduled lift?
   a. Discovering the load rigging does not have inspection tags.
   b. The load weight is greater than planned.
   c. Clouds blocking the sun.
   d. The crane's configuration is different from the lift plan.

10. Participants should conduct post-lift reviews to _____.
    a. establish blame for any mistakes that occurred during the lift
    b. obtain lessons learned for planning future operations
    c. approve the lift plan
    d. critique the work of the lift planner

Trade Terms Introduced in This Module

Ground bearing pressure (GBP): The maximum force per unit area (pressure), measured in pounds per square foot, that can be exerted on the ground's surface without causing deformation of the surface. Can only be determined by testing.

Lift data sheet (LDS): The essential page or pages of a written lift plan that organize and summarize the important information required to develop a lift plan.

Building Architectural Plans

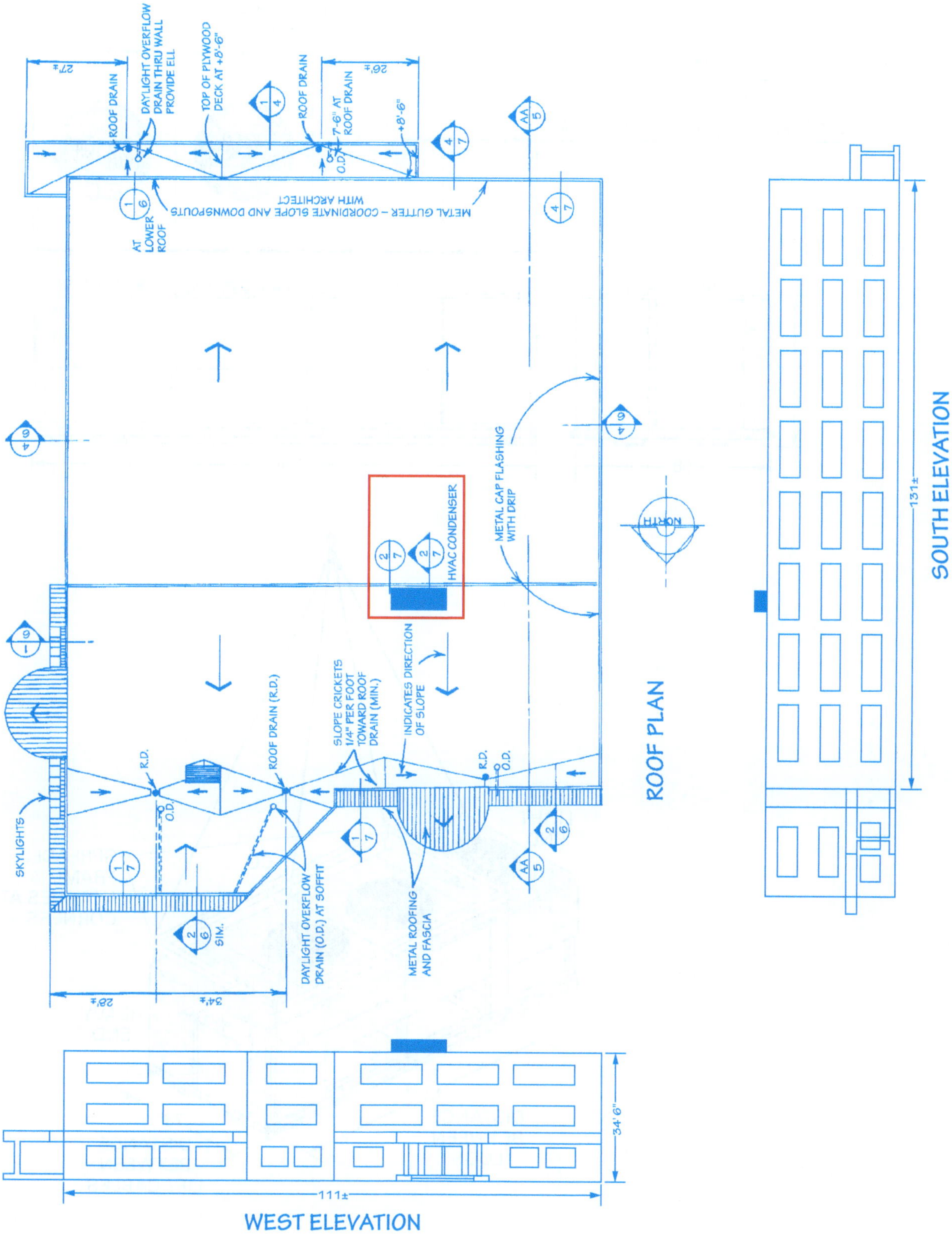

HVAC Condenser Schematics

WEIGHT DATA

SHIPPING	2797 LB
INSTALLED	2437 LB

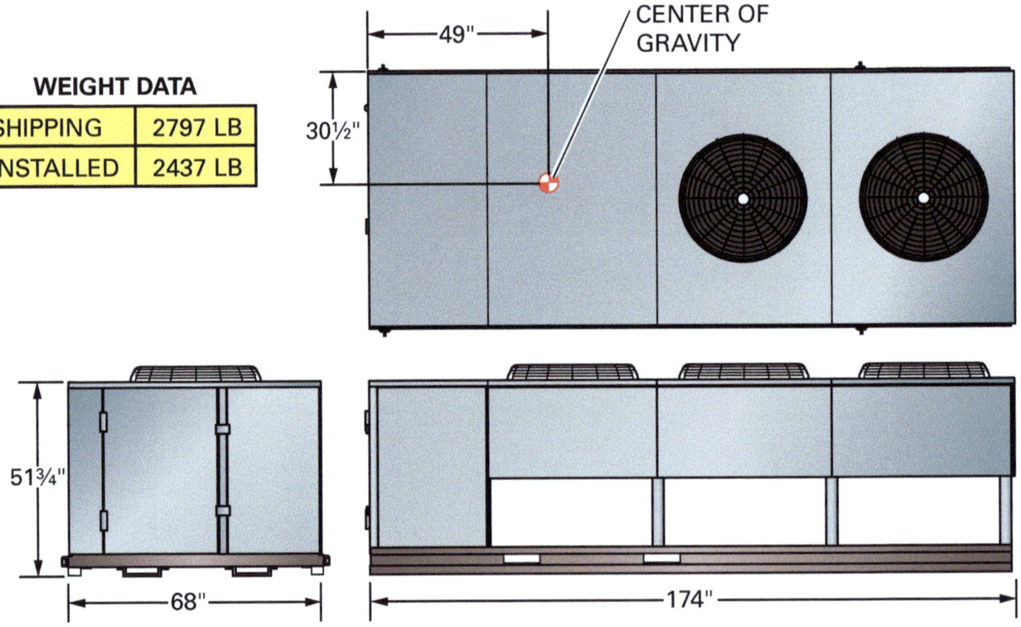

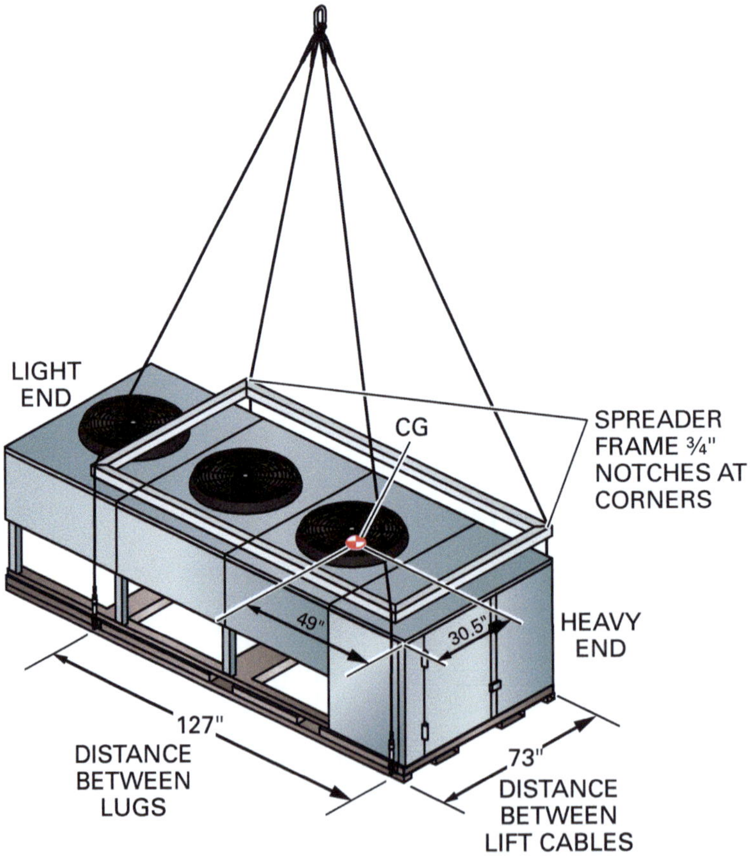

GROVE TM1500 LOAD CHART

GROVE CRANE
A GROVE WORLDWIDE COMPANY

TM1500

Carrier-mounted hydraulic crane /85% Domestic
46 ft. – 173 ft. aerial power pinned boom

WORKING RANGE DIAGRAM
(BOOM DEFLECTION NOT SHOWN)

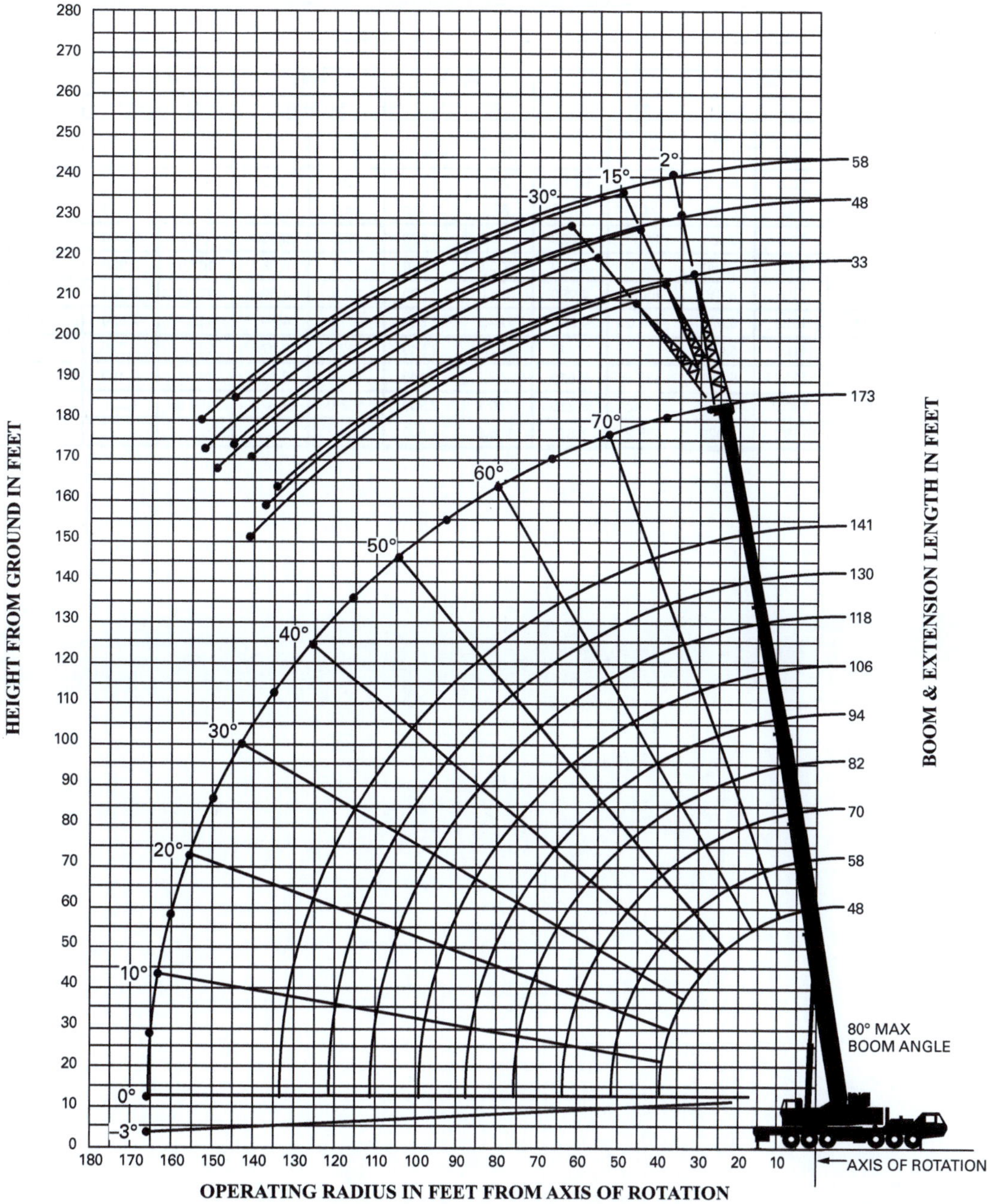

46 FT. - 88 FT. FIXED JIBS ON OUTRIGGERS - 360°

Main Boom Angle (Deg.)	46 FT. JIB						60 FT. JIB					
	#0071		#0072		#0073		#0074		#0075		#0076	
	5° OFFSET		17° OFFSET		30° OFFSET		5° OFFSET		17° OFFSET		30° OFFSET	
	Rad. Ref. (ft.)*	Cap. lbs.**	Rad. Ref. (ft.)*	Cap. lbs.**	Rad. Ref. (ft.)*	Cap. lbs.**	Rad. Ref. (ft.)*	Cap. lbs.**	Rad. Ref. (ft.)*	Cap. lbs.**	Rad. Ref. (ft.)*	Cap. lbs.**
80	44.4	14,900	52.4	12,600	59.8	8,120	47.1	12,500	58.6	8,530	68.7	5,350
77.5	53.7	14,450	61.6	11,650	68.8	7,750	57.1	11,650	68.2	7,850	78.1	5,070
75	62.9	14,000	70.6	10,850	77.6	7,410	67.0	10,800	77.9	7,260	87.3	4,830
72.5	72.0	13,600	79.5	10,150	86.2	7,120	76.8	10,000	87.2	6,760	96.5	4,610
70	81.0	12,470	88.3	9,550	94.8	6,860	86.5	9,230	96.5	6,330	105.3	4,420
67.5	89.8	11,250	96.9	8,980	103.2	6,630	96.0	8,500	105.6	5,950	114.2	4,250
65	98.4	8,850	105.3	7,940	113.2	6,430	105.2	7,850	114.5	5,620	122.7	4,100
62.5	106.9	6,900	113.5	6,550	119.2	6,210	114.3	6,430	123.2	5,270	130.9	3,970
60	115.1	5,370	121.5	5,080	126.9	4,800	123.2	5,010	131.6	4,500	138.9	3,580
55	131.0	3,050	136.8	2,870	141.6	2,700	140.2	2,980	147.7	2,670	154.1	2,370
50	145.9	1,550	151.1	1,460	155.2	1,370	156.2	1,430	162.7	1,260	168.2	1,100

Main Boom Angle (Deg.)	74 FT. JIB						88 FT. JIB					
	#0077		#0078		#0079		#0080		#0081		#0082	
	5° OFFSET		17° OFFSET		30° OFFSET		5° OFFSET		17° OFFSET		30° OFFSET	
	Rad. Ref. (ft.)*	Cap. lbs.**	Rad. Ref. (ft.)*	Cap. lbs.**	Rad. Ref. (ft.)*	Cap. lbs.**	Rad. Ref. (ft.)*	Cap. lbs.**	Rad. Ref. (ft.)*	Cap. lbs.**	Rad. Ref. (ft.)*	Cap. lbs.**
80	50.8	9,470	64.3	6,000	78.0	3,530	52.9	7,300	70.1	3,970	86.0	2,330
77.5	61.2	8,790	74.6	5,470	87.8	3,310	64.1	6,580	80.9	3,590	96.2	2,060
75	71.6	7,980	84.7	5,020	97.3	3,120	75.2	5,950	91.5	3,260	106.3	1,820
72.5	81.9	7,280	94.7	4,630	106.8	2,950	86.1	5,380	102.0	2,980	116.2	1,610
70	92.0	6,610	104.3	4,290	116.0	2,800	96.9	4,800	112.2	2,740	126.0	1,450
67.5	101.9	6,030	114.0	4,000	124.9	2,660	107.5	4,300	122.2	2,530	135.3	1,330
65	111.6	5,500	123.3	3,740	133.7	2,540	117.9	3,870	132.1	2,340	144.6	1,220
62.5	121.1	5,040	132.5	3,510	142.1	2,440	128.0	3,510	141.7	2,180	153.5	1,150
60	130.2	4,050	141.2	3,260	150.2	2,350	138.0	2,870	150.9	1,970	162.1	1,080
55	148.1	2,020	158.2	1,810	165.8	1,600						

A-829-007220

GROVE CRANE

A GROVE WORLDWIDE COMPANY

TM1500

Carrier-mounted hydraulic crane /85% Domestic
46 ft. – 173 ft. aerial power pinned boom

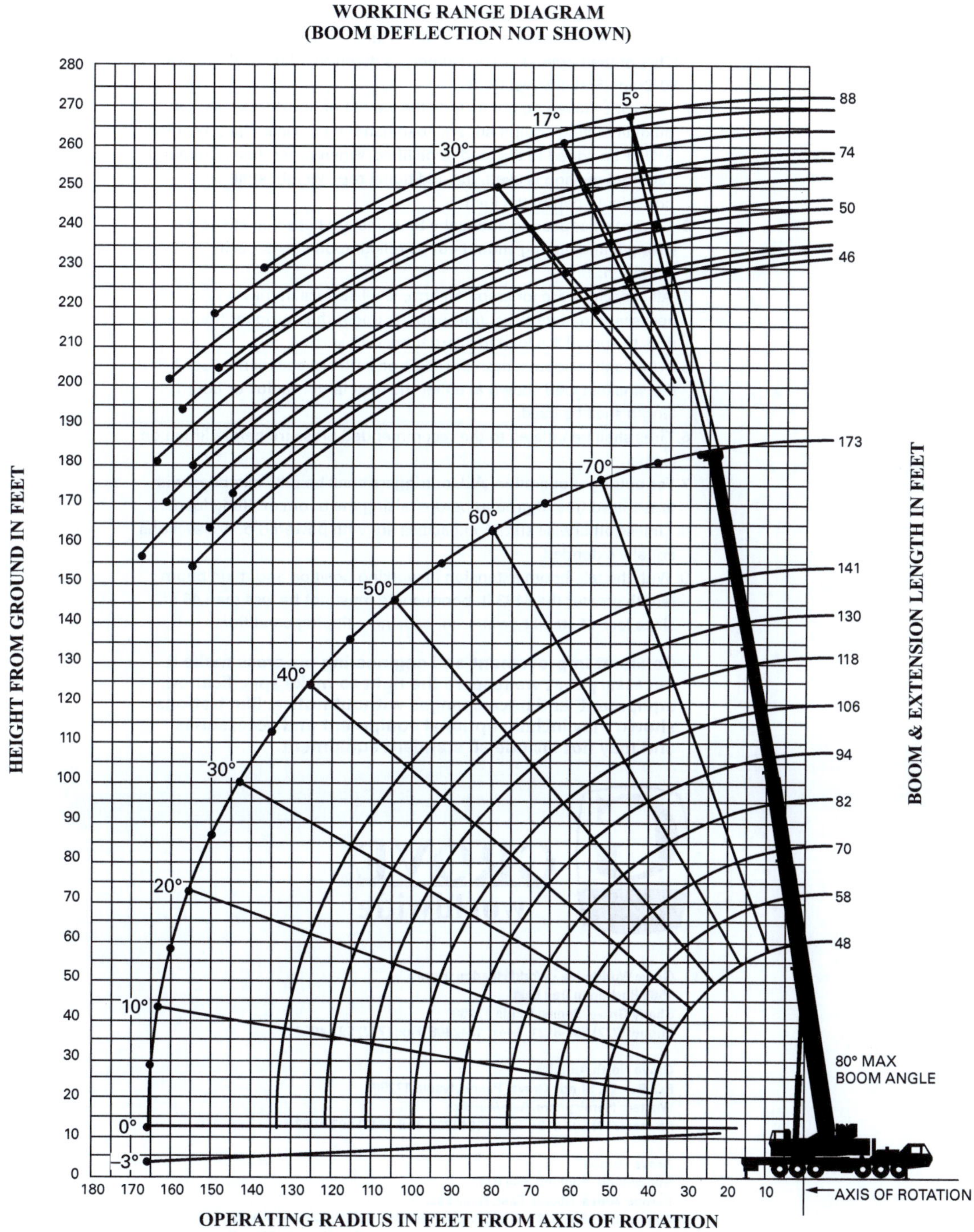

NOTES FOR LIFTING CAPACITIES

WARNING: THIS CHART IS ONLY A GUIDE. The Notes below are for illustration only and should not be relied upon to operate the crane. The individual crane's load chart, operating instructions and other instruction plates must be read and understood prior to operating the crane.

1. All rated loads have been tested to and meet minimum requirements of SAE J1063 OCT80 - Cantilevered Boom Crane Structures - Method of Test, and do not exceed 85% of the tipping load on outriggers (75% of the tipping load on rubber) as determined by SAE J765 OCT 80 Crane Stability Test Code.
2. Capacities given do not include the weight of hookblocks, slings, auxiliary lifting equipment and load handling devices. Their weights MUST be added to the load to be lifted. When more than minimum required reeving is used, the additional rope weight shall be considered part of the load.
3. Capacities appearing above the bold line are based on structural strength and tipping should not be relied upon as a capacity limitation.
4. All capacities are for crane on firm, level surface. It may be necessary to have structural supports under the outrigger floats or tires to spread the load to a larger bearing surface.
5. When either boom length or radius or both are between values listed, the smallest load shown at either the next larger radius or boom length shall be used.
6. Tires shall be inflated to the recommended pressure before lifting on rubber.
7. For outrigger operation, ALL outriggers shall be fully extended with tires raised free of ground before raising the boom or lifting loads.
8. Unless otherwise stated, capacities are with powered boom sections equally extended.
9. With the tele boom extension in working position and main boom length greater than 133 ft. boom angle must not be less than 46.5°, since loss of stability will occur causing a tipping condition.

Constant improvement and engineering progress make it necessary that we reserve the right to make specification, equipment, and price changes without notice. Illustrations shown may include optional equipment and accessories and may not include all standard equipment.

GROVE® worldwide

Grove Worldwide - World Headquarters
1565 Buchanan Trail East
Shady Grove, Pennsylvania 17256
Phone: (717) 597-8121 Telex: 1842308 Fax: (717) 597-4062

Grove North America
P.O. Box 21, Shady Grove, Pennsylvania 17256
Western Hemisphere, Asia/Pacific
Phone: (717) 597-8121 Telex: 1842308 Fax: (717) 597-4062

Grove Europe*
Sunderland, England SR4 6TT
Europe, Africa, Middle East, Near East
Phone: (091) 565-6281 Telex: 53484 CRANESG FAX: (091) 564-0442
***Grove Europe Limited, Registered in England, Number 1845128, Registered office, Crown Works, Pallion, Sunderland, Tyne & Wear, England SR4 6TT.**

FORM NO: LC-TMI500-Dom. P/N 3-416 793-3M PRINTED IN U.S.A.

RATED LIFTING CAPACITIES IN POUNDS
46 FT. - 173 FT. BOOM ON OUTRIGGERS - OVER REAR

Radius in Feet	#0001 Main Boom Length in Feet (Power Pinned Fly Retracted)									#0002 Power Pin Fly Ext. & 141 ft.
	46	58	70	82	94	106	118	130	141	173
10	300,000 (74.5)									
12	280,000 (72)	143,500 (76)	142,000 (79)							
15	235,000 (67.5)	143,500 (72.5)	141,500 (76.5)	130,000 (78.5)						
20	173,500 (60.5)	143,500 (67.5)	123,500 (72)	112,000 (75)	102,000 (77.5)	90,300 (79.5)				
25	136,500 (52)	131,500 (61.5)	110,500 (67.5)	98,650 (71)	89,250 (74)	78,550 (76.5)	73,700 (78.5)	69,300 (80)		
30	106,000 (43)	106,000 (55.5)	98,000 (63)	88,350 (67.5)	78,750 (71)	69,250 (73.5)	65,100 (75)	61,000 (77.5)	60,000 (79.5)	
35	84,700 (30.5)	84,700 (49)	84,700 (58)	80,150 (63.5)	69,000 (67.5)	60,750 (70.5)	57,150 (73)	54,000 (75.5)	52,150 (77.5)	
40		70,500 (41)	70,500 (52.5)	70,500 (59.5)	51,300 (64)	54,000 (67.5)	50,600 (70.5)	48,300 (73)	45,850 (75)	38,000 (79)
45		58,850 (32)	58,850 (47)	58,850 (55)	55,000 (60.5)	48,500 (64.5)	46,200 (68)	43,050 (71)	40,400 (73)	35,750 (77)
50		49,600 (17.5)	49,600 (40.5)	49,600 (50.5)	48,750 (57)	43,050 (61.5)	40,700 (65)	38,250 (68.5)	35,750 (71)	32,100 (75.5)
60			36,200 (22.5)	36,200 (38.5)	36,200 (48.5)	34,300 (55)	33,600 (58.5)	30,750 (63.5)	28,500 (66.5)	26,350 (72)
70				26,050 (25)	26,050 (39.5)	26,050 (47.5)	26,050 (53)	24,750 (58)	23,100 (61.5)	22,000 (68.5)
80					18,850 (27)	18,850 (39)	18,850 (46.5)	18,850 (52.5)	18,700 (56.5)	18,500 (64.5)
90						13,500 (28)	13,500 (38.5)	13,500 (46.5)	13,500 (51.5)	15,250 (60.5)
100							9,390 (29)	9,390 (39)	9,390 (45.5)	12,600 (56.5)
110							6,080 (12.5)	6,080 (30.5)	6,080 (39)	10,100 (52)
120								3,390 (17.5)	3,390 (31)	7,530 (47.5)
130									1,150 (19.5)	5,390 (42.5)
140										3,610 (36.5)
150										2,100 (30)
Minimum boom angle (deg.) for indicated length (no load)									10	19
Maximum boom length (ft) at 0 deg boom angle (no load)									140	167

NOTE: ( ) Boom angles are in degrees.
#LMI operating code. Refer to LMI manual for instructions.

A6-829-007136 & -007143

WEIGHT REDUCTIONS FOR LOAD HANDLING DEVICES

33 ft. Extension	
* Stowed -	892 lbs.
* Erected -	5,704 lbs.

33 ft. - 58 ft. Extension	
* Stowed -	1,246 lbs.
* Erected (Ret.) -	8,412 lbs.
* Erected (Ext.) -	11,406 lbs.

46 ft. - 173 ft. Boom with	
* 46 ft. Jib Erected -	9,613 lbs.
* 60 ft. Jib Erected -	14,571 lbs.
* 74 ft. Jib Erected -	20,443 lbs.
* 88 ft. Jib Erected -	27,199 lbs.
* Fixed Jib Accessories -	327 lbs.

* Reduction of main boom capacities

HOOKBLOCKS:	
30 Ton, 1 Sheave	1,022 lbs.
150 Ton, 8 Sheave	5,254 lbs.
Auxiliary Boom Head	261 lbs.
10 Ton Headache Ball	560 lbs.
15 Ton Headache Ball	803 lbs.

RATED LIFTING CAPACITIES IN POUNDS
46 FT. - 173 FT. BOOM ON OUTRIGGERS - 360°

Radius in Feet	#0001									#0002
	Main Boom Length in Feet (Power Pinned Fly Retracted)									Power Pin Fly Ext. & 141 ft.
	46	58	70	82	94	106	118	130	141	173
10	300,000 (74.5)									
12	280,000 (72)	143,500 (76)	142,000 (79)							
15	235,000 (67.5)	143,500 (72.5)	141,500 (76.5)	130,000 (78.5)						
20	173,500 (60.5)	143,500 (67.5)	123,500 (72)	112,000 (75)	102,000 (77.5)	90,300 (79.5)				
25	135,500 (52)	131,500 (61.5)	110,500 (67.5)	98,650 (71)	89,250 (74)	78,550 (76.5)	73,700 (78.5)	69,300 (80)		
30	106,000 (43)	106,000 (55.5)	98,000 (63)	88,350 (67.5)	78,750 (71)	69,250 (73.5)	65,100 (76)	61,000 (77.5)	60,000 (79.5)	
35	84,700 (30.5)	84,700 (49)	84,700 (58)	80,150 (63.5)	69,000 (67.5)	60,750 (70.5)	57,150 (73)	54,000 (75.5)	52,150 (77.5)	
40		70,500 (41)	70,500 (52.5)	70,500 (59.5)	61,300 (64)	54,000 (67.5)	50,600 (70.5)	48,300 (73)	45,850 (75)	38,000 (79)
45		57,250 (32)	57,250 (47)	57,250 (55)	55,000 (60.5)	48,500 (64.5)	45,200 (68)	43,050 (71)	40,400 (73)	35,750 (77)
50		46,550 (17.5)	46,550 (40.5)	46,550 (50.5)	46,550 (57)	43,050 (61.5)	40,700 (65)	38,250 (68.5)	35,750 (71)	32,100 (75.5)
60			32,000 (22.5)	32,000 (39.5)	32,000 (48.5)	32,000 (55)	32,600 (58.5)	30,750 (63.5)	28,500 (66.5)	26,350 (72)
70				22,550 (25)	22,550 (39.5)	22,550 (47.5)	22,550 (53)	22,550 (58)	22,550 (61.5)	22,000 (68.5)
80					15,900 (27)	15,900 (39)	15,900 (46.5)	15,900 (52.5)	15,900 (56.5)	18,500 (64.5)
90						11,000 (28)	11,000 (38.5)	11,000 (46.5)	11,000 (51.5)	15,250 (60.5)
100							7,280 (29)	7,280 (39)	7,260 (45.5)	11,750 (56.5)
110							4,260 (12.5)	4,260 (30.5)	4,260 (39)	8,520 (52)
120								1,830 (17.5)	1,830 (31)	5,850 (47.5)
130										3,610 (42.5)
140										1,710 (36.5)
Minimum boom angle (deg.) for indicated length (no load)									20	30
Maximum boom length (ft) at 0 dog boom angle (no load)									133	166

NOTE: ( ) Boom angles are in degrees.
#LMI operating code. Refer to LMI manual for instructions.

A6-829-007136 & -007143

33 FT. FIXED LENGTH EXTENSION ON OUTRIGGERS - 360°

Main Boom Angle (Deg.)	#0051		#0052		#0053	
	2° OFFSET		15° OFFSET		30° OFFSET	
	Rad. Ref. (ft.)	Cap. lbs.	Rad. Ref. (ft.)	Cap. lbs.	Rad. Ref. (ft.)	Cap. lbs.
80	39.1	22,850	45.9	20,200	52.0	14,750
75	56.6	17,400	63.0	16,100	68.7	12,750
70	73.7	14,000	79.6	12,950	84.8	10,700
65	90.2	11,850	95.5	10,750	100.3	9,210
60	106.0	10,750	110.8	9,140	115.0	8,050
55	121.1	7,240	125.2	6,160	128.8	5,320
50	135.2	4,330	138.6	3,700	141.8	3,150

A6-829-008157

DIMENSIONS ARE FOR LARGEST GROVE FURNISHED HOOK BLOCK AND HEADACHE BALL, WITH ANTI-TWO BLOCK ACTIVATED.

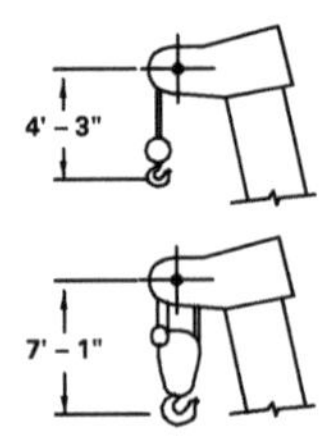

33 FT. - 58 FT. TELE EXTENSION ON OUTRIGGERS - 360°

Main Boom Angle (Deg.)	33 ft. EXTENSION						48 ft. EXTENSION						58 ft. EXTENSION					
	#0021		#0022		#0023		#0031		#0032		#0033		#0041		#0042		#0043	
	2° OFFSET		15° OFFSET		30° OFFSET		2° OFFSET		15° OFFSET		30° OFFSET		2° OFFSET		15° OFFSET		30° OFFSET	
	Rad. Ref. (ft.)	Cap. lbs.	Rad. Ref. (ft.)	Cap. lbs.	Rad. Ref. (ft.)	Cap. lbs.	Rad. Ref. (ft.)	Cap. lbs.	Rad. Ref. (ft.)	Cap. lbs.	Rad. Ref. (ft.)	Cap. lbs.	Rad. Ref. (ft.)	Cap. lbs.	Rad. Ref. (ft.)	Cap. lbs.	Rad. Ref. (ft.)	Cap. lbs.
80	39.1	22,300	45.9	19,650	52.0	14,200	43.9	14,800	55.8	12,700	65.0	9,040	45.1	9,600	59.1	8,300	69.6	7,000
75	56.6	16,850	63.0	15,550	68.7	12,200	62.5	13,700	73.4	10,900	81.8	8,320	65.0	9,150	77.9	7,950	87.8	6,440
70	73.7	13,450	78.6	12,400	84.8	10,150	80.7	11,600	90.3	8,840	97.9	7,110	84.4	8,550	96.2	7,200	105.5	5,760
65	90.2	11,300	95.5	10,200	100.3	8,870	98.2	9,230	106.7	7,330	113.3	6,120	103.2	7,630	113.8	6,000	122.3	4,970
60	106.0	10,200	110.8	8,600	115.0	7,450	115.1	7,550	122.1	8,210	127.9	5,340	121.2	6,260	130.5	5,100	138.2	4,380
55	121.1	6,400	125.2	5,370	128.8	4,540	131.1	5,200	136.7	3,860	141.5	2,950	138.2	4,650	146.2	3,265	153.1	2,380
50	135.2	3,520	138.6	2,810	141.8	2,360	146.1	2,580	150.2	1,790			154.3	2,235				

#LMI operating code. Refer to LMI manual for instructions.

A6-829-007158

Lift Data Sheet

Lift Data Sheet (Single Mobile Crane Lift)				Date:
Load Name:		**Lift Description:**		
Project:			**Units:**	U.S. (ft, lb)
Crane Details	**Manufacturer:**		**Model Number:**	
Configuration:		Base Mount Type:	Track/Outrigger Ext:	ft
Boom Type:		Boom Length Used: ft	Boom/Jib Angle:	deg.
Jib Type:		Jib Length Used: ft	Tail Swing:	ft
Crane Ballast:	lb	Aux Counterweigh lb		
Block Capacity:	ton	Line Si in		
Single Line Pull:	lb	Parts Line Use	Reeved Capacity:	lb
Heavy Lift Attachments:		Load Radius: ft	Superlift Weight:	ton

Load Details

	Quantity	Wt. ea.	Total Weight
Basic Weight of Load/Item		lb	lb
		lb	lb
		lb	lb
		lb	lb
		lb	lb
		Total Weight of Load/Item:	lb

Rigging Data (size, type, and capacity)

		lb	lb
		lb	lb
		lb	lb
		lb	lb
		lb	lb
		lb	lb
		lb	lb
		Total Rigging Weight:	lb

Additional Weight Items

Main Hook		lb	lb
Wire Rope		lb	lb
Other Suspended Hooks		lb	lb
Aux Boom Sheaves		lb	lb
Jib		lb	lb
		lb	lb
		lb	lb
		Total Additional Weight:	lb
		GROSS LOAD (sum of all items above):	8,991 lb

Crane Capacities

Condition:	Placing Condenser		
Total Load to Crane:	8,991 lb	lb	lb
Planned Radius:	73 ft	ft	ft
Chart Radius Used:	80 ft	ft	ft
Chart Capacity:	18,850 lb	lb	lb
% of Chart Capacity:	47.7 %	%	%

MAXIMUM PERCENTAGE OF CHART CAPACITY TO BE USED IS 47.7 %

Total Suspended Load: 3,476 lb   Reeved Capacity: 51,800 lb

MAXIMUM PERCENTAGE OF REEVED CAPACITY PLANNED TO BE USED IS 6.7 %

Document Attachments

○ Crane Layout	○ Lift Plan Category	○ GBP Source/Calculations
○ Rigging Hookup Diagram	○ AHA	○ Project Scope
○ Crane Load Chart	○ Risk Evaluation	○ Load Weight/CG Source
○ Foundation Details	○ Weather Forecast	○ Load/Crane Clearances
○ Crane Cribbing Arrangement	○ Load Schematic	○

Notes

	Name (Print)	Signature	Title	Date
Prepared by:				
Checked by:				
Approved:				

Critical Lift Plan

CRITICAL LIFT PLAN

DESCRIPTION OF WORK — **SECTION 1A**

Date:	Time:	
Associated Work Permit Numbers:		
JOB Number:		
Location Name:		
Street Address/Building Number:		
City / State:		
Weather:	Wind Speed:	Visibility:
Load Description:		
Task:		
Estimated Duration for Task:		
Crane Make & Model:	Crane Operator:	
Lift Supervisor Name [print]: ______________________	Phone: (______)______-________	

HAZARDS IDENTIFIED

- ☐ OVERHEAD UTILITIES / OBSTRUCTION
- ☐ GROUND CONDITIONS
- ☐ UNDERGROUND UTILITIES
- ☐ VEHICLE & PEDESTRIAN TRAFFIC
- ☐ PINCH POINTS
- ☐ OTHER EQUIPMENT
- ☐ FALL HAZARDS
- ☐ OTHER: ______________________

____

PRECAUTIONS

- ☐ SPOTTER
- ☐ ELEVATED WARNING SIGNS
- ☐ CRIBBING / SHORING / TRENCH PLATES
- ☐ K-RAILS / JERSEY BARRIER
- ☐ CAUTION TAPE / BARRICADE
- ☐ FALL PROTECTION
- ☐ OTHER: ______________________

____

____

____

Version 2c: 10/16/12

RIGGING EQUIPMENT — SECTION 1B

SAFETY

☐ Slings & Rigging Inspected **PRIOR TO USE**

☐ Objects in eyelets of synthetic slings **DO NOT** exceed 1/3 the length of the eyelet

☐ Objects in eyelets of wire rope slings **DO NOT** exceed 1/2 the length of the eyelet

☐ The D/d Ratio of any attached rigging equipment to the Wire Rope is 25:1

☐ The Included Angle of the slings attaching to the hook **DOES NOT** exceed 90°

☐ Sharp Edge Protection

☐ Shield and protect slings and rigging against chemical exposure and extreme heat

☐ Safety notified 5 days prior to all lifts

SLINGS

☐ SYNTHETIC ☐ WIRE ROPE
☐ CHAIN ☐ WIRE MESH

SLING CONFIGURATION & CAPACITY

☐ Vertical: ______________________
[Vertical is within 0° to 5° to be 100%]

☐ Basket: ______________________
[Wire Rope Slings require D/d of 25:1]

☐ Choker: ______________________
[Starting @ approximately 75% of Vertical]

Number of Slings: ______________________

☐ 2 Leg Bridle
☐ 3 Leg Bridle
☐ 4 Leg Bridle
☐ More than 4 Leg Bridle
☐ Tag Line

SHACKLES

☐ Shackles are at 100% of capacity when slings are from 0° to 5° of vertical
☐ Shackles are at 70% of capacity when the included sling angles are from 12° to 90° of vertical
☐ Shackles are at 50% of capacity when the included sling angles are from 91° to 180°

Number of Shackles: ______________________

Shackle Capacity: ______________________

EYEBOLTS

☐ Non-shouldered eyebolts are at 100% of capacity when the slings are from 0° to 5° of vertical
☐ Non-shouldered eyebolts are **NOT** recommended with sling angles more than 5° from vertical
☐ Shouldered eyebolts are at 100% of capacity when the slings are from 0° to 5° of vertical
☐ Shouldered eyebolts are at 70% of capacity when the sling angles are from 6° to 45° of vertical
☐ Shouldered eyebolts are at 50% of capacity when he sling angles are from 46° to 90° of vertical

☐ Shouldered ☐ Non-shouldered

Eyebolt Capacity: ______________________ (lbs.)

HOOK

Included angle to the hook **DOES NOT** exceed 90°

☐ Latch required
☐ Latch moused in the open position
☐ Latch moused in the closed position

MASTERLINK

Masterlink Capacity: ______________________

SWIVEL HOIST RING

Torque value: ______________________

Swivel hoist ring capacity:

Version 2c: 10/16/12

ADDITIONAL RIGGING EQUIPMENT	SECTION 1C

LIST ALL ADDITIONAL EQUIPMENT AND INFORMATION NOT COVERED IN SECTION 1B

- ☐ ______________________________
- ☐ ______________________________
- ☐ ______________________________
- ☐ ______________________________
- ☐ ______________________________
- ☐ ______________________________
- ☐ ______________________________
- ☐ ______________________________
- ☐ ______________________________
- ☐ ______________________________
- ☐ ______________________________

Version 2c: 10/16/12

LOAD CALCULATIONS — SECTION 1D

LOAD #1

Width: __________ Length: __________

Height: __________ Weight: __________

Number of lift points: __________

Number of slings: _____ Length of Sling Leg: _____

Distance Between Opposing Sling: ______________

Height Between Hook and Load Plane: ___________

Sling Tension:

LOAD #2

Width: __________ Length: __________

Height: __________ Weight: __________

Number of lift points: __________

Number of slings: _____ Length of Sling Leg: _____

Distance Between Opposing Sling: ______________

Height Between Hook and Load Plane: ___________

Sling Tension:

LOAD #3

Width: __________ Length: __________

Height: __________ Weight: __________

Number of lift points: __________

Number of slings: _____ Length of Sling Leg: _____

Distance Between Opposing Sling: ______________

Height Between Hook and Load Plane: ___________

Sling Tension:

LOAD #4

Width: __________ Length: __________

Height: __________ Weight: __________

Number of lift points: __________

Number of slings: _____ Length of Sling Leg: _____

Distance Between Opposing Sling: ______________

Height Between Hook and Load Plane: ___________

Sling Tension:

LOAD #5

Width: __________ Length: __________

Height: __________ Weight: __________

Number of lift points: __________

Number of slings: _____ Length of Sling Leg: _____

Distance Between Opposing Sling: ______________

Height Between Hook and Load Plane: ___________

Sling Tension:

SLING TENSION FORMULA

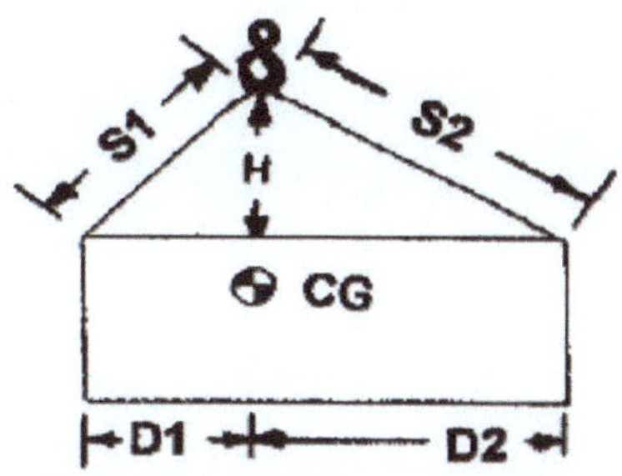

LOAD ON SLING CALCULATED
TENSION 1 = LOAD X D2 X S1/(H(D1+D2))
TENSION 2 = LOAD X D1 X S2/(H(D1+D2))

USE ADDITIONAL PAGES AS NEEDED

Version 2c: 10/16/12

LOAD INFORMATION	SECTION 1E
DIAGRAM OF LOAD [INCLUDE MEASUREMENTS]	
DIAGRAM OF PROPOSED RIGGING & SLINGS [INCLUDE MEASUREMENTS]	

Version 2c: 10/16/12

MAN BASKET / PERSONNEL PLATFORM — SECTION 1F

OSHA 29 CFR 1926.1400 ; NASA-STD-8719.9

SAFETY

Types of Communication: ☐ Hand ☐ Voice ☐ Radio ☐ Hardline

☐ **CRANE / WIRE ROPE CAPACITY DE-RATED 50%** ☐ **WINDS UNDER 20MPH**

☐ Over Water
Depth of Water: ____________________ Height to Water: ______________
☐ Personal Floatation Device Required ☐ Fall Harness Required

☐ Basket & Rigging Inspected ☐ 125% Capacity Test Pick at each radius

CRANE INFORMATION

Crane Operator Name: ____________________ Crane Make & Model: ____________________

Outrigger / Crawler Configuration:
☐ On Rubber ☐ 0% Extended ☐ 50% Extended ☐ 100% Extended

CAPACITY INFORMATION

☐ **CRANE CAPACITY DE-RATED 50%** ☐ **WIRE ROPE CAPACITY DE-RATED 50%**

PICK #1

Main Boom Length & Configuration: ____________ Jib Length & Configuration: ______________

Radius: _____ Crane Cap: ________ Wire Rope Cap: ________ **Capacity LESS 50%:** ________

PICK #2

Main Boom Length & Configuration: ____________ Jib Length & Configuration: ______________

Radius: _____ Crane Cap: ________ Wire Rope Cap: ________ **Capacity LESS 50%:** ________

NAMES & WEIGHTS OF PERSONNEL IN BASKET

Name: ____________________ WT: _____ Name: ____________________ WT: _____

Name: ____________________ WT: _____ Name: ____________________ WT: _____

Rated Capacity of the Man Basket: ________ Total Weight of Men in Basket: __________
Weight of Rigging: __________

☐ 125% Capacity Test Pick at each radius

TOTAL WEIGHT:

Version 2c: 10/16/12

WORKING NEAR POWER LINES	SECTION 1G

OSHA 29 CFR 1926.1400 ; NASA-STD-8719.9

Utility Owner: ______________________________ Utility Contact: ________________________

Contact Phone Number: _____________________________

Verified Voltage of Power Lines: ________________ Verified Elevation of Power Lines: _____

Required **MINIMUM** Clearance (see table below): ______________________

Cranes Working Radius: _____________________________

☐ DE-ENERGIZED & GROUNDED

☐ LOCKED OUT / TAGGED OUT
☐ Verified By: _______________________________________
[printed name]

☐ ENERGIZED

☐ Insulating Link / Device
☐ Non-Conductive Slings
☐ Non-Conductive Tag Line
☐ Distance from Crane to Power Line: ____________________
☐ Distance from Crane Boom Tip to Power Line: ___________

WARNING DEVICES

☐ Proximity Alarm
☐ Elevated Warning Signs
☐ Spotter

REQUIRED MINIMUM CLEARANCES

0—50kV..................10'
50—200kV...............15"
200—350kV.............20'
350—500kV.............25'
500—750kV.............35'
750—1,000kV...........45'
1,000kV +.................Determined by Registered Engineer / Utility Owner

Version 2c: 10/16/12

MULTIPLE CRANE LIFT

SECTION 1H

OSHA 29 CFR 1926.1400; NASA-STD-8719.9

SAFETY

- ☐ **SAFETY LIFT OBSERVER PRESENT**
- ☐ **HOIST LINES VERTICAL**
- ☐ **NO SIDE LOADING**
- ☐ **CRANES DE-RATED 25%**
- ☐ **ALL CRANES ARE LEVEL**

NAME OF OBSERVER: ____________________

CRANE # 1 INFORMATION

Crane Operator Name: ____________________ NCCCO License #: ____________________

Crane Make & Model: ____

Outrigger / Crawler Configurations:
☐ On Rubber ☐ 0% Extended ☐ 50% Extended ☐ 100% Extended

Main Boom Length & Configuration: ________ Jib Length & Configuration: ____________________

Furthest Operating Radius: ____________ Rated Capacity at Furthest Radius **LESS 25%:** ________

CRANE #2 INFORMATION

Crane Operator Name: ____________________ NCCCO License #: ____________________

Crane Make & Model: ____

Outrigger / Crawler Configurations:
☐ On Rubber ☐ 0% Extended ☐ 50% Extended ☐ 100% Extended

Main Boom Length & Configuration: ________ Jib Length & Configuration: ____________________

Furthest Operating Radius: ____________ Rated Capacity at Furthest Radius **LESS 25%:** ________

CRANE #3 INFORMATION

Crane Operator Name: ____________________ NCCCO License #: ____________________

Crane Make & Model: ____

Outrigger / Crawler Configurations:
☐ On Rubber ☐ 0% Extended ☐ 50% Extended ☐ 100% Extended

Main Boom Length & Configuration: ________ Jib Length & Configuration: ____________________

Furthest Operating Radius: ____________ Rated Capacity at Furthest Radius **LESS 25%:** ________

Version 2c: 10/16/12

OVER 75% RATED CAPACITY	SECTION 1I

Crane Operator Name: ______________________ Crane Make & Model: ______________________

Outrigger / Crawler Configuration:
☐ On Rubber ☐ 0% Extended ☐ 50% Extended ☐ 100% Extended

PICK #1

Main Boom Length & Configuration: ____________ Jib Length & Configuration: ________________

Radius: _____ Crane Cap: ________ Wire Rope Cap: ________ Load Wt: ________

PICK #2

Main Boom Length & Configuration: ____________ Jib Length & Configuration: ________________

Radius: _____ Crane Cap: ________ Wire Rope Cap: ________ Load Wt: ________

PICK #3

Main Boom Length & Configuration: ____________ Jib Length & Configuration: ________________

Radius: _____ Crane Cap: ________ Wire Rope Cap: ________ Load Wt: ________

PICK #4

Main Boom Length & Configuration: ____________ Jib Length & Configuration: ________________

Radius: _____ Crane Cap: ________ Wire Rope Cap: ________ Load Wt: ________

PICK #5

Main Boom Length & Configuration: ____________ Jib Length & Configuration: ________________

Radius: _____ Crane Cap: ________ Wire Rope Cap: ________ Load Wt: ________

PICK #6

Main Boom Length & Configuration: ____________ Jib Length & Configuration: ________________

Radius: _____ Crane Cap: ________ Wire Rope Cap: ________ Load Wt: ________

Version 2c: 10/16/12

CRITICAL LIFT PLAN

COMMENTS	SECTION 2

Lifting Device and Equipment Manger

Safety Concerns/Comments:

Version 2c: 10/16/12

CRITICAL LIFT PLAN

APPROVALS — SECTION 3

Notifications: If lift is to occur in or adjacent to a facility, notify the appropriate FSM (or functional equivalent) at the lift site and nearby facilities.

	Name (Printed)	Name (Signature)	Date
LDEM or Alternate			
Project Lift Manager Approval			
Safety Lift Observer (SLO)			
FSM (if lift is in facility)			
Engineer			
Operator			
Rigger			

Version 2c: 10/16/12

CRITICAL LIFT PLAN

CHECKLIST SECTION 4

1. ☐ Verify crane and equipment are within inspection and testing intervals for operation. Review crane history for recent discrepancies, as well as required repairs or modifications that could affect the planned lift. Such history review should typically assess the 12 months prior to the lift.

2. ☐ Verify that rigging equipment to be used is within inspection and test intervals for operation (Include all load-bearing elements such as shackles, turnbuckles, and eyebolts.) and that it has a thirteen digit tag.

3. ☐ Verify operator(s) has current certifications for operation and has received specific operational training instructions on the devices and equipment to be used for the lift.

4. ☐ Identify approved variance (deviation or waiver) if applicable.

5. ☐ Establish safety zones and install appropriate barriers.

6. ☐ Designate cognizant individual to be positioned at the crane's main electrical disconnect (for facility cranes & hoists), to provide alternate emergency stop (required for Critical Lifts).

7. ☐ Assemble operations crew and conduct pre-lift briefing. The briefing should emphasize such details as lift signals, emergency procedures, as well as that emergency stop signal intended to immediately stop all crane operations and crane motions. It should be reiterated that only personnel with identified tasks are allowed in the lift zone, and then only with required personal protective equipment (PPE), proper training, etc.

8. The Operator is to conduct the following "daily inspection": (NOTE: Document all deficiencies. Correct all hazardous deficiencies before using the equipment.)
   a. ☐ Check functional operation of all control mechanisms for proper action and/or response.
   b. ☐ Without disassembling anything, visually inspect lines, tanks, valves, drain pumps, gear casings, and other components of fluid systems for deterioration and leaks. This applies to components that you can see from ground level or safely check from crane inspection walkways.
   c. ☐ Without disassembling anything, visually inspect functional operating and control mechanisms for excessive wear and contamination by excessive lubricants or foreign matter.
   d. ☐ Visually inspect the hook, latch and swivel for proper operation, cracks, and deformities.
   e. ☐ Visually (without climbing up to bridge) inspect rope reeving for proper travel and drum lay, and inspect wire rope for obvious kinks, deformations, or damage.
   f. ☐ Visually inspect hoist chain (where applicable) for excessive wear or distortion.
   g. ☐ Verify operation of the upper limit switch with no load on the hook.
   h. ☐ Verify that any multiple lines (ropes) are not twisted around each other, and that the hook is centered over the load before lifting.
   i. ☐ Verify communications.
   j. ☐ Verify that all clearances, especially from electrical equipment and power lines, as well as weather conditions, etc., are acceptable.
   k. ☐ Ensure the load is properly rigged and securely attached to the hook. Use tag lines when required to minimize and control unwanted motion or rotation of the load.
   l. ☐ Slowly lift the load just an inch or two and hold in suspension for a few moments. Allow the load dynamics to dampen and make sure the load is in balance. Verify that the brake is holding.

9. If off-site travel is required, verify that necessary actions and preparations have been made to travel the roadway, such as obtaining permission, permits, arranging for escorts, etc.

WARNING: KNOW AND OBEY ALL EMERGENCY STOP SIGNALS

Version 2c: 10/16/12

Additional Resources

This module is intended as a thorough resource for task training. The following reference materials are recommended for further study.

Planning for Load Handling Activities,*ASME P30.1*. New York: The American Society of Mechanical Engineers.

"Load Charts for Use With Written Examinations—Grove TM1500." Manitowoc Crane Group and NCCER. **www.nccer.org/docs/default-source/Load-Charts/grovetm1500.pdf?sfvrsn=853e0e4f_0**

29 *CFR* 1926, Subpart CC,*Cranes and Derricks in Construction*. **www.ecfr.gov**

Figure Credits

Courtesy of DOE/NREL, Credit - Pat Corkery, Module Opener

© iStock.com/BanksPhotos, Figure 1

© iStock.com/Knaupe, Figure 2

SCE&G® A SCANA Company, Figure 3

© iStock.com/Paul Vasarhelyi, Figure 4

© iStock.com/pictafolio, Figure 5

The Manitowoc Company, Inc., Figures 8, 9, Appendix C

© iStock.com/MichaelKnudsen, Figure 13

© Gary Whitton/Shutterstock.com, Figure 15

National Oceanic and Atmospheric Administration/Department of Commerce/NOAA Weather in Focus Photo Contest 2015/Photographer: Jeff Stillman, Figure 17

NASA, Appendix E

Section Review Answer Key

Answer Section One	Section Reference	Objective
1. b	1.1.1	1a
2. d	1.2.2	1b
3. c	1.3.1	1c

NCCER CURRICULA — USER UPDATE

NCCER makes every effort to keep its textbooks up-to-date and free of technical errors. We appreciate your help in this process. If you find an error, a typographical mistake, or an inaccuracy in NCCER's curricula, please fill out this form (or a photocopy), or complete the online form at **www.nccer.org/olf**. Be sure to include the exact module ID number, page number, a detailed description, and your recommended correction. Your input will be brought to the attention of the Authoring Team. Thank you for your assistance.

Instructors – If you have an idea for improving this textbook, or have found that additional materials were necessary to teach this module effectively, please let us know so that we may present your suggestions to the Authoring Team.

NCCER Product Development and Revision

13614 Progress Blvd., Alachua, FL 32615

Email: curriculum@nccer.org

Online: www.nccer.org/olf

❑ Trainee Guide ❑ Lesson Plans ❑ Exam ❑ PowerPoints Other ____________

Craft / Level: Copyright Date:

Module ID Number / Title:

Section Number(s):

Description:

Recommended Correction:

Your Name:

Address:

Email: Phone:

Telescopic-Boom Attachment Setup and Assembly

OVERVIEW

Manufacturers build and sell cranes in various models with specific options. However, operators can greatly increase the utility of cranes by installing add-on components provided by manufacturers. This module describes the assembly and disassembly of several types of telescopic-boom extensions and accessories.

Module 21302

Trainees with successful module completions may be eligible for credentialing through the NCCER Registry. To learn more, go to **www.nccer.org** or contact us at 1.888.622.3720. Our website has information on the latest product releases and training, as well as online versions of our *Cornerstone* magazine and Pearson's product catalog.

Your feedback is welcome. You may email your comments to **curriculum@nccer.org**, send general comments and inquiries to **info@nccer.org**, or fill in the User Update form at the back of this module.

This information is general in nature and intended for training purposes only. Actual performance of activities described in this manual requires compliance with all applicable operating, service, maintenance, and safety procedures under the direction of qualified personnel. References in this manual to patented or proprietary devices do not constitute a recommendation of their use.

21302 V3

21302

Telescopic-Boom Attachment Setup and Assembly

Objective

When you have completed this module, you will be able to do the following:

1. Describe the process of setting up, assembling, and disassembling telescopic-boom components.
   a. Explain how to set up and stow a swing-away lattice extension.
   b. Explain how to set up an auxiliary-sheave boom head.
   c. Identify pre- and post-assembly considerations.
   d. Explain how to install and remove intermediate boom sections.
   e. Discuss the design and operations of manual and power luffing jibs.

Performance Tasks

Under the supervision of your instructor, you should be able to do the following:

1. Analyze the site for adequate assembly/disassembly space.
2. Analyze and determine counterweight requirements needed for boom assembly.
3. Determine the ground stability and ground support needed for assembly and disassembly.

Trade Terms

A/D director
Boom-extend mode
Connector pin
Fly
Level luffing
Luffing
Manual luffing jib
Power luffing jib
Stop rail

Industry Recognized Credentials

If you are training through an NCCER-accredited sponsor, you may be eligible for credentials from NCCER's Registry. The ID number for this module is 21302. Note that this module may have been used in other NCCER curricula and may apply to other level completions. Contact NCCER's Registry at 888.622.3720 or go to **www.nccer.org** for more information.

Contents

1.0.0 Telescopic-Boom Attachment Setup and Assembly 1
1.1.0 Swing-Away Lattice Extension 1
1.1.1 Installing a Swing-Away Lattice Extension 2
1.1.2 Stowing a Swing-Away Lattice Extension 8
1.2.0 Auxiliary Boom Head 8
1.3.0 Pre- and Post-Assembly Considerations 9
1.3.1 Counterweight Considerations 10
1.4.0 Intermediate Boom Sections 10
1.4.1 Assembling Intermediate Boom Sections 11
1.4.2 Removing Intermediate Boom Sections 12
1.5.0 Luffing Extensions 13
1.5.1 Manual Luffing Jibs 13
1.5.2 Power Luffing Jibs 14

Figures

Figure 1 Stowed jib extension. 2
Figure 2 Two-piece adjustable fly assembled at an offset. 2
Figure 3 Stop rail viewed from under stowed extension. 3
Figure 4 Extension-boom stop rail in the raised position. 3
Figure 5 Main boom, swing-away, and support brackets. 4
Figure 6 Positioning the swing-away. 5
Figure 7 Securing the swing-away on the boom nose. 6
Figure 8 Reeving the hoist line to the swing-away. 7
Figure 9 Auxiliary boom head with a hoist block. 8
Figure 10 Boom-extend mode on an LMI. 10
Figure 11 Intermediate boom sections. 11
Figure 12 Links used to set a manual luffing jib offset. 13
Figure 13 Mobile crane equipped with a long power luffing jib. 14

SECTION ONE

1.0.0 TELESCOPIC-BOOM ATTACHMENT SETUP AND ASSEMBLY

Objective

Describe the process of assembling and disassembling telescopic-boom components.

a. Explain how to set up and stow a swing-away lattice extension.
b. Explain how to set up an auxiliary-sheave boom head.
c. Identify pre- and post-assembly considerations.
d. Explain how to install and remove intermediate boom sections.
e. Discuss the design and operations of manual and power luffing jibs.

Performance Tasks

1. Analyze the site for assembly/disassembly room.
2. Analyze and determine counterweight requirements needed for boom assembly.
3. Determine the ground stability and ground support needed for assembly and disassembly.

Trade Terms

A/D director: The individual specified by OSHA standards to be responsible for the assembly and disassembly of cranes.

Boom-extend mode: Refers to the order, and percentage, of extension telescopic-boom sections. The standard mode is to extend all segments the same percentage. Other modes may selectively extend certain sections to provide different capacities for a given load radius.

Connector pin: Any solid, cylindrical, metal part used to fasten, support, or attach the various components and members of a crane's boom and other hoisting gear.

Fly: With reference to mobile cranes, a boom extension carried on the side of the main boom when not in use, that can swing into position to extend the main boom; also called a swing-away.

Level luffing: The process of luffing the boom of a crane while adjusting the hoist cable and boom extension offset so that the load travels in a horizontal direction.

Luffing: Raising or lowering a crane's boom. A luffing jib can luff independently of the main boom.

Manual luffing jib: A semi-fixed jib extension whose offset must be manually set between lifts.

Power luffing jib: A jib extension that has the means to change its offset during a lift by remote control of the operator.

Stop rail: A heavy metal guiderail used by some crane manufacturers to help align a swing-away to its transport location as it is being stowed.

Many mobile crane accidents result from human errors made during crane assembly. For this reason, operators and owners must make sure workers set up, assemble, and disassemble boom attachments properly. Specific guidelines for installing various attachments are included in the manufacturer's manual for each telescopic-boom crane. This module presents considerations for setting up and taking down jib extensions, and describes the assembly and disassembly of intermediate sections and powered jibs.

Deploying a simple folding jib is considered a setup activity rather than an assembly/disassembly operation. More challenging tasks, such as assembling intermediate sections of a significant boom extension, do fall into the latter category. In such cases, OSHA requires that an assembly/disassembly supervisor, referred to as the **A/D director**, be involved to ensure proper and safe assembly.

This module begins by presenting information on component deployments that are classified as setup tasks. More complex tasks that require an A/D director are discussed later in the module.

1.1.0 Swing-Away Lattice Extension

A swing-away lattice extension, sometimes called a **fly** (or *swing-away*) is a jib extension that stows in brackets on the base section of the telescopic boom. The extension and its parts are built for each model crane and are serial-number specific. The operator should only operate the crane using those parts of the extension with the same serial number or other identifier as the crane. If the crane is equipped with a load moment indicator (LMI), operators must program it for this specific extension only. If any other extension is used, the LMI cannot function properly and damage may occur.

Figure 1 shows a stowed jib extension, while *Figure 2* shows a two-piece adjustable fly in use. Operators can assemble a jib as a straight extension of the boom or they can set it at various offsets to the boom as a **luffing** jib.

> **NOTE**
>
> 29 *CFR* 1926.1403–1406 outline safety requirements that apply when assembling or disassembling a crane. However, OSHA's interpretation of the standards indicates that preparing a swing-away jib is not an assembly process, but is instead part of setting up an assembled crane. Therefore, an A/D director is not required for this task.

1.1.1 Installing a Swing-Away Lattice Extension

The length of nonadjustable swing-away extensions can vary depending on the type of crane and the manufacturer. Some older cranes may be equipped with both a swing-away and an A-frame jib on the side of their main booms, but this is much less common today. *Figure 1* is an example of a rough-terrain crane with only a swing-away extension.

Figure 1 Stowed jib extension.

Figure 2 Two-piece adjustable fly assembled at an offset.

Operators must always follow the manufacturer's setup instructions and procedures for their crane. Note that the procedure described here applies to a specific crane model only. Other cranes may require different steps, but overall, the process will be similar. The following is the general procedure for installing a nonadjustable swing-away extension on a telescoping-boom crane.

> **WARNING!**
>
> The operator and any other personnel involved in boom assembly should observe the proper warning and safety information provided in the crane manual or site safety manual.

Steps 1 through 5 prepare the crane to set up the swing-away extension:

Step 1 Extend and set the outriggers or crawlers according to the manufacturer's recommendations. Install proper counterweights as necessary for the intended load and boom length.

Step 2 Fully retract all boom sections, and lower the boom to minimum elevation to permit access to the boom nose.

Step 3 Rig either the main hoist or optional auxiliary hoist cable for a single part of line. Lay the cable off to the side opposite of the boom from the swing-away (normally the left side) while erecting the swing-away.

Step 4 To deploy the **stop rail** of the swing-away support bracket, remove the retaining clip from the stop rail **connector pin**, and remove the pin (*Figure 3*). Hold the free end of the stop rail with your hand to prevent the stop rail from swinging open.

> **CAUTION**
>
> Connector pins used on cranes come equipped with some form of retainer to prevent them from falling out. These devices are typically a reusable spring-wire retaining clip inserted into a hole drilled through the pin, or a cotter key that is disposed of after each use. Operators should replace missing retainers immediately. Never use connector pins without the proper retainer in place.

Step 5 Rotate the stop rail until it is in a horizontal position, and insert the stop rail connector pin into the outer pin hole (*Figure 4*). Secure the stop rail pin in place with its retaining clip. Check that the section is secure with the stop rail pin and a pivot pin in the bearing point on the stop rail.

Steps 6 through 9 separate the swing-away from the A-frame jib (if present). These steps also prepare the swing-away for setup on the boom nose. The part numbers referenced are shown in *Figure 5*.

Step 6 Ensure that the A-frame is properly secured. At the front support bracket (6), check engagement of the lock pin (1) and its lock plate (3); at the rear bracket (7), check the lock pin (2) and its lock plate (4).

Step 7 Detach the swing-away from the A-frame jib. At the rear end of the A-frame (5), remove the pin (8), which secures the swing-away (9) to the A-frame at that end. Stow the pin in the A-frame attaching link (13).

Step 8 At the front end of the A-frame (5), remove the pin (10), which secures the swing-away to the A-frame at that end. Lower the attaching link (11) to its stowed-position bracket (12) on the A-frame, and secure it with the pin (10).

Step 9 Unlock the swing-away. At the swing-away rear support bracket (7), pull the lock pin (15), and swing the latch plate (16) out of the way.

Figure 3 Stop rail viewed from under stowed extension.

Figure 4 Extension-boom stop rail in the raised position.

> **NOTE**
>
> Steps 10 through 15 position the swing-away on the boom nose; refer to *Figure 6.*

Step 10 With the front support bracket (6) acting as a pivot, pull outward on the swing-away tip to rotate the swing-away anchor points (18) onto the attach fittings (17) on the right side of the boom nose.

Step 11 Remove the retaining clips and pull the two attach pins (14) from their stowage brackets in the base of the swing-away. Insert the pins through the attach and anchor fittings on the right side of the boom nose. Reinstall the retaining clips in the pins.

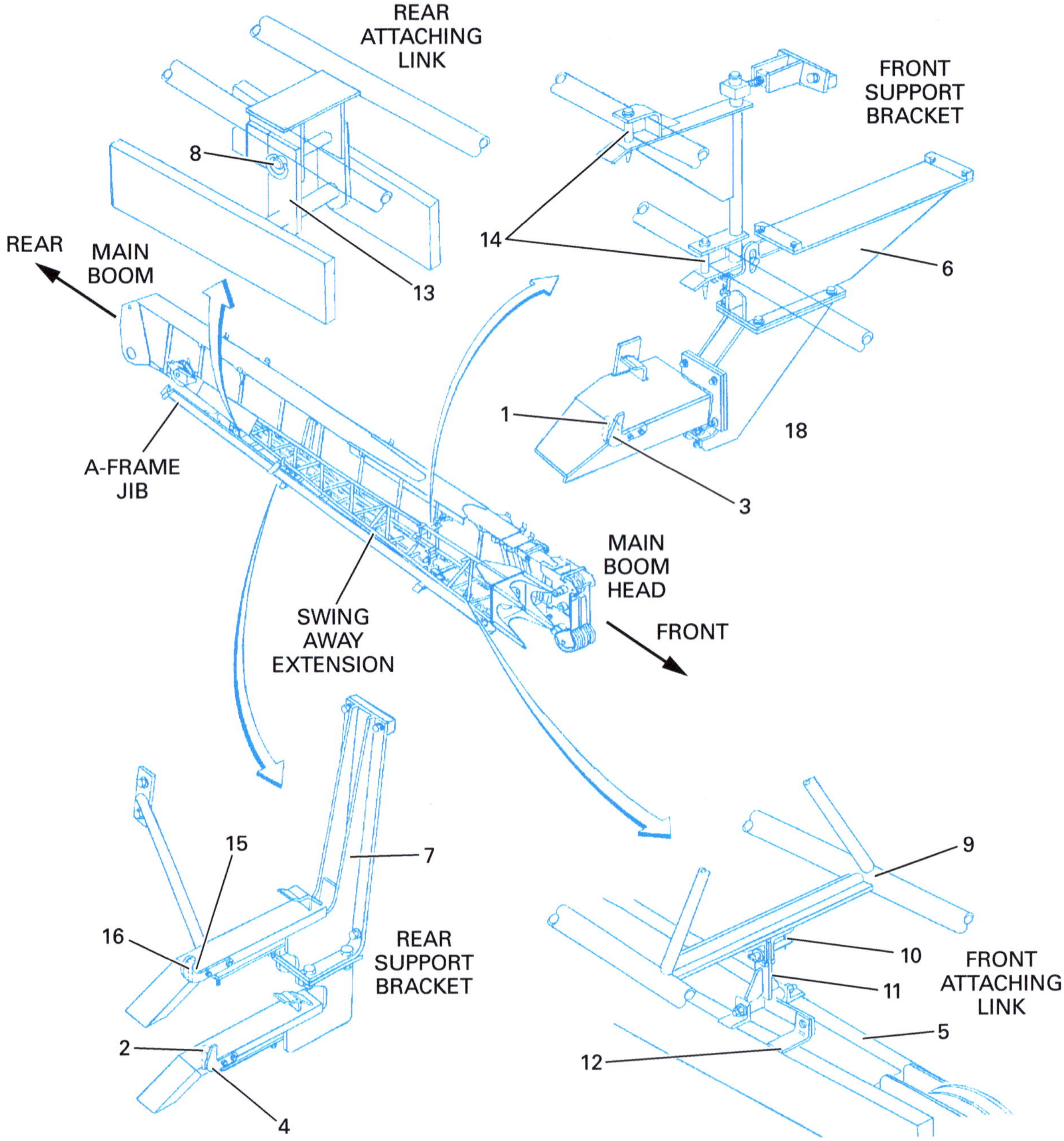

Figure 5 Main boom, swing-away, and support brackets.

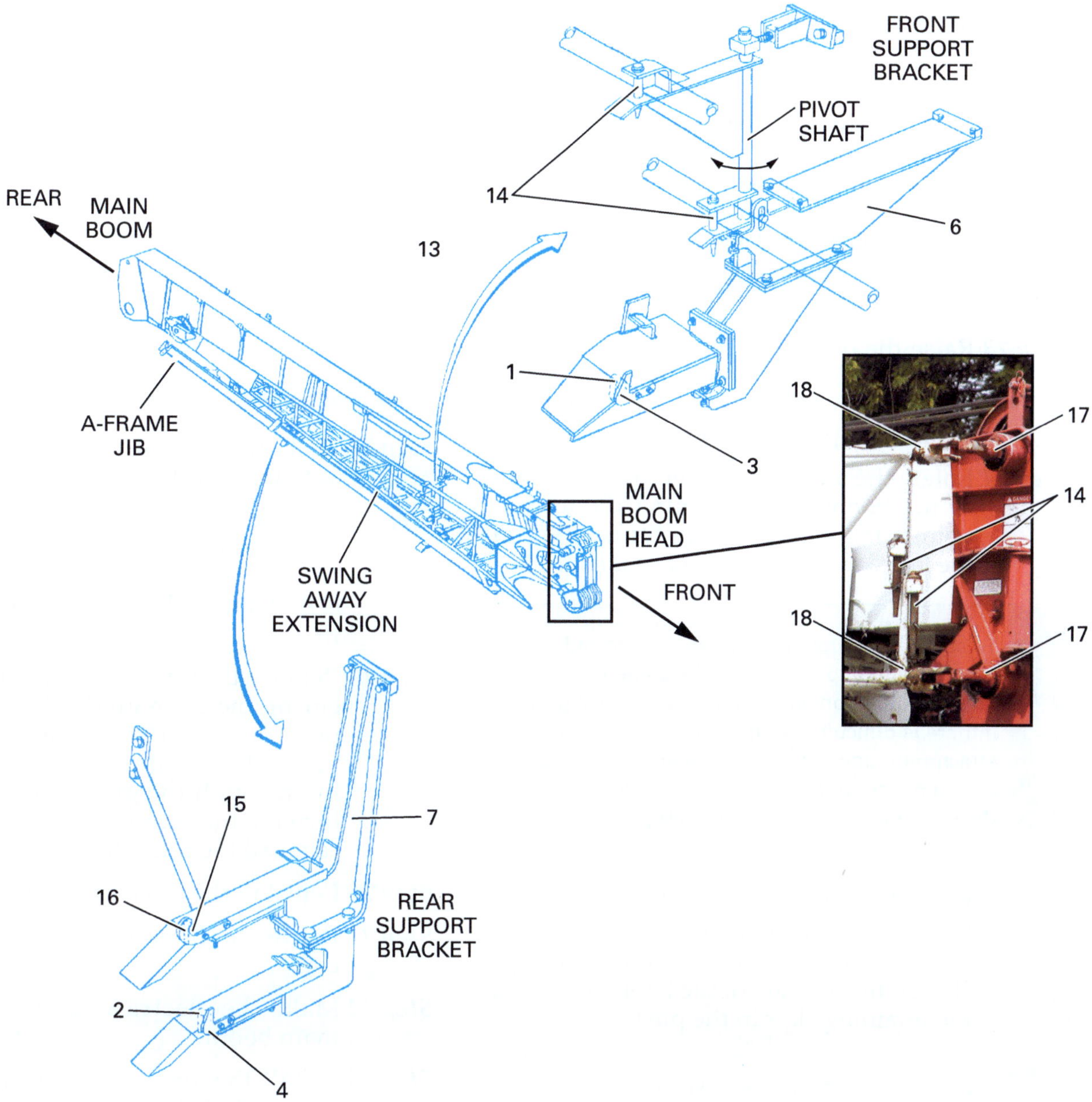

Figure 6 Positioning the swing-away.

Step 12 Attach a tag line to the swing-away. This ensures workers can safely control the motion of the swing-away extension while positioning it.

> **WARNING!**
>
> The boom must be raised to horizontal to prevent the swing-away from swinging around after removing the connector pins from the front support bracket. When erecting the swing-away, make sure all personnel and equipment remain clear of the swing path.

Step 13 Raise the boom to a level, horizontal position. Then, remove the two connector pins (14) and their retaining clips from the front support bracket (6).

Step 14 Rotate the swing-away into place in front of the boom nose, engaging the anchor points (18) with the attach fittings (17) on the left side of the boom nose (*Figure 7*).

> **NOTE**
>
> To assist in swinging the extension around, use the outriggers to slightly tilt the crane into a diagonal position and incline the boom slightly. If there is difficulty in aligning the extension's attachment points to the boom nose, assembly personnel should place blocking under the tip of the swing-away to relieve the weight.

Step 15 Lock the swing-away to the main boom by inserting the two connector pins (14) into the anchor (18) and attach fittings (17) on the left side of the boom. Reinstall the retaining clips in the pins.

> **NOTE**
>
> Steps 16 through 25 describe reeving the hoist cables and completing the electrical connections to the swing-away extension. Refer to *Figure 8.*

Step 16 Remove a pin (19) and retaining clip from the cable-roller sheave (20). Move the cable-roller sheave (20) into working position in accordance with the prescribed inclination. Lock it into place with a pin (19), and secure the pin with its retaining clip.

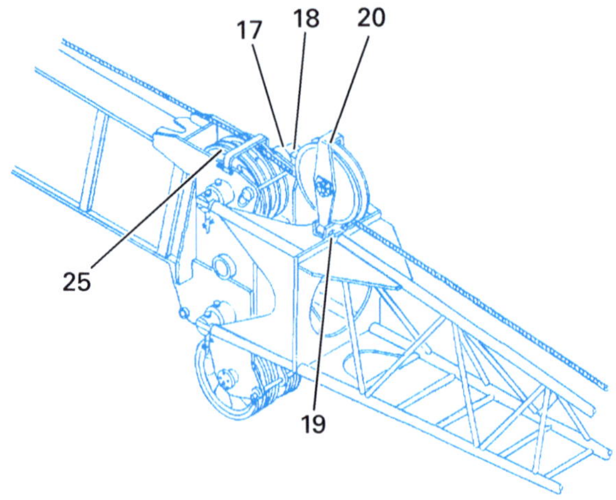

Figure 7 Securing the swing-away on the boom nose.

Step 17 Reel off the hoist rope from the main or auxiliary hoist.

Step 18 Lay the hoist cable over the boom-head sheave (25), under the cable-roller sheave (20), and over the extension tip sheave (24). Remove the cable retaining pins (21) first.

Step 19 Remove the pin (22) and retaining clip securing the cable guide (23) in the stowed position. Position the cable guide in the vertical position, with the cable passing through the guide. Reinstall the pin through the second hole in the guide, and reinstall the retaining clip.

Step 20 Reinstall the cable-retaining pins (21).

Step 21 Reeve the hoist cable on the headache ball.

Step 22 Make the electrical connections to the main boom.

Step 23 Install the anemometer (if equipped) at the head of the swing-away, and electrically connect it.

Step 24 Attach the anti-two-block switch weight. Feed the hoist rope through the weight.

Step 25 Stow the stop rail in the transport position by reversing Steps 4 and 5.

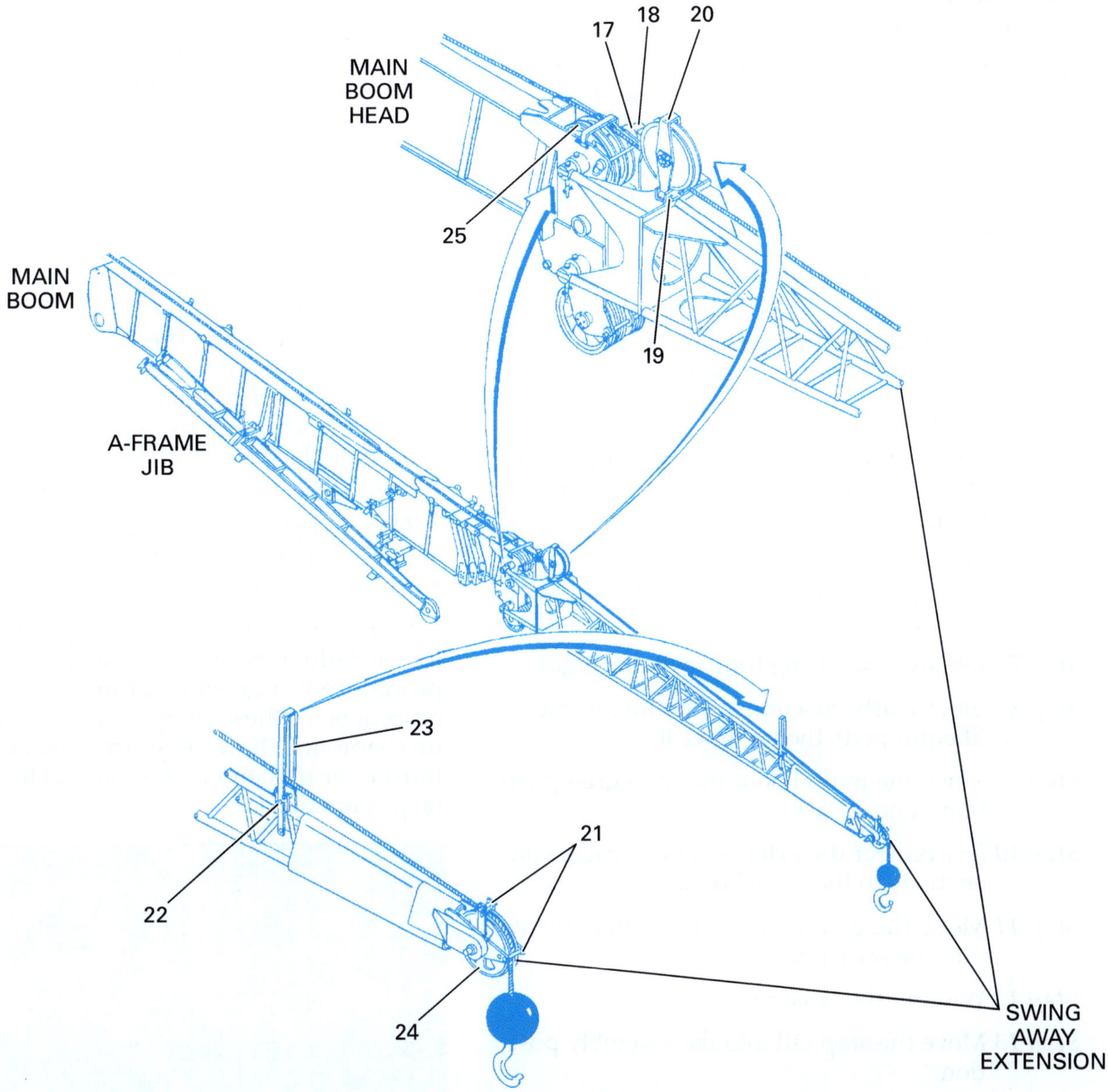

Figure 8 Reeving the hoist line to the swing-away.

1.1.2 Stowing a Swing-Away Lattice Extension

Always follow the manufacturer's instructions for stowing a swing-away lattice extension. For general guidelines, refer to the following abbreviated procedure:

Step 1 Fully retract the telescopic sections of the boom.

Step 2 Turn the superstructure to the back or the side.

Step 3 Lower the main boom until the load moment indicator (LMI) switches off.

Step 4 Override the LMI.

Step 5 Lower the main boom until the head sheave is approximately five feet above the ground.

Step 6 Unreeve the hoist rope from the load block, and remove it from the head sheave.

Step 7 Remove the lifting limit switch weight.

Step 8 Electrically disconnect the anemometer (if equipped), then remove it.

Step 9 Move the main boom into 0-degree position, if necessary.

Step 10 Disconnect the extension's electrical connections to the main boom.

Step 11 Move the extension's cable-roller sheave into the transport position.

Step 12 Wind in the hoist rope.

Step 13 Move the stop rail into the assembly position.

Step 14 Attach a tag line to the swing-away.

Step 15 Remove the connector pins between the swing-away and main boom on the left-hand side of the boom head. Place them in their storage brackets on the extension and reinsert their retaining clips.

Step 16 Rotate the swing-away on the main boom until the pivot points on the pivot bearing align.

Step 17 Remove the pins between the swing-away and main boom on the right-hand side of the boom head.

Step 18 Insert the pins in the boom clips at the pivot bearing and reinsert their retaining clips.

Step 19 Rotate the swing-away to the transport position on the pivot bearing. At the rear support bracket, install the latch lock pin to secure the head of the swing-away.

Step 20 Stow the stop rail in transport position.

1.2.0 Auxiliary Boom Head

Crane booms may require an additional piece of equipment called the auxiliary boom head (*Figure 9*). Auxiliary hoists allow lifting relatively light loads without the need to use the main hoist block, which involves extra wear and tear on the main hoist system. Auxiliary boom heads on modern cranes may be one-, two-, and three-sheave models. To attach the auxiliary head to the boom, configure the crane as follows:

- On outriggers or crawlers
- Superstructure swung to a suitable position
- Main boom retracted and set horizontal

Setup requires the same preparations as those for installing boom extensions. The following procedure is a general summary of the steps required. Since these components are quite variable in shape and how they are attached, operators must refer to their crane's manual to install auxiliary boom heads.

Figure 9 Auxiliary boom head with a hoist block.

Step 1 Install the auxiliary boom head in position. Insert the required connector pins and secure them with their retaining clips.

- This step may involve swinging up the component from its transport position below the main boom head to in front of it. In other cranes, you may have to unpin the auxiliary head on one side of the main boom and rotate it so that it is in front of the main boom sheaves. Some cranes do not transport their auxiliary boom head at all. In those cases, an auxiliary crane must lift the component into position to install it.

Step 2 Install the auxiliary head's signal cable(s) into the appropriate socket(s) on the main boom head.

Step 3 Remove the hoist rope guide bars from the head of the main boom and the auxiliary boom head.

Step 4 Lay the hoist rope over the auxiliary boom head and over the appropriate main boom sheave(s).

Step 5 Reinsert the rope guide bars, and secure them with their retaining pins.

Step 6 Install an anti-two-blocking weight on the auxiliary boom head (and on the lifting unit switch on the main boom, if not already installed).

Step 7 Guide the hoist rope through the anti-two-blocking weight. Note that, depending on the anti-two-blocking weight configuration, operators may be able to complete Step 7 after attaching the headache ball.

Step 8 Attach the headache ball on the auxiliary hoist rope.

Step 9 For attaching hook blocks, reeve the auxiliary hoist rope through the hook sheave and attach it to the auxiliary boom head per manufacturer's instructions.

1.3.0 Pre- and Post-Assembly Considerations

While the previous procedures are considered setup, the installation of intermediate jib sections and luffing jibs is considered assembly and disassembly. These tasks have additional requirements.

Company or project supervision must designate a responsible person, called the A/D director, to assemble or disassemble a crane. This individual has legal responsibilities described in 29 *CFR* 1926.1404, *Assembly/Disassembly—General Requirements*. An operator working alone may act as the A/D director, but that individual must be both competent and qualified as defined by OSHA standards.

Workers should rope off an erection area large enough to allow ingress and egress of trucks carrying crane components. Its size should allow layout of the components in proper order for assembly. The perimeter should prevent interference or hazards to other site personnel during crane assembly. In addition, as the crane swings, there needs to be at least two feet of clearance (0.6 m) between the superstructure counterweight and the nearest obstacle.

Before assembly begins, the operator and the person responsible for the lift must consider the following:

- Is there enough room for the crane to maneuver into position on the site?
- Is there a large enough area in which to erect or extend the boom?
- Can the trucks hauling boom sections get into place and be unloaded safely?
- Is there enough timber blocking to support the boom when it is being assembled and disassembled?
- Will crane operations occur within the vicinity of overhead power lines as described in 29 *CFR* 1926.1407–1411?

WARNING!

Personnel involved in crane A/D work must always follow the manufacturer's guidelines. Never take shortcuts that could lead to injury when climbing or handling boom attachment components. Most are heavy enough to require another crane. Pinching and crushing are common sources of injury.

Self-Assembling Cranes

Some of today's mobile cranes are designed for self-assembly. Of course, not every task can be automated, such as inserting all the connector pins and their retaining hardware. But cranes such as the Liebherr LR1250 crawler crane can self-assemble the main boom along with a luffing jib after it has been delivered to the jobsite. Videos of the process can be found on sites such as YouTube. Any design innovation that makes the task of assembly easier and safer is welcomed by the industry.

1.3.1 Counterweight Considerations

Everyone involved in crane component assembly and disassembly must be aware of counterweight requirements. When attaching boom extensions, the mounted counterweight places restrictions on the **boom-extend mode** and superstructure position with some cranes. The crane may overturn if the leverage of the boom-extend mode or position of the superstructure is not within the crane's load chart parameters. Many LMIs can display boom extend-mode information (*Figure 10*).

> **WARNING!**
>
> Workers must always wear the appropriate PPE and fall protection equipment, and observe safe working practices when assembling and disassembling cranes.

Operators must correctly enter the crane's counterweight configuration into the LMI for the system to function properly. All LMI tipping alerts and protective actions rely on accurate

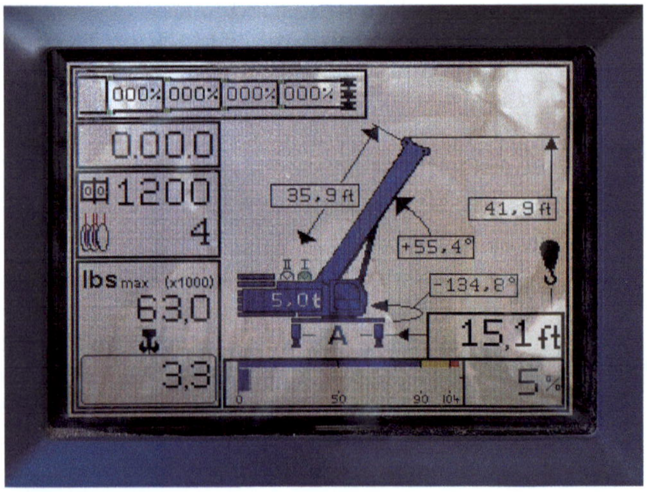

Figure 10 Boom-extend mode on an LMI.

configuration inputs. Remember to reprogram the LMI with revised configurations when removing counterweights or changing boom components. LMIs are equipped with an override button to allow assembly operations that would otherwise be prevented by the LMI's programming. The override button must only be used as the manufacturer specifically allows, and never as a matter of convenience to defeat safety programming.

1.4.0 Intermediate Boom Sections

The boom extension of a telescopic crane may include intermediate boom sections (*Figure 11*). This section covers intermediate boom sections and the procedures for installing them on one crane model. Installation on other models is similar, but will differ in the details.

Before installing the boom extensions, determine the boom-extend mode of the crane required for the lift. As with the setup of a swing-away, the operator must extend and set the outriggers, verify the appropriate amount of counterweight, retract the main boom completely, and lower it into the horizontal position.

> **WARNING!**
>
> Only the combinations of boom-extension spans and counterweights indicated in the crane's load chart are permissible. Insufficient counterweight creates the risk of overturning the crane during assembly.
>
> Assembly and disassembly personnel must follow the manufacturer's instructions when installing and removing boom extensions. Use only original equipment manufacturer (OEM) parts, or authorized substitutions with sufficient loadbearing capacity, when assembling boom extensions.

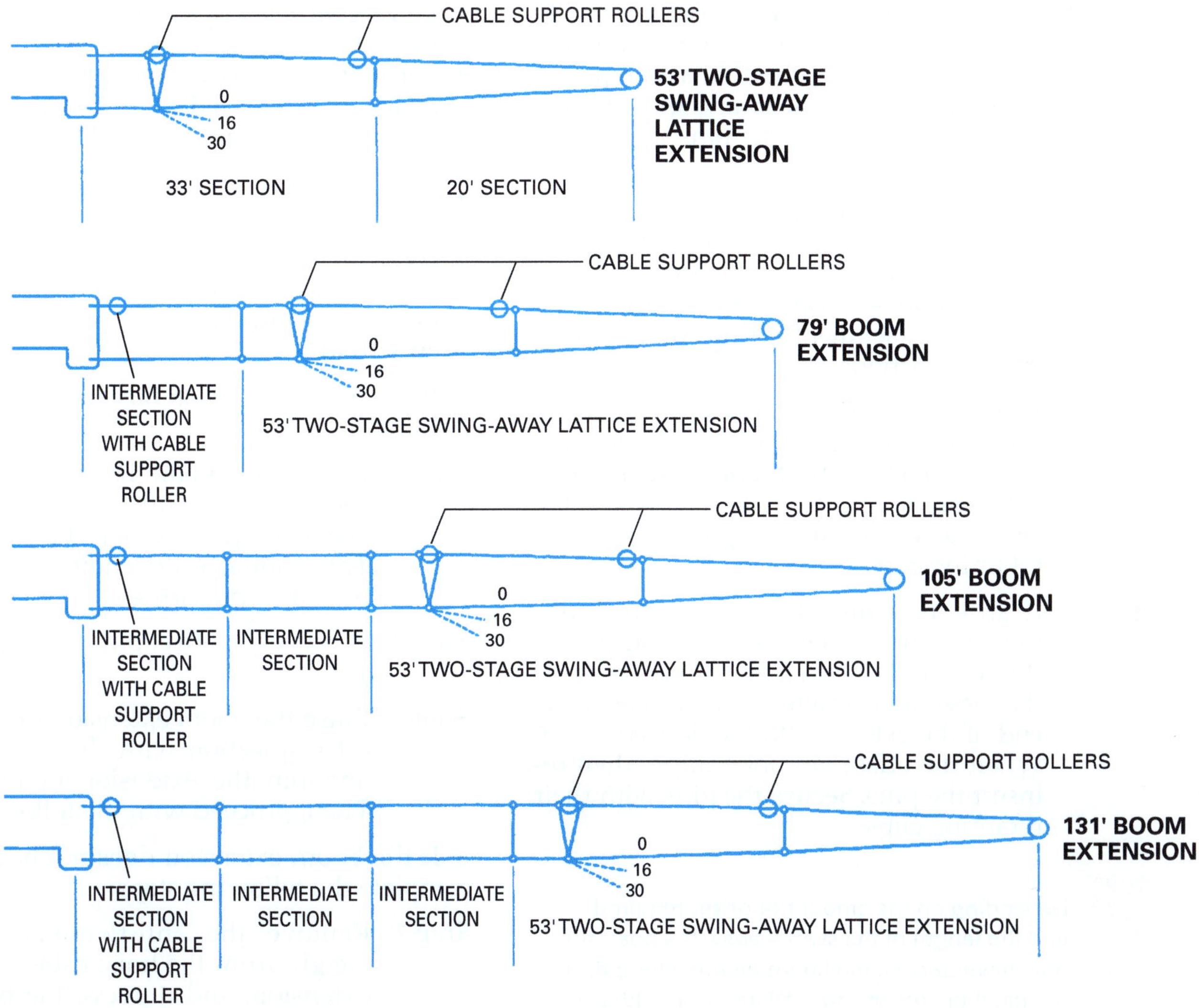

Figure 11 Intermediate boom sections.

1.4.1 Assembling Intermediate Boom Sections

A/D operations usually require an auxiliary crane when attaching intermediate boom sections to the main boom. The following abbreviated procedure is for one specific crane, but the process is similar for other cranes. Always consult the manufacturer's guidelines for the crane in use.

Step 1 Before mounting the intermediate boom sections, remove the swing-away from the main boom. Hook an auxiliary crane to the extension per manufacturer's instructions, hold the swing-away, remove and stow the swing-away connector pins, and place it on the ground.

Step 2 The first intermediate section should have a cable support roller. Attach the lift rigging to the intermediate section. Boom sections include specific attachment points for this purpose.

Step 3 The end of the intermediate section with the cable support roller should nearest the boom head. Use the auxiliary crane to position the intermediate section so that its anchor points align with the attachment fittings on the boom head.

CAUTION

Always assemble the intermediate section with the cable support roller directly in front of the main boom head. Damage to the hoist rope by the main boom head may occur if the first intermediate section does not include the cable support roller.

WARNING!

Pinch and crush injuries are too common during crane assembly and disassembly. Keep hands clear of attachment points while mating up intermediate boom sections.

Step 1 Remove the pins from the stowage brackets on the intermediate section, and insert them at the top and bottom attach points between the boom head and the intermediate section. Secure the pins using their retaining clips.

Step 2 Use the auxiliary crane to raise and position the swing-away in front of the intermediate section. Align the anchor points on the foot section of the swing-away with the attachment fittings on the intermediate section.

Step 3 Remove the pins from the stowage brackets on the swing-away and insert them at the top and bottom attachment points between the intermediate section and the swing-away. Secure the pins using their retaining clips.

Step 4 If an extension offset is required, use the auxiliary crane to support the end of the assembled extension. Then remove the offset adjustment pins and lower the end of the extension to the desired offset. When the offset pin holes align, then re-insert the pins. Secure the pins with their retaining clips.

> **WARNING!**
> Depending on the amount of offset required and the length of the swing-away, the operator may have to raise the boom before attempting to adjust the offset angle. Workers should use fall protection if they need to climb to the offset hinge in such cases.

Step 5 Restore all electrical connections and devices. These may include the anti-two-blocking switch, anemometer, and high-voltage proximity sensors.

Step 6 Place the hoist cable through the anti-two-blocking switch weight, and reeve the hook ball or load block.

1.4.2 Removing Intermediate Boom Sections

When the intermediate boom sections are no longer needed, their removal requires a relatively simple procedure. An auxiliary crane is typically required to disassemble the intermediate sections and make any adjustments to the swing-away offset. Configure the crane as follows for disassembly:

- Retract all telescopic sections completely with the boom raised.
- Turn the superstructure to an appropriate quadrant permitted by the crane's load chart. The amount of space available may be a deciding factor in this choice.
- Lower the main boom until the LMI triggers a shutdown.
- Override the LMI, and lower the main boom into the horizontal position (or until the offset extension can touch the ground, whichever occurs first).

If the boom extension has an offset, proceed as follows:

Step 1 Lower the main boom until the head sheave on the offset boom extension is just resting on a board on the ground. This should exert enough force to leave the offset pins loose.

Step 2 Remove the offset pins from their bearing points.

Step 3 Raise the boom extension to the 0-degree offset position with the auxiliary crane, and pin the extension in this position. Then, proceed with the following steps.

If the boom extension does not have an offset, complete the following steps:

Step 1 Remove the anti-two-blocking switch weight from the hoist cable on the boom extension, and unreeve the ball or load-block hook.

Step 2 Remove the hoist cable from all sheaves on the boom extension.

Step 3 Remove the retaining pin locking the anti-two-block switch from the swing-away.

Step 4 Release the anti-two-block switch lock on the main boom.

Step 5 Disconnect all electrical connections between the distribution boxes at the main boom and the swing-away.

Step 6 Spool the cable onto the appropriate devices.

Step 7 Screw the plugs of the connecting cable onto the dummy sockets of the individual parts of the boom extension.

Step 8 Screw the shorting plugs onto the distribution boxes on the main boom and on the swing-away.

Step 9 Screw the dummy plugs onto the dummy sockets on the intermediate boom sections.

Step 10 Attach the appropriate lifting gear to the swing-away, and hold it with the auxiliary crane.

Step 11 Remove the connector pins between the swing-away and the intermediate boom section, and insert the pins in their stowage brackets on the swing-away. Secure the pins with their retaining clips.

Step 12 Place the swing-away on the ground.

Step 13 Attach the lifting tackle to the intermediate boom section, and hold the section with the auxiliary crane.

Step 14 Remove the pins securing the intermediate boom section, and insert the pins into their stowage brackets on the intermediate boom section. Secure the pins with their retaining clips.

Step 15 Place the intermediate boom section on a transport vehicle.

1.5.0 Luffing Extensions

When the construction site permits, a standard telescopic-crane boom configuration with the swing-away jib may reach all areas of need from one location. Operators can change the load radius within the limitations of the crane's capacity by increasing the boom angle. However, an increase in load radius (and boom tip height) is possible by inserting intermediate sections to extend the jib, lengthening the boom as described earlier. This provides a greater load radius with a smaller boom angle.

In many situations, however, obstructions near the crane site can limit the load radius of a crane using straight-boom configurations. For example, lifting to the roof of a tall structure or hoisting a tall load requires high boom angles. Placing the load may be impossible if the intended location is farther from the crane than the load radius it can achieve at that boom angle.

1.5.1 Manual Luffing Jibs

Cranes can overcome load radius limitations to a certain extent by using a relatively lightweight boom extension at some fixed offset angle. At larger offsets, the crane achieves a greater load radius for a given main boom angle, but at the cost of reduced capacity. Many crane manufacturers provide boom extensions with different lengths to accommodate longer load radiuses.

Setting the offset of a fixed jib—also called a **manual luffing jib**—requires lowering the boom to the ground. The method for setting the offset varies with the type of extension and the crane manufacturer. Some luffing jibs use offset pins inserted into different pairs of holes to fix the angle of the jib at its base, as described earlier. Other manual jibs consist of the extension hinged-to-base section. Workers set the boom angle by inserting or removing heavy metal links between the two parts opposite the offset hinge (*Figure 12*).

Manually luffing a jib is labor intensive, and requires space to lower the crane boom to make the change. At larger offset angles, the hinge point at the head of the main boom may be many feet off the ground, even with the jib-head sheaves resting on the ground. This increases the fall hazard to workers involved, who will either require a lift or must climb to the jib base to remove and insert the offset pins while making the angle adjustment.

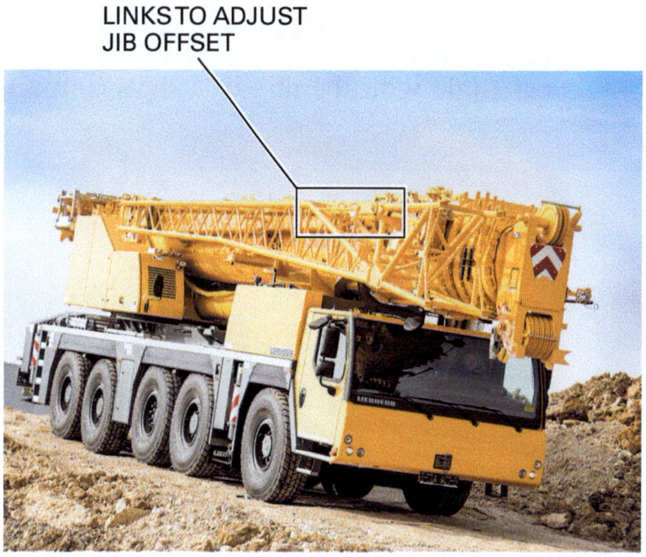

Figure 12 Links used to set a manual luffing jib offset.

1.5.2 Power Luffing Jibs

Several factors have prompted manufacturers to address cost and delay issues related to manual adjustment. Crane users have requested mobile cranes with longer load radiuses at smaller boom angles, but without the need to lay the boom down to adjust the jib offset. Many times, there simply isn't any room to do this. The other factor involves simple physics. As boom angles become smaller and luffing reaches increase, cranes reach both structural-strength and stability limits.

Manufacturers solved the first problem in some luffing jibs by inserting a hydraulic actuator at a hinge point in the jib extension. This feature allows the operator to luff the jib from the cab. The term **power luffing jib** applies to such boom extensions. Modern crane boom control systems can coordinate the offset with the main boom angle to produce a **level luffing** action. Since the crane's capacity changes dynamically with such a system, the LMI must be able to integrate the luffing jib's position with the other boom and load parameters.

Solving the second problem required adding one or more jib masts, forestays, and backstays to support longer luffing jibs. Extra-long lattice jibs can more than double the reach of a telescopic crane. The system of stays reduces the tendency of the main boom to buckle by allowing it to operate at high angles. These jibs are luffed by adjusting the lengths of the cable stays and pivoting the jib mast supports (*Figure 13*). In extreme cases, the main boom can be elevated to its highest angle with the jib's backstays connected to a separate set of counterweights behind the crane. In this arrangement, the operator can luff the extension closer to a horizontal position. Operations with long luffing jibs require reference to special load charts and the use of LMIs programmed with these special boom configurations.

Figure 13 Mobile crane equipped with a long power luffing jib.

Cranes and Wind Farms

To get the best value from the investment, wind turbines have grown to incredible heights and rotor sizes. The crane in *Figure 13* is setting a portion of the wind-turbine tower in place. Eventually the nacelle will be installed, shown here in an exploded view. The rotor assembly, with all blades attached to a hub, is often lifted into place fully assembled.

Many crane manufacturers have developed and built cranes specifically designed to accommodate the extremely high lifts of complex structures, while still being able to get in and out of these remote jobsites. Many are large telescopic-boom cranes fitted with complex power luffing jib systems and other accessories to improve ground stability and make the cranes footprint as large as possible. This type of crane takes less time to assemble on site than a comparable lattice-boom model, and fewer trucks are required to transport the boom components.

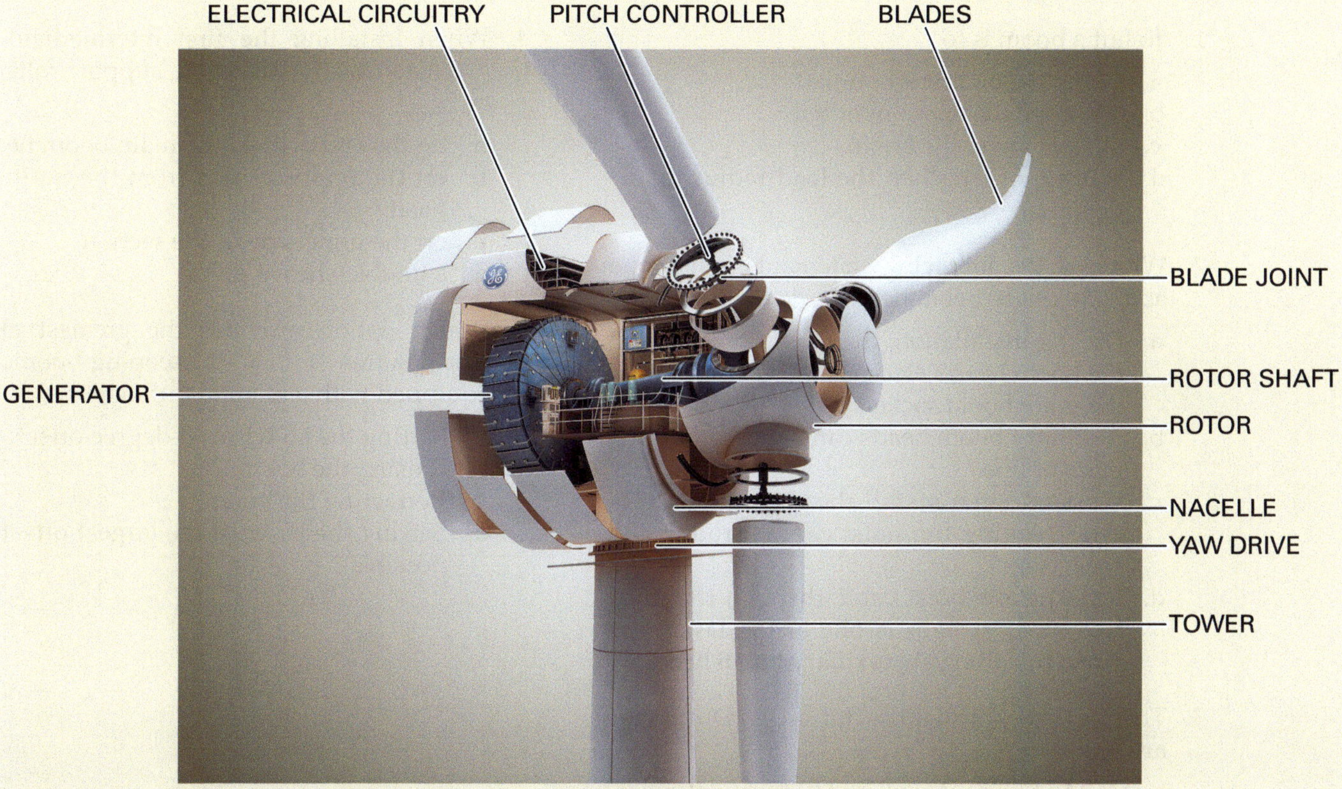

Figure Credit: Nick Kaloterakis from kollected.com <http://kollected.com>

Additional Resources

ASME Standard B30.5, Mobile and Locomotive Cranes. Current edition. New York, NY: American Society of Mechanical Engineers.

Cranes: Design, Practice, and Maintenance. Second edition. 2002. Ing J. Verschoof. Hoboken, NJ: John Wiley and Sons, Inc.

Crane Handbook. D. E. Dickie. 2013. Waltham, MA: Butterworth-Heinemann.

Mobile Crane Safety Manual. 2014. Milwaukee, WI: Association of Equipment Manufacturers.

29 *CFR* 1926, Subpart CC,*Cranes and Derricks in Construction.* **www.ecfr.gov**

1.0.0 Section Review

1. To luff a boom is to _____.
   a. swing the boom right or left
   b. telescope the boom in or out
   c. raise or lower the boom
   d. shorten or lengthen the load radius by articulation

2. Which of the following statements is *not* true about auxiliary boom heads?
   a. During installation, the workers must plug the boom head's electrical signal cables into the main boom sockets.
   b. Auxiliary boom heads can handle up to ten parts of line.
   c. Workers must install the rope guide bars after reeving the hoist cable through the sheave(s).
   d. Leading the hoist cable through the anti-two-block weight is one of the last steps when installing an auxiliary boom head.

3. The legal responsibilities for an A/D director are found in _____.
   a. *ASME B50.5—Mobile and Locomotive Cranes*
   b. *ASME P30.1—Planning for Load Handling Activities*
   c. 29 *CFR* 1926.32
   d. 29 *CFR* 1926.1404

4. When installing the first intermediate boom extension section, its cable support roller must be _____.
   a. on the end nearest the main boom head
   b. on the opposite end from the main boom head
   c. on the underside of the section
   d. a double-sheave roller

5. Which option provides the greatest allowed load radius for a telescoping-boom crane equipped with a manual luffing jib?
   a. Setting the jib with a 0-degree offset
   b. Raising the boom
   c. Retracting the boom
   d. Setting the jib with the largest offset angle possible

SUMMARY

Lifting operations depend on the proper setup and assembly of the crane boom. On telescopic cranes, this can involve the attachment of jibs or intermediate boom extensions to the main boom. While some tasks, such as deploying a fixed fold-away jib, are considered crane setup, other tasks fall into the category of assembly and disassembly. These tasks require an A/D director to ensure they are performed safely.

Workers must be aware of clearance requirements on the jobsite and follow the safety requirements associated with setup and assembly/disassembly work. The procedures presented in this module should be considered general guidelines; always follow the manufacturer's instructions when performing setup and assembly for a specific crane.

Review Questions

1. A significant factor contributing to accidents during mobile crane assembly/disassembly is _____.
   a. incomplete manufacturer's instructions
   b. using worn parts
   c. human error
   d. weather

2. During setup of a swing-away extension, the purpose of the tag line is to _____.
   a. secure the swing-away until ready to position it
   b. mark on the ground the extent of the extension's swing to keep spectators clear
   c. string together all the connector pins to prevent losing them
   d. control the motion of the swing-away during setup

3. When stowing a swing-away extension, the step listed below that should be completed first is to _____.
   a. disconnect and secure all extension electrical connections
   b. remove the connector pins from the extension base
   c. stow the stop rail in the transport position
   d. wind in the hoist rope

4. When a crane does not have an attached auxiliary boom head that can be deployed at the end of the main boom, _____.
   a. an auxiliary boom head cannot be used at all
   b. an auxiliary crane must lift the component into position
   c. the boom will have to be significantly field-modified
   d. the additional weight of adding one will cause instability

5. The amount of counterweight required for a given boom-extend mode is determined by the _____.
   a. A/D director
   b. load moment indicator (LMI)
   c. crane operator
   d. crane's load chart

6. Usually, the swing-away for a telescoping-boom crane is set up on the last intermediate extension section by _____.
   a. rotating it into position from its stowed location
   b. lowering the boom tip to the ground and attaching the swing-away there
   c. using an auxiliary crane to lift the swing-away into position
   d. having a crew of workers lift the swing-away into position

7. When disassembling a boom extension that includes intermediate sections and a swing-away with an offset, workers must first _____.
   a. remove the connector pins from the swing-away and place it on the ground
   b. use an auxiliary crane to set the swing-away offset to 0 degrees
   c. place the intermediate sections on transport vehicles
   d. remove the connector pins from the first intermediate section

8. Before removing any type of boom extension from a telescoping-boom, it is not necessary to _____.
   a. reprogram the LMI
   b. remove the hoist cable from all sheaves in the extension
   c. disconnect all electrical connections between the boom and swing-away
   d. remove the anti-two-blocking switch and weight from the swing-away

9. Which limitation in crane operations led to the development of longer power luffing jibs?
   a. Low ground-bearing pressures
   b. Too-small outrigger footprints
   c. Crane instability at long load radiuses and small boom angles
   d. Hoist ropes with too-low single-line pull capacities

10. Which feature do all power luffing jibs have?
    a. Extension offsets are set with pins.
    b. Offset angles are controlled remotely by the crane operator.
    c. Luffing angles are set by moving the main boom.
    d. They are all the telescoping type.

Trade Terms Introduced in This Module

A/D director: The individual specified by OSHA standards to be responsible for the assembly and disassembly of cranes.

Boom-extend mode: Refers to the order, and percentage, of extension telescopic-boom sections. The standard mode is to extend all segments the same percentage. Other modes may selectively extend certain sections to provide different capacities for a given load radius.

Connector pin: Any solid, cylindrical, metal part used to fasten, support, or attach the various components and members of a crane's boom and other hoisting gear.

Fly: With reference to mobile cranes, a boom extension carried on the side of the main boom when not in use, that can swing into position to extend the main boom; also called a swing-away.

Level luffing: The process of luffing the boom of a crane while adjusting the hoist cable and boom extension offset so that the load travels in a horizontal direction.

Luffing: Raising or lowering a crane's boom. A luffing jib can luff independently of the main boom.

Manual luffing jib: A semi-fixed jib extension whose offset must be manually set between lifts.

Power luffing jib: A jib extension that has the means to change its offset during a lift by remote control of the operator.

Stop rail: A heavy metal guiderail used by some crane manufacturers to help align a swing-away to its transport location as it is being stowed.

Additional Resources

This module is intended as a thorough resource for task training. The following reference materials are recommended for further study.

ASME Standard B30.5, Mobile and Locomotive Cranes. Current edition. New York, NY: American Society of Mechanical Engineers.

Cranes: Design, Practice, and Maintenance. Second edition. 2002. Ing J. Verschoof. Hoboken, NJ: John Wiley and Sons, Inc.

Crane Handbook. D. E. Dickie. 2013. Waltham, MA: Butterworth-Heinemann.

Mobile Crane Safety Manual. 2014. Milwaukee, WI: Association of Equipment Manufacturers.

29 *CFR* 1926, Subpart CC,*Cranes and Derricks in Construction*. **www.ecfr.gov**

Figure Credits

© Eugenesergeev/Dreamstime.com, Module Opener

Link-Belt Construction Equipment Company, Figures 1, 2

The Manitowoc Company, Inc., Figures 3–8, 11

© iStock.com/gmalandra, Figure 9

Liebherr USA, Co., Figure 12

© iStock.com/axel-ellerhorst, Figure 13

Nick Kaloterakis from kollected.com <**http://kollected.com**>, SA01

Section Review Answer Key

Answer	Section Reference	Objective
Section One		
1. c	1.1.0	1a
2. b	1.2.0	1b
3. d	1.3.0	1c
4. a	1.4.1	1d
5. d	1.5.1	1e

NCCER CURRICULA — USER UPDATE

NCCER makes every effort to keep its textbooks up-to-date and free of technical errors. We appreciate your help in this process. If you find an error, a typographical mistake, or an inaccuracy in NCCER's curricula, please fill out this form (or a photocopy), or complete the online form at **www.nccer.org/olf**. Be sure to include the exact module ID number, page number, a detailed description, and your recommended correction. Your input will be brought to the attention of the Authoring Team. Thank you for your assistance.

Instructors – If you have an idea for improving this textbook, or have found that additional materials were necessary to teach this module effectively, please let us know so that we may present your suggestions to the Authoring Team.

NCCER Product Development and Revision

13614 Progress Blvd., Alachua, FL 32615

Email: curriculum@nccer.org
Online: www.nccer.org/olf

❑ Trainee Guide ❑ Lesson Plans ❑ Exam ❑ PowerPoints Other ____________________

Craft / Level: Copyright Date:

Module ID Number / Title:

Section Number(s):

Description:

Recommended Correction:

Your Name:

Address:

Email: Phone:

Lattice Boom Assembly and Disassembly

Overview

The long length of some lattice booms makes it impossible for them to be transported when assembled. Therefore, many lattice-boom cranes require assembly at the work site. Boom assembly and disassembly can be extremely hazardous. Due to the weight and size of boom components, workers must take care when handling them and remain safety-conscious at all times. This module presents the general procedures for lattice-boom assembly and disassembly.

Module 21306

Trainees with successful module completions may be eligible for credentialing through the NCCER Registry. To learn more, go to **www.nccer.org** or contact us at 1.888.622.3720. Our website has information on the latest product releases and training, as well as online versions of our *Cornerstone* magazine and Pearson's product catalog.

Your feedback is welcome. You may email your comments to **curriculum@nccer.org**, send general comments and inquiries to **info@nccer.org**, or fill in the User Update form at the back of this module.

This information is general in nature and intended for training purposes only. Actual performance of activities described in this manual requires compliance with all applicable operating, service, maintenance, and safety procedures under the direction of qualified personnel. References in this manual to patented or proprietary devices do not constitute a recommendation of their use.

21306 V3

21306
Lattice Boom Assembly and Disassembly

Objectives

When you have completed this module, you will be able to do the following:

1. Identify boom components and primary considerations in the assembly and disassembly of lattice booms.
   a. Identify primary lattice-boom components.
   b. Identify pre- and post-assembly considerations.
2. Explain how to assemble and disassemble lattice booms and jibs.
   a. Explain the general method for assembling lattice booms.
   b. Explain the precautions and considerations for assembling and raising long lattice booms.
   c. Explain how to assemble jibs.
   d. Explain how to disassemble lattice booms.

Performance Tasks

Under the supervision of your instructor, you should be able to do the following:

1. Identify lattice-boom components.
2. Determine the amount of counterweight required for lattice boom erection.
3. Determine the proper boom-section count and their layout for a specific lattice boom.

Trade Terms

Back hitch
Belly lines
Boom stops
Cantilever length
Diagonal struts
Gantry
Hoist kick-out
Lattice lacing
Live boom mast
Main chords
Pendant
Suspension rope

Industry Recognized Credentials

If you are training through an NCCER-accredited sponsor, you may be eligible for credentials from NCCER's Registry. The ID number for this module is 21306. Note that this module may have been used in other NCCER curricula and may apply to other level completions. Contact NCCER's Registry at 888.622.3720 or go to **www.nccer.org** for more information.

Contents

1.0.0 Lattice Booms and Assembly Considerations 1
1.1.0 Boom Sections 3
1.2.0 Pre- and Post-Assembly Considerations 6
1.2.1 Counterweight Considerations 7
2.0.0 Lattice Boom and Jib Assembly 9
2.1.0 Assembling Lattice Booms 9
2.1.1 Initial Crane Setup 9
2.1.2 Boom Assembly 10
2.1.3 Raising the Boom 12
2.2.0 Assembling and Raising Long Lattice Booms 14
2.2.1 Preparing for Boom Assembly 14
2.2.2 Boom Assembly 14
2.2.3 Setting Up Hoist Line Components 15
2.2.4 Raising the Boom After Assembly 15
2.3.0 Jib Assembly 17
2.4.0 Disassembly of Lattice Booms 18
2.4.1 Crane Preparation 18
2.4.2 Boom Disassembly 19

Figures

Figure 1 Crane assembly process 2
Figure 2 Never remove pendant pins when the pendant is under tension 3
Figure 3 Never remove boom connector pins between the pendant under tension and the boom base 3
Figure 4 Never remove boom connector pins when the boom is supported only at its end or the connectors are under tension 4
Figure 5 Crane assembly procedure table of contents 5
Figure 6 Construction of a lattice-boom section 6
Figure 7 Connector pin secured with a reusable retaining clip 6
Figure 8 A boom lifting lug 7
Figure 9 Assembly area for a large crane 7
Figure 10 Installing counterweights 7
Figure 11 Extending the live boom mast 10
Figure 12 Installing the boom base 11
Figure 13 Install the upper connector pins 11
Figure 14 Installing lower pins 12
Figure 15 Boom head supported by blocking 13
Figure 16 Cable reel 13
Figure 17 Wheeled counterweights 14
Figure 18 A cantilevered boom 15
Figure 19 Securing the lower boom connector pins 15
Figure 20 Move the pendants to the first boom insert 15
Figure 21 Pendant spreader bar 15

Figure 22 Extending the pendants to the last attached insert 16
Figure 23 Completing the assembly of the boom inserts............................ 16
Figure 24 Extending the pendants to the boom tip..................................... 16
Figure 25 Belly lines suspending a long lattice boom 17
Figure 26 Long live mast for operations with small boom angles 17
Figure 27 Boom tip with jib bracket.. 17
Figure 28 Jib assembly with forestay and backstay pendants.................... 18
Figure 29 Boom pin properly stowed.. 19

SECTION ONE

1.0.0 LATTICE BOOMS AND ASSEMBLY CONSIDERATIONS

Objective

Identify boom components and primary considerations in the assembly and disassembly of lattice booms.

a. Identify primary lattice-boom components.
b. Identify pre- and post-assembly considerations.

Performance Tasks

1. Identify lattice-boom components.
2. Determine the amount of counterweight required for lattice boom erection.
3. Determine the proper boom-section count and their layout for a specific lattice boom.

Trade Terms

Diagonal struts: In a lattice boom with a rectangular cross section, pieces of structural pipe or angle iron welded to diagonally opposite chords of the boom to stiffen the assembly.

Gantry: A heavy frame with roller sheaves mounted on the upperworks of a crane, used for providing a mechanical advantage to the boom hoist ropes; also refers to a jib strut or mast.

Lattice lacing: In a lattice boom, the zig-zag pattern of structural metal supporting the main chords along the sides, top, and bottom of the boom sections.

Main chords: The heavy structural pipe or angle iron members running the length of a boom section.

Pendant: A length of heavy wire rope or flexibly connected sections of flat metal bars, equipped with connectors at both ends, used to hoist and/or support sections of a crane's boom.

Suspension rope: The rope reeved through a crane's gantry and mast sheaves to the sheaves suspending the boom pendants.

Depending on a crane's size and weight, assembly and disassembly (A/D) workers must disassemble cranes for transportation and reassemble them at new work sites. For most lattice-boom cranes, A/D workers remove the boom and counterweights before transport. For larger cranes, they also remove the crawler treads, because the carbody with crawlers is too wide for standard road lanes. Once the crane arrives at the site, A/D personnel and operators must unload and assemble the crane and its boom. *Figure 1* illustrates the general assembly process for one model of crane, though the steps are similar for most lattice-boom cranes.

Assembling cranes can be hazardous. The results of a two-decade study ending in 2006 indicate that over half of all fatalities from impact by falling crane boom components were the result of unskilled workers setting up cranes. OSHA has addressed the particularly hazardous nature of crane assembly and disassembly in 29 *CFR* 1926.1403 through 1926.1406. Some of these requirements include the following:

- An individual called the A/D director, who meets OSHA's criteria for being both a competent and qualified person, or a competent person assisted by qualified persons, must supervise all A/D operations.
- A/D personnel must understand their tasks and the hazards associated with crane assembly.
- The operator and A/D personnel must communicate and coordinate actions to avoid movement of equipment when workers are out of the operator's sight.
- The A/D director must address crane ground support, load characteristics, and other planning considerations associated with handling crane components.
- Before use, operators must test any hoist or swing brake they intend to rely upon to prevent crane movement. Otherwise, the operator must engage a pawl or install a locking device on the drum or gear, or workers must use blocking or an assist crane to prevent movement.
- A/D personnel must know the weights of all components they will handle to ensure the stability of the crane they are assembling, and to remain within the capacities of assist cranes.

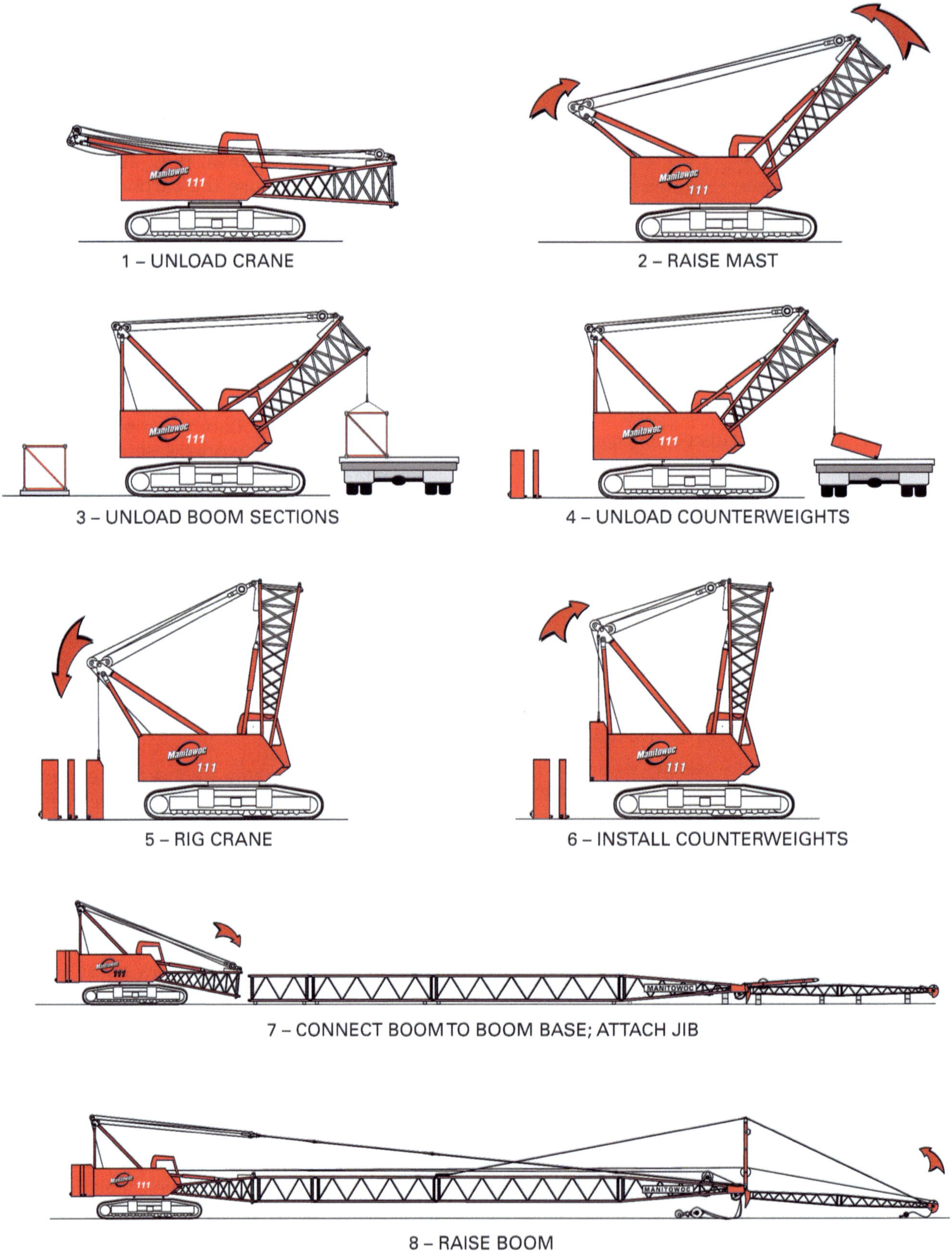

Figure 1 Crane assembly process.

- A qualified person must conduct a post-assembly inspection before lifting operations may begin, in accordance with 29 *CFR* 1926.1412, *Inspections*.
- Qualified riggers must perform all A/D rigging work.
- Workers shall not partially or completely remove pendant pins when the pendants are in tension (*Figure 2*).
- Workers shall not remove boom connector pins between the pendant attachment points and the crane body when the pendants are in tension (*Figure 3*).
- Workers shall not remove any boom connector pins when only blocking under the boom tip supports the boom, or the upper pins in the cantilevered section of the boom (*Figure 4*).

Crane operators must help to ensure crane assembly is done correctly according to the manufacturer's instructions. The project engineer, site superintendent, contractor, and the user of the crane are also responsible for safety.

Specific guidelines and instructions for assembling the crane and its boom are provided in the manufacturer's documentation for each crane. Some manufacturers may provide a separate A/D manual. *Figure 5* shows the table of contents for a typical set of assembly and disassembly procedures.

1.1.0 Boom Sections

Lattice booms, especially large ones with various jibs added, can be comprised of many independent sections. After lift planners have determined the boom model and configuration needed to accomplish the lift, the necessary pieces must be collected from inventory and prepared for shipment. The pieces required can be determined from a crane's boom combination chart in the documentation. An example of such a chart for the Manitowoc 18000 can be reviewed in the *Appendix*. The components should be inspected before they are loaded for shipment to avoid hauling damaged components to the site.

After the boom sections are unloaded at the site, they should be inventoried and inspected again. The crane's rigging drawings shows the arrangement of boom sections and their names or identification numbers. The A/D workers should inspect the boom sections for damage or wear. Qualified crane repair workers must repair or replace damaged components according to the manufacturer's specifications.

Damaged or worn components can cause serious injuries. The boom sections should be checked for bent**main chords**, **lattice lacing** or **diagonal struts** (*Figure 6*), as well as any failing welds. A/D workers can't fit the boom sections together if there are bent or damaged fittings. Components that are damaged or deformed in some way should be rejected. When alignment problems occur consistently, it often indicates that one or more components are damaged. Check that the **gantry** or mast **suspension rope**, **pendant** sections, and boom pendant pins are all in good condition. Connector pins must have spring-steel retaining clips or cotter pins, as shown in *Figure 7*. Cotter pins should never be reused, but retaining clips and pins provided by the manufacturer are intended for reuse.

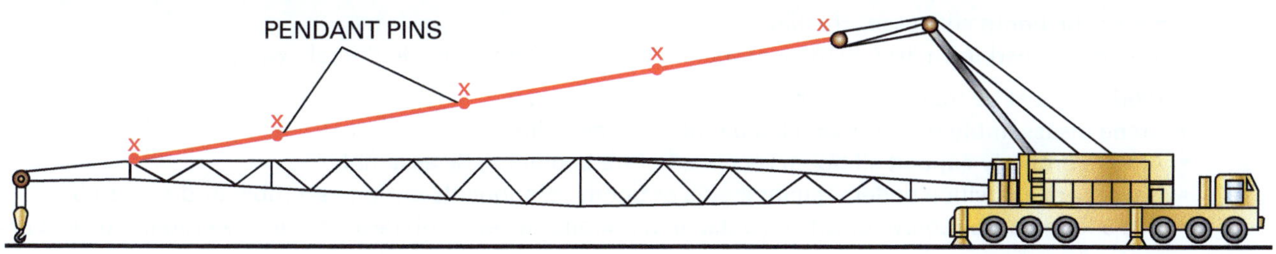

Figure 2 Never remove pendant pins when the pendant is under tension.

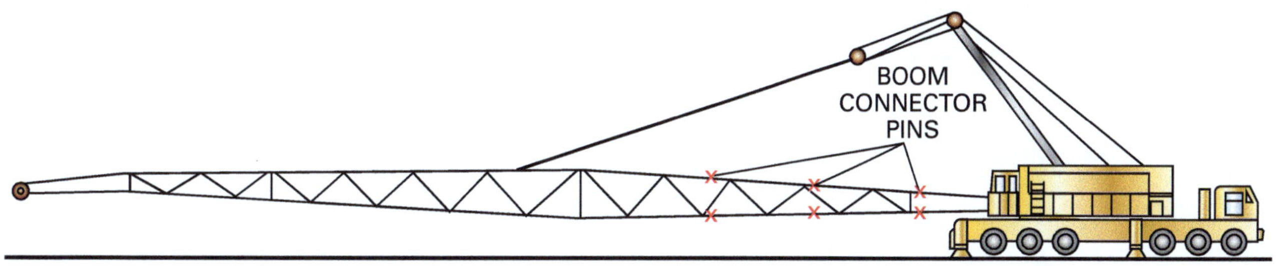

Figure 3 Never remove boom connector pins between the pendant under tension and the boom base.

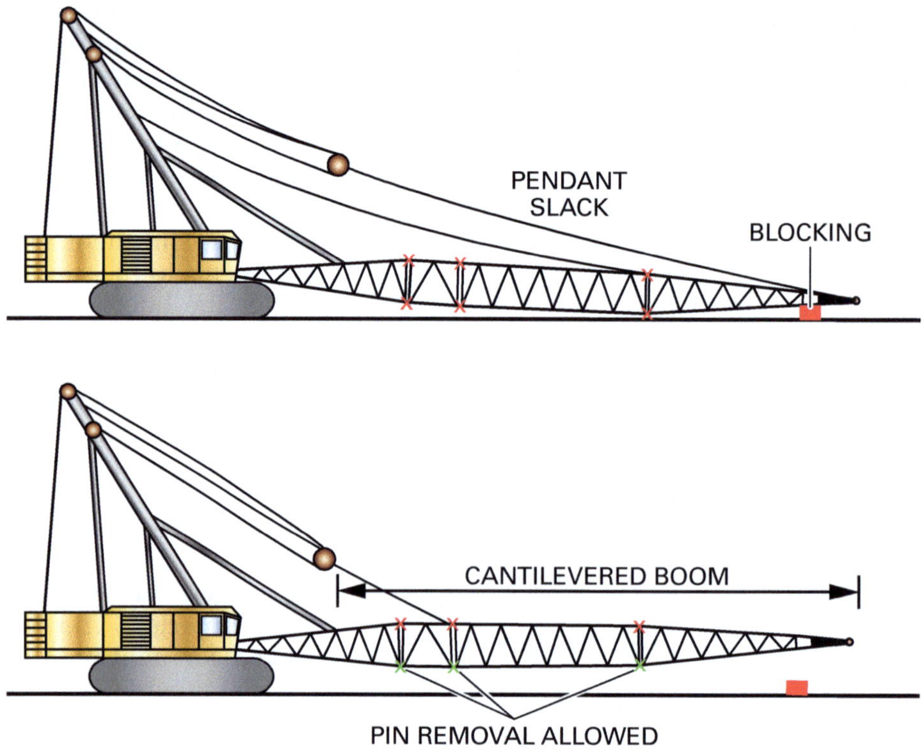

Figure 4 Never remove boom connector pins when the boom is supported only at its end or the connectors are under tension.

Case History

Lattice Boom Collapse During Disassembly

In 2008, an adult male died when crushed while assisting in the disassembly of a lattice boom. The boom consisted of a 15-foot base section, two intermediate boom extensions (20 feet and 10 feet), and a 10-foot tip that included the suspended hook block. Four connector pins and eight bolts secured adjacent intermediate sections.

The worker was a friend of the owner and had no formal training with assembling cranes. The owner had purchased the used crane, and he also had no formal crane A/D training. After positioning the boom near the horizontal with an auxiliary crane, the two men proceeded to remove the bolts and pins for the outermost sections. The worker reported that two of the pins seemed stuck. When he struck one of the pins with a sledge hammer, the boom collapsed, crushing him.

Lessons learned from the OSHA investigation of this accident include the following:

- Workers must be trained for crane disassembly operations.
- Crane disassembly procedures must be followed as written.
- Always properly support all boom sections before beginning disassembly.
- Never take a position underneath or inside of boom components during assembly or disassembly.
- Use a long bar to remove pins from outside the boom; never climb inside to remove retaining clips or to remove connector pins.
- Workers should be trained to insert boom assembly pins from the inside of the boom so the retaining clips are on the outside for easy access.

CRANE ASSEMBLY AND DISASSEMBLY GUIDE

MODEL 999

Table of Contents

Safety 2
Operating Limitations (Table 1) 2
General 3
Crane Weights 3
Operating Controls 3
Assembly Personnel 3
Assembly/Disassembly Area 3
Crane Orientation 3
Retaining Connecting Pins 3
Pre-Start Checks 3
- Engine 3
- Planetaries/Gear Cases 3
- Hydraulic System 3

Assembly Procedure
- A — Remove Crane from Trailer 4
- B — Raise Mast To Operating Position 6
- C — Install Assembly Block 8
- D — Install Crawlers 11
- E — Install Counterweight 14
- F — Assemble Boom 18
- G — Remove Boom Butt From Trailer 18
- H — Remove Assembly Block 18
- I-1 — Install #22E or #22EL Boom Butt 20
- I-2 — Install #82 Boom Butt 22
- J-1 — Connect #22E or #22EL Boom Butt To Boom 26
- J-2 — Connect #82 Boom Butt To Boom 28

Pre-Raising Checks 31

Disassembly Procedure
- A — Disconnect Boom Butt From Boom 31
  - General 31
  - Disconnect #22E or #22EL Boom Butt From Boom . 32
  - Disconnect #82 Boom Butt From Boom 34
- B-1 — Remove #22E or #22EL Boom Butt 36
- B-2 — Remove #82 Boom Butt 38
- C — Install Assembly Block 42
- D — Install Boom Butt On Trailer 42
- E — Disassemble Boom 42
- F — Remove Counterweight 42
- G — Remove Crawlers 44
- H — Remove Assembly Block 48
- I — Lower Mast To Transport Position 48
- J — Install Crane On Trailer 48

Component Identification

NOTE: Except for procedures specific to #82 boom, mast for #22E boom is shown throughout this folio. Mast for #22EL and #82 booms is identical except that pendants (#22EL boom) or straps (#82 and #82LR boom) are suspended from mast.

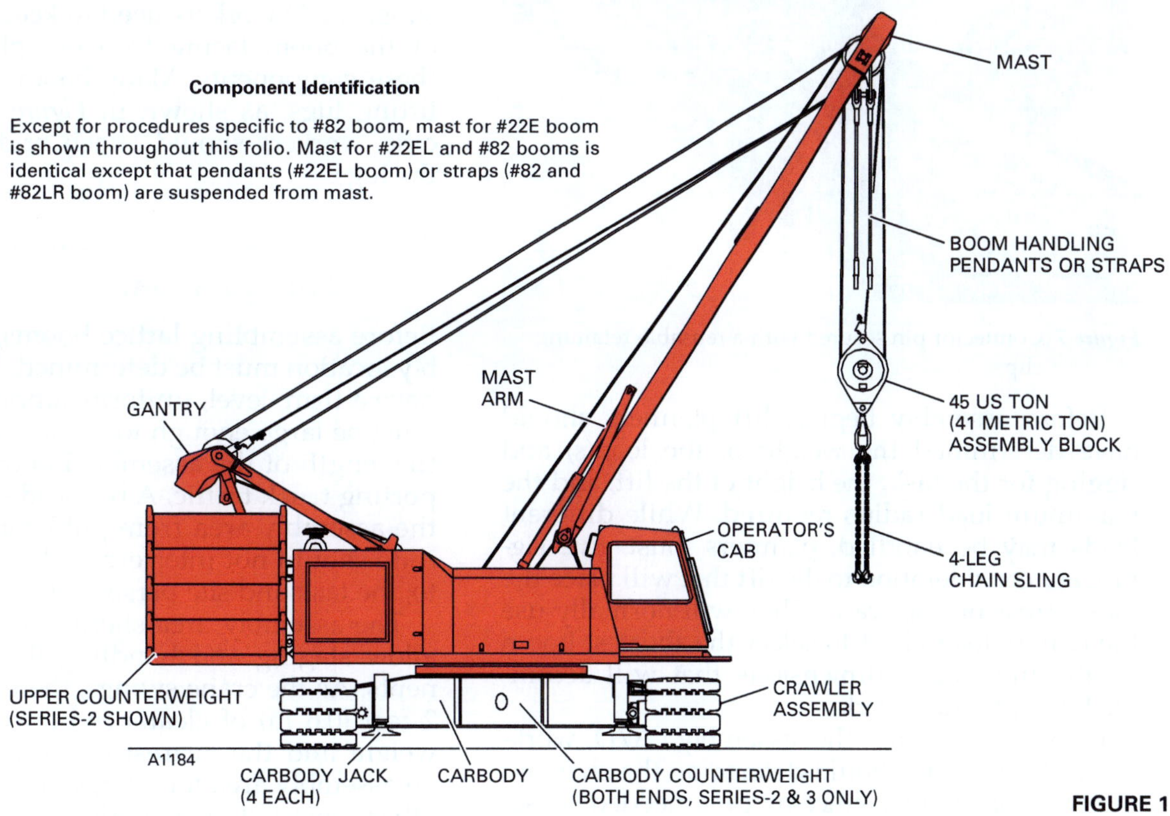

FIGURE 1

Figure 5 Crane assembly procedure table of contents.

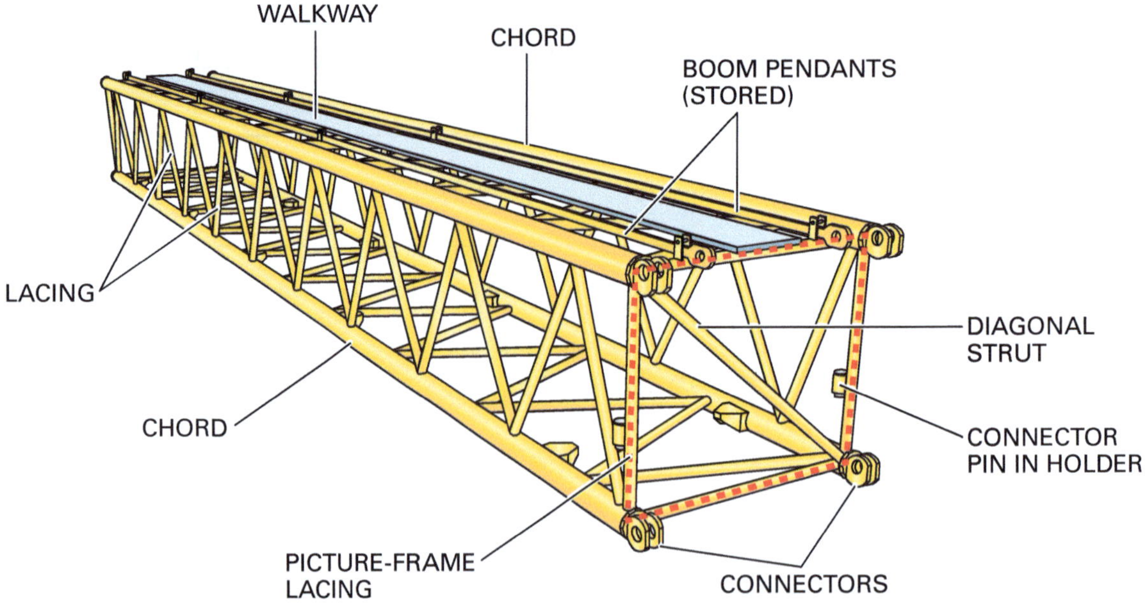

Figure 6 Construction of a lattice-boom section.

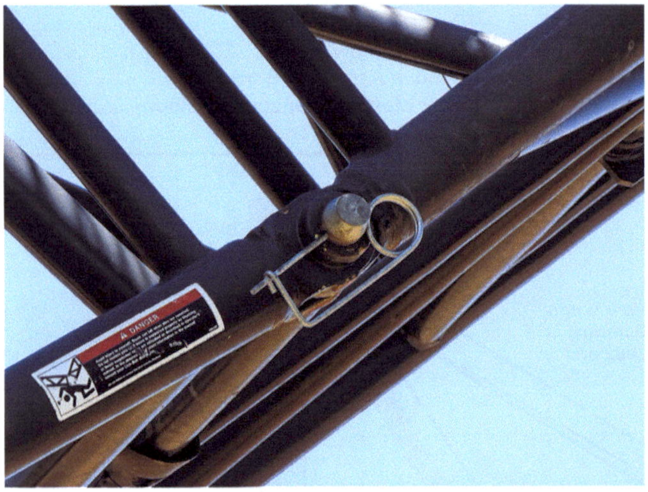

Figure 7 Connector pin secured with a reusable retaining clip.

Before assembly begins, lift planners should have determined the weight of the load(s) and rigging for the task, the height of the lift, and the maximum load radius required. While different loads may be handled, planners must give significant consideration to the lift that will place the most stress on the crane. They will normally use the crane's load chart to select the shortest boom height and radius dimensions that will accomplish the planned work.

Before beginning the assembly, A/D workers and operators should determine the size and quantity of boom connector pins needed to secure the sections. They will need to inventory the required components and replace any damaged or missing items with components meeting the original manufacturer's specifications.

Boom sections must be handled with care to prevent damage to chords, lacing, and connectors. It is important to pay close attention to the type of slings used for the sections. Synthetic slings are generally best, although workers may use other rigging. When using wire or chain slings, A/D workers need to keep the slings clear of the boom lacing to avoid placing stress on these components. Many boom sections include lifting lugs, as shown in *Figure 8*. If this is the case, workers should attach bridles or slings to the lugs.

1.2.0 Pre- and Post-Assembly Considerations

Before assembling lattice booms, a good assembly location must be determined. The area should have a firm, level, uniform supporting surface. It must be large enough to accommodate the crane, the length of the assembled boom, and the supporting truck traffic. A barricade should separate the assembly area from public areas. The barricade should not interfere with, or pose a hazard to, the task and site personnel.

The assembly area should be large enough to allow staging and handling the largest components. As the crane swings, there must be at least 2 feet (0.6 m) of clearance between the counterweight and the nearest obstacle. *Figure 9* shows an assembly area for a large crane, set apart from other construction activities.

The A/D director, operator, lift planners, and persons responsible for the intended lifts must take the following logistics into consideration:

- Is there enough room for the crane to maneuver into position?
- Is there sufficient room to assemble the boom?
- Can the trucks hauling boom sections get into place and be safely unloaded?
- Is there enough timber blocking to support the boom during assembly and disassembly?

Figure 8 A boom lifting lug.

Figure 9 Assembly area for a large crane.

- Is there an effective system for the crane operator to communicate with the A/D workers?

1.2.1 Counterweight Considerations

Crane operators and any personnel involved in the assembly of lattice-boom cranes must be aware of counterweight requirements. Before assembly begins, verify that the correct counterweight sections required to lift the boom after assembly are available and installed. Lift planners should also verify that the counterweight is sufficient for the lifting operation(s).

Never add or remove a counterweight from a carrier-mounted crane without first positioning the machine on firm, level ground, and extending all outriggers, as shown in *Figure 10*. To lift longer booms off the ground, it may be necessary to install additional counterweight sections. Workers may add these to the standard counterweight on the upperworks, on the sides of the crawler assemblies, or in the bumper areas of the carrier. For some cranes, workers must remove this extra weight after the boom is raised; too much weight behind the point of rotation can reduce backward stability. Always refer to the crane's load chart for allowable counterweight configurations.

WARNING!

To avoid tipping the crane, install counterweight sections in the exact sequence shown in the manufacturer's documentation. Ensure that all counterweight components are properly secured; if they move or fall unexpectedly, they can cause serious injury or death.

Figure 10 Installing counterweights.

Additional Resources

ASME P30.1, Planning for Load Handling Activities. Current edition. New York, NY: American Society of Mechanical Engineers.

Mobile Crane Safety Manual. 2014. Milwaukee, WI: Association of Equipment Manufacturers.

29 *CFR* 1926, Subpart CC,*Cranes and Derricks in Construction*. **www.ecfr.gov**

1.0.0 Section Review

1. Connector pins in pendants should never be removed _____.
   a. unless the pendant is under light tension
   b. without first removing the pendant from the crane
   c. until the proper lubricant has been applied
   d. when the pendant is under any tension

2. Which of the following features is necessary for a crane assembly area?
   a. Soft, sandy soil for cushioning the boom sections
   b. Easy entry and egress for transport trucks
   c. Gently sloping area for assembling the boom
   d. Minimum of 220 VAC electrical service

Section Two

2.0.0 Lattice Boom and Jib Assembly

Objective

Explain how to assemble and disassemble lattice booms and jibs.

a. Explain the general method for assembling lattice booms.
b. Explain the precautions and considerations for assembling and raising long lattice booms.
c. Explain how to assemble jibs.
d. Explain how to disassemble lattice booms.

Trade Terms

Back hitch: The mechanism that permits adjusting the position of the boom gantry to maximize its leverage on the boom hoist cables.

Belly lines: Additional pendants attached to the boom hoisting bail and secured at a location near the midpoint of a long boom to minimize sagging.

Boom stops: Devices that mechanically prevent the boom from exceeding a maximum boom angle, reducing the potential for backward tipping or structural damage.

Cantilever length: The length of a horizontal boom that projects beyond the support of the boom pendants, leaving that portion of the boom supported at only one end, like a cantilever.

Hoist kick-out: A term for any safety device that automatically stops a hoist drum when a limit is reached.

Live boom mast: A heavy framework or lattice boom with sheaves, hinged at the lower end of its base on the crane's upperworks, whose purpose is to control the angle of the main boom through attached boom pendants. Boom hoist cables reeved through the boom gantry sheaves or hydraulic actuators control its position.

NOTE

The information presented in this module is general in nature because there are numerous crane models and manufacturers. Always refer to the manufacturer's documentation for the specific crane in use before beginning any assembly work.

Manufacturers provide a boom rigging drawing with each boom ordered. The drawings can include details about boom, boom gantry, back hitch, boom-point, and pendant assembly. The operator should study and understand the content of these drawings before beginning the assembly.

WARNING!

Boom section connection points can be potentially hazardous pinch points. Avoid placing the fingers and limbs into these areas when an unstable condition exists.

2.1.0 Assembling Lattice Booms

This section presents a general procedure for assembling lattice booms. Always consult the manufacturer's documentation before any assembly or disassembly takes place.

2.1.1 Initial Crane Setup

Before beginning assembly, the assembly team must review the planned crane configuration. It is essential to know the model and length of the boom to be assembled. With this information, the operator and A/D director can verify, from the crane's load chart, the amount of counterweight required. Ideally, there needs to be enough counterweight to safely lift the boom off the ground unassisted. However, this is not always the case.

Depending on the type of carrier, assembly may include installation of the crawler assemblies. This step generally involves erecting the gantry system and often a live boom mast as specified by the manufacturer. Initial setup of a basic crawler crane will generally proceed in the following order:

1. Deploy outriggers for supporting the carbody/upperworks
2. Erect the gantry and boom mast rigging
3. Install crawler assemblies
4. Set down on crawlers
5. Install the required counterweight sections

An attempt should be made to position the crane so that the upperworks align with the crane's most stable quadrant pointing in the direction of boom assembly. The boom sections can then be laid out. This may not be possible, however, due to space restrictions. Every jobsite is unique. It may be easier to lay out the boom sections first, then travel the crane into position.

Some crane models have an extendable counterweight that can increase the crane's leverage. Manufacturers may require extending a movable counterweight for boom assembly and disassembly.

2.1.2 Boom Assembly

After positioning the crane over the boom base section according to the documentation, set the swing lock to secure the upperworks. Extend the gantry or mast (*Figure 11*), if applicable, to install the boom base. A mast with special pendants, as shown in the figure, permits the crane to self-assemble the boom base section without the need for an assist crane.

> **NOTE**
>
> Crane load charts or manuals may specifically address the use of the gantry or live boom mast for assembly operations. If required, workers must set up these components first before attempting to assemble and raise a boom. Crane manufacturers will specify the maximum length of boom a crane can raise using its standard equipment. Longer booms require the use of special equipment.

Using pendants or slings as required, raise the hinge connectors of the boom base into position and insert the boom hinge pins. On many models, the pins are hydraulically actuated due to their size. After connecting the boom base hinge, attach the initial set of boom pendants (*Figure 12*). Then install the **boom stops**, if applicable.

> **WARNING!**
>
> A/D workers must install boom sections and components using the sequence and methods specified by the manufacturer. A boom can collapse if not properly assembled. Do not mix boom sections from different crane models or manufacturers.

With the boom base installed, workers can proceed with assembling the boom sections in the order specified by the manufacturer's manual. The boom number may be stamped on the side of the boom joint connectors or on the diagonal struts of adjacent boom sections. The boom may also have an identification plate. Boom identification is done to prevent mixing boom sections of one type with those of another. Boom identification numbers can be in a variety of formats.

The A/D director must ensure that workers lay down the boom sections in the order of assembly. Generally, manufacturers recommend placing the shorter boom sections closest to the boom base. For shorter booms, workers may pre-assemble the boom and boom head as a unit before attaching it to the boom base.

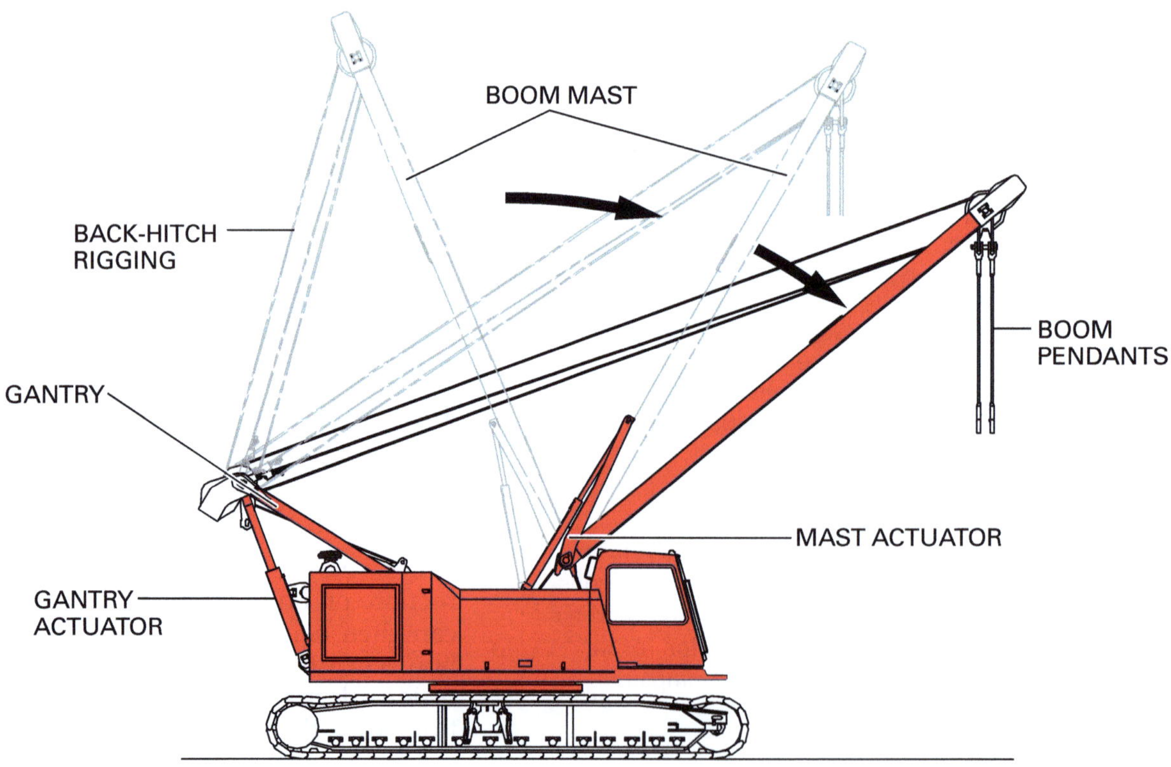

Figure 11 Extending the live boom mast.

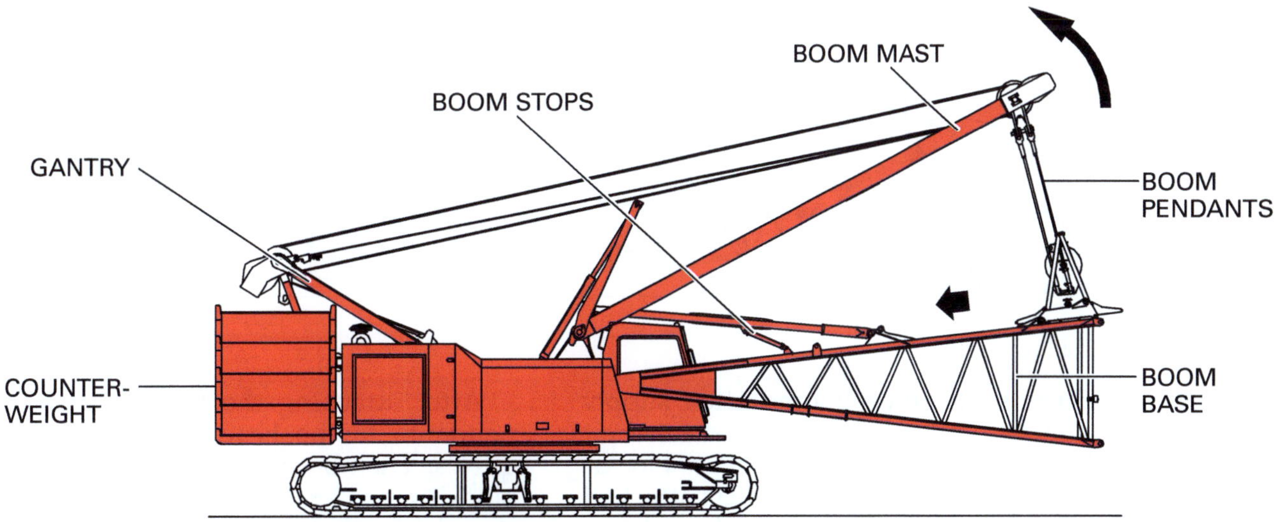

Figure 12 Installing the boom base.

Travel the crane to position the end of the boom base above the first intermediate boom section, also called an *insert*. Lower the boom base to the level of the first boom section. Align the upper connector holes between the two sections being pinned together. Insert the two upper connector pins and their pin retainers or cotter keys (*Figure 13*).

> **WARNING!**
>
> Do not stand inside or under the boom while installing or removing connector pins. Always stand outside the boom unless the employer determines that there is no other way to accomplish the task. In such a situation, the A/D director must implement positive measures to prevent equipment movement when performing these tasks. *Appendix B* of 29 *CFR* 1926, Subpart CC, *Assembly/Disassembly: Sample Procedures for Minimizing the Risk of Unintended Dangerous Boom Movement,* covers these situations.

Boom up slightly until the lower pin connection points align. Engage the boom or gantry/mast hoist pawl. Install the pins and retainers (*Figure 14*).

Typically, assembly procedures require attaching pendants to the outermost section of the assembled boom that will connect to the next section lying on the ground. This approach provides the maximum leverage for lifting the boom without assistance. Pendants come in matched sets of either two or four. Builders usually stamp identification numbers into both ends of each pendant to aid in keeping them in sets. Always install pendants in the sequence noted on the rigging drawing.

> **WARNING!**
>
> A set of pendants must be exactly the same length. If they are not, the boom will twist, creating the potential for catastrophic boom failure.

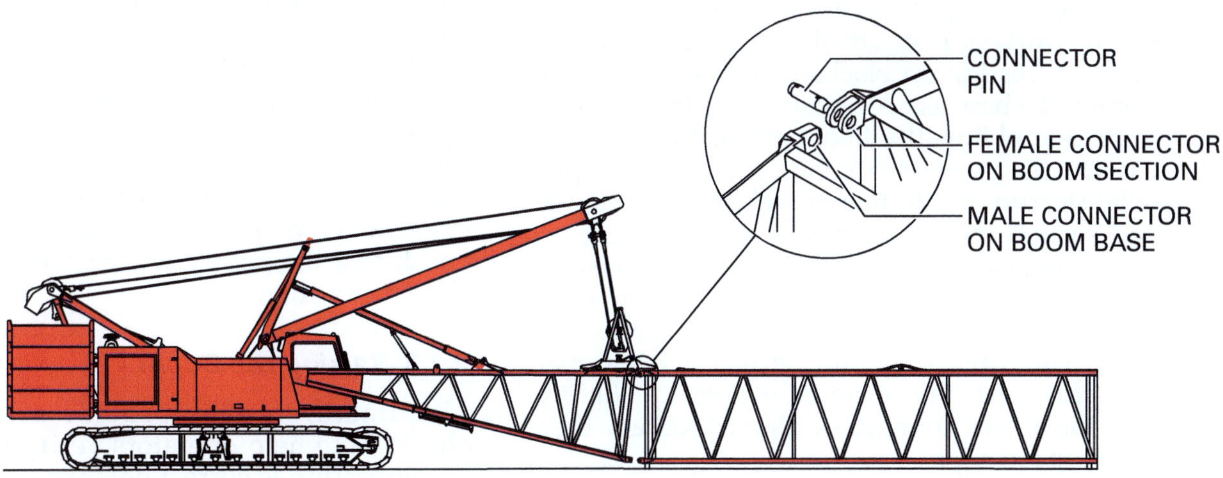

Figure 13 Install the upper connector pins.

Figure 14 Installing lower pins.

Manufacturers of lattice booms often supply pendants attached to each boom insert for their cranes. With the boom end resting on blocks, workers extend the pendants to the last attached section by detaching the previous set of pendants. The workers then attach these to the free ends of the next set of pendants using pendant assembly pins and retainer clips.

CAUTION

When assembling the boom, make sure that slack suspension ropes or pendants do not snag the connector pins and/or their retainers.

WARNING!

Check the crane's load chart before raising the boom. It may be necessary to block the crawlers, if applicable, to maintain stability.

Assembly of the boom continues in the same manner—join the upper connectors of the next insert, slightly raise the boom to join the lower connectors, lower the boom onto blocks, and move the pendants to the last assembled insert. Repeat until workers have attached all the inserts.

Installing the boom head section is similar to attaching an insert. If the crane's instructions permit, workers may attach the head section to the boom resting on blocks before attaching the boom to the base as previously described. If separately assembling the boom head, the operator first lifts the assembled boom, sets the appropriate hoist drum pawls, then travels the crane to the head section resting on blocks. Alternatively, an assist crane can place the head section at the end of the assembled boom on blocking. The operator adjusts the boom so that the upper connectors align. After inserting and securing the upper connector pins, the operator booms up slightly until the lower connectors align. Workers then insert the lower pins.

After attaching the head section, extend the pendants to the head, if the manufacturer requires. Then boom up so that workers can place blocking under the boom head (*Figure 15*) for completing the remainder of the assembly. Many boom heads include a special support bracket or foot for this purpose.

CAUTION

If the pendants are too long for a crane equipped with a shorter boom and a live boom mast, the mast may tip backwards when approaching maximum boom angle. The weight of the tipping mast can pull an unloaded boom into the boom stops. The operator will have to pull the boom away from the stops by loading the hook.

The remainder of the boom assembly process involves preparing the boom for lifting operations. Workers should make up any electrical connections needed between the crane base and the boom head (*e.g.*, the anti-two-blocking circuit). The cabling required is typically stored on a reel to accommodate varying boom lengths (*Figure 16*). Then the main hoist line is rigged as well as the whip line, if needed. Depending on the length of the boom, workers may have to set up cable roller sheaves and rope guides at various points along the boom.

Check for proper installation and adjustment of the boom angle indicator. If necessary, boom up slightly and engage the boom hoist pawl. Then reeve the block and/or ball. Verify that the hoist wire rope is spooled tightly onto the drum and engaged with the proper sheaves. If applicable, check that the gantry wedge socket securely anchors the wire rope.

2.1.3 Raising the Boom

Manufacturers may provide a boom raising checklist to help operators verify that the crane configuration is correct before raising the boom.

Figure 15 Boom head supported by blocking.

Boom leverage on the crane's components is most extreme when raising the boom from blocking on the ground, in a horizontal position. Examples of items that may be included on such a checklist include the following:

- Crane is on a firm, level surface.
- Boom hinge pins are fully engaged and the pin locking cover is in the Up position.
- Crawler connecting pins are engaged and locking pins installed.
- Carbody jack pads are stored.
- Carbody jacks are fully retracted and pinned in stored position.
- Boom and jib inserts are installed in proper sequence per the rigging drawing.
- Spreader (if required) is installed at the proper location and spread to the proper position.
- Intermediate suspension is installed, if required (for longer booms).
- Links are properly connected between all straps.
- All insert and pendant connector pins are installed. Cotter pins are installed and spread.
- The gantry is fully raised.
- Mast arms are fully lowered.
- The mast arms bypass handle is turned off.
- Setup mode is turned off (select and confirm desired LMI operating mode).
- Boom hoist wire rope is spooled tightly onto the drum and engaged with proper sheaves. Wire rope is securely anchored to the wedge socket at the gantry.
- Load lines are spooled tightly onto the drums and engaged with proper sheaves. Load lines are securely anchored to the wedge sockets at boom and jib point or at load block and weight ball.

Figure 16 Cable reel.

- All blocking, tools, and other items have been removed from the boom and jib.
- Automatic boom stop is properly installed and adjusted.
- Cable from the crane control system is connected to the cable reel and boom butt.
- Block-up limit control is properly installed and operational; LMI is properly installed and operational.
- Crane and attachments are properly lubricated.
- Crawlers are blocked if required per load chart.
- Wind is within allowable limits for the operation given in load chart.

When ready to raise the boom to the operating position with the standard hoisting attachments, confirm that the crane's configuration agrees with the load chart for this operation. Do not lift the load block or ball when raising the boom; there is no need to place an even greater load on the crane. Pay out the hoist line to keep the additional weight on the ground. Raise the boom smoothly in one continuous lift until reaching the minimum allowed operating boom angle. Once the boom is in the air, check that the boom hoist limiting device (**hoist kick-out**) is installed and in working order.

> **CAUTION**
>
> Always follow the manufacturer's recommendations when initially raising the boom to avoid damaging the equipment.

2.2.0 Assembling and Raising Long Lattice Booms

The safe assembly of longer lattice booms requires some additional considerations. There is no absolute definition for a long lattice boom compared to a short one. The main differences are as follows:

- The tendency of long lattice booms to sag as they come closer to a horizontal position
- The effect of a long boom's leverage on crane stability

This section provides general guidelines for assembling longer lattice booms. Operators and A/D personnel must always follow the crane manufacturer's specific instructions.

2.2.1 Preparing for Boom Assembly

The preparations for assembling a longer boom are essentially the same as those for a shorter boom. Consult the crane's load chart and documentation for the counterweight required for raising the assembled boom and for the planned lifts. The manufacturer may require the installation of counterweights at other locations on the crane for this purpose. For very long booms, some manufacturers provide a separate wheeled counterweight that attaches like a trailer to the main crane carbody (*Figure 17*). Boom hoist cables anchored to the counterweight trailer pass through an extra lattice-boom mast, mounted behind the main boom.

Always verify that the installed counterweight is sufficient to lift the boom off the ground. If so equipped and if required, extend the extendible counterweight. Due to the larger moment forces exerted when raising long booms, some crane manufacturers recommend blocking the front idlers on the crawler ends.

Figure 17 Wheeled counterweights.

WARNING!

The crane must be equipped with the appropriate counterweights or the crane may tip when lifting the assembled boom.

2.2.2 Boom Assembly

Install the assembly slings and/or pendants to the boom mast and assemble the base section of the boom to the upperworks. Then proceed to assemble the boom inserts in the order specified by the manufacturer. Generally, manufacturers require placing the shorter inserts closest to the base section. Assembly workers should lay out and draw together some, but not all, of the sections. The total length of the assembled boom attached beyond the boom pendants must not exceed the maximum **cantilever length** specified by the manufacturer (*Figure 18*). The cantilever length is the length of a horizontal boom that projects beyond the support of the boom pendants. The portion of the boom beyond the pendants is being supported at only one end, like a cantilever. The maximum cantilever length is usually a structural limitation of the boom, rather than a crane stability issue. Align the upper pin connection points and insert those pins only, securing them with pin retainers.

With the base section attached by the upper connector pins, boom up slightly until the lower pin connections align. Engage the boom hoist pawl, and install the lower pins and pin retainers (*Figure 19*).

Before moving to assemble the next boom section, workers need to extend the boom pendants to the next position along the boom, according to the documentation. Lower the boom onto blocking until the pendants are slack, then disconnect the pendants from the boom. Connect their free ends to the pendants leading to the end of the last attached insert (*Figure 20*) using pendant assembly pins and retainers. Remember to watch for boom twist. Use pendant spreader bars (*Figure 21*) as required by the manufacturer.

Assemble the next inserts in the same manner, following the manufacturer's procedure. Lower the boom onto blocking until the pendants are slack. Move the pendants out to the end of the last attached boom section (*Figure 22*). Note how the shorter inserts are near the base of the boom. Repeat for the remaining inserts (*Figure 23*).

Repeat the assembly procedure for the boom tip if not already completed. Lower the boom onto the blocks again and move the boom pendants to their operating position near the end of the boom (*Figure 24*).

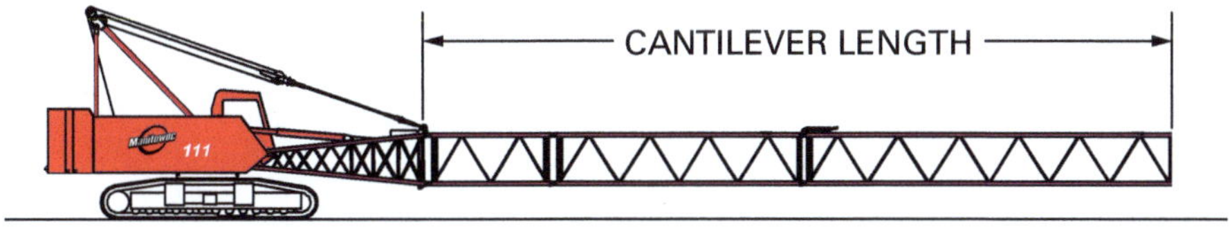

Figure 18 A cantilevered boom.

Figure 19 Securing the lower boom connector pins.

> **WARNING!**
> If the pendants are too short on a crane with a longer boom, it will overload the boom, mast, pendants, and boom hoist systems when handling near-capacity loads.

Operators should verify that the length of the final set of pendants is the length specified in the documentation for assembly. Depending on the boom manufacturer and the electrical features provided, workers will also need to make the electrical connections along the boom between the boom head and the crane junction box on the upperworks. Verify the proper routing and securement of the electrical cable(s) to avoid pinching during boom movement. The documentation will identify the cables and the point in the procedure for completing this part of the assembly.

Figure 21 Pendant spreader bar.

2.2.3 Setting Up Hoist Line Components

After assembly of the lattice boom is complete, assembly workers must turn their attention to rigging the hoist line, reeving the sheaves, and attaching the hoist block or ball. Rig the main hoist line and, if necessary, the whip line along the boom. Generally, this process requires attaching or erecting boom roller sheaves and guides on the designated boom sections to support the lines between the hoist drum(s) and boom head. If necessary, boom up slightly and engage the boom hoist pawl. Reeve the boom head sheaves and block.

> **NOTE**
> Information regarding the installation of wire rope and reeving the load block can be found in Module 21204 of this curriculum.

2.2.4 Raising the Boom After Assembly

Some manufacturers recommend that assembly workers and operators perform certain checks before raising the boom. Refer to the checklist provided earlier for raising the boom.

> **CAUTION**
> When initially lifting a long lattice boom following assembly, some manufacturers require an assist crane to suspend and/or lift the boom.

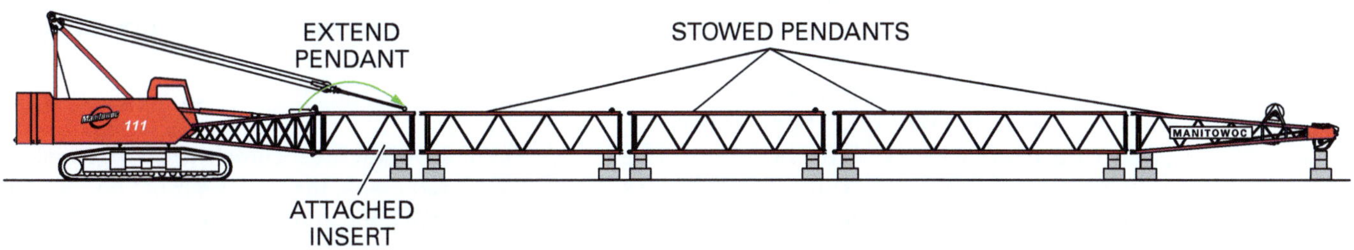

Figure 20 Move the pendants to the first boom insert.

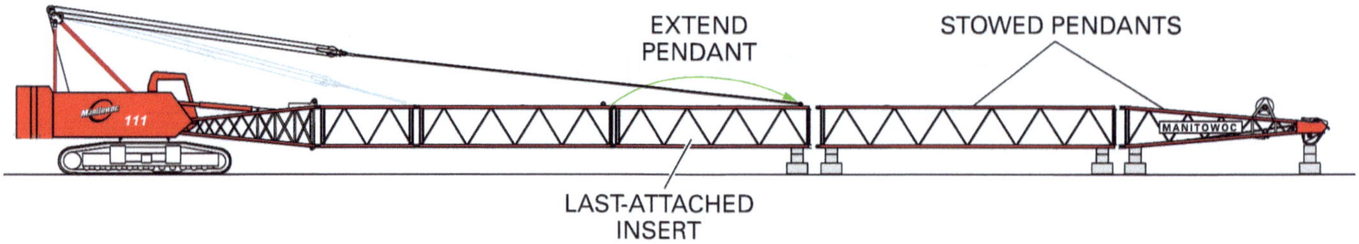

Figure 22 Extending the pendants to the last attached insert.

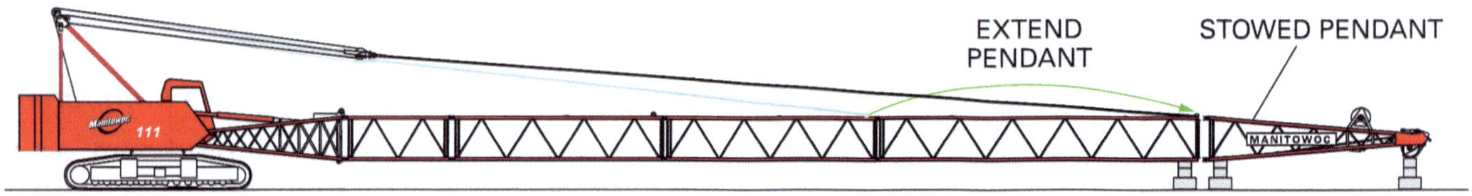

Figure 23 Completing the assembly of the boom inserts.

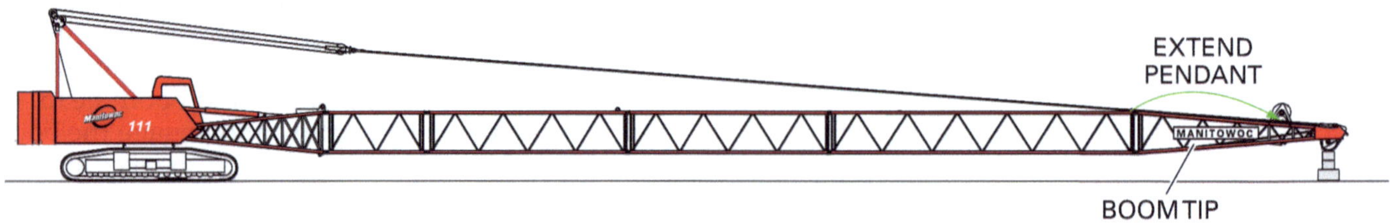

Figure 24 Extending the pendants to the boom tip.

Long lattice booms can be several hundred feet long. If the boom hoisting pendants connect to just the boom head, the boom structure between the base and head can visibly sag as it approaches horizontal. Sagging places excessive stress on the boom chords and lacing, leading to structural damage. Crane manufacturers can minimize sagging by requiring the use of boom suspension rigging, often called **belly lines**. These may be separate pendants attached to the boom hoisting bail and secured to the boom at a location near its midpoint. Belly lines may also be wire rope pendants reeved through sheaves at the boom hoisting bail and attached to the boom pendants, as shown in *Figure 25*. This configuration equalizes the tension between the main hoist pendants and belly lines, which can vary depending on the boom angle.

The other challenge with long booms is the excessive leverage the boom imposes on the crane when it is at or near horizontal. There are methods that can be used to solve this problem. The crane may require additional counterweight. As noted earlier, this may include counterweights on a separate carrier attached to the crane base. Some manufacturers provide for a longer live mast (*Figure 26*), which supplies the needed leverage to lift the boom. Finally, an assist crane may lift the boom tip into the permitted operating region of the crane's load chart. The manufacturer's documentation will provide directions on which method to use.

When the A/D director gives permission, raise the boom slowly. To gain extra stability when raising a boom without assistance, do not lift the load block. Pay out the hoist line to keep the block on the ground. If an assist crane is involved, coordinate with the other operator and maintain tension in your crane's boom hoist rigging as the assist crane lifts the boom. If the crane model has a manual LMI bypass feature, be sure to set the proper LMI mode when the boom is within normal operating limits. Detach the assist crane when the boom is within the operating limits of the crane.

WARNING!

If a worker must climb the boom to release the assist crane, that person must be trained for and use the appropriate fall arrest PPE.

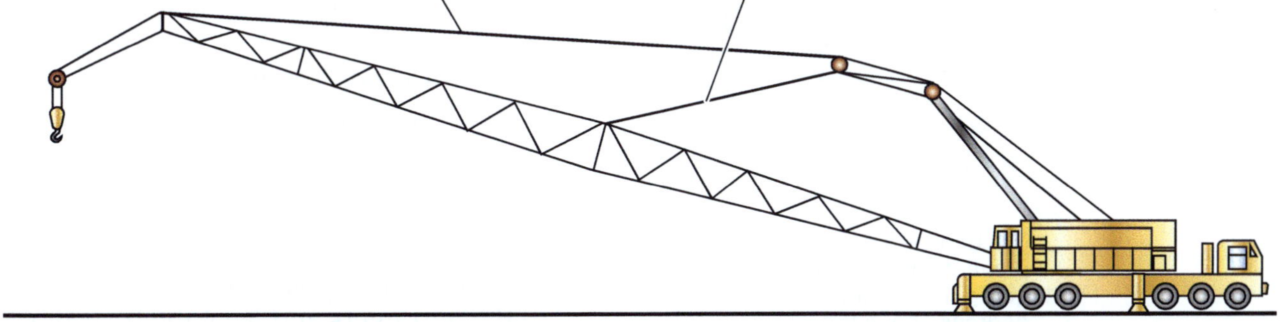

Figure 25 Belly lines suspending a long lattice boom.

Figure 26 Long live mast for operations with small boom angles.

2.3.0 Jib Assembly

After assembling the boom, workers may assemble and attach a jib extension as part of the process. The advantage of installing a jib extension during boom assembly is that the main boom is already at ground level on blocks. Adding an extension to an operating crane requires lowering the boom to ground level, which often involves significant effort.

If necessary, lower the boom onto blocking supporting the boom head, but clear of the sheaves. Pay out the main or auxiliary hoist line as required, and lay it on the ground off to the appropriate side of the boom per the documentation. Some boom tips have welded brackets to attach a jib, as shown in *Figure 27*. If a jib adapter is required for the boom, install the jib adapter to the boom tip.

Assemble the jib resting on blocking on the ground. Jibs may be a single section or they may have multiple sections. Next, workers attach the base of the jib to the boom head. They can do this by having the operator lower the main boom to the jib or having an assist crane raise the jib to the boom head. Workers then pin the jib base to the boom tip.

Figure 27 Boom tip with jib bracket.

The process of rigging the jib's stays or operating gear depends entirely on the type of jib and whether it has remote luffing capability. A jib that can be offset manually may have nothing but the baseplate attachment points where assembly workers insert pins to establish the required offset angle. If the jib is of the fixed, A-frame type, workers must install the jib mast (*Figure 28*) first, then secure the A-frame by installing the appropriate fore- and backstay pendants for the required offset. Pin the forestay and backstays to the strut tip using the pins furnished with the pendants.

If the jib extension is a luffing jib, workers must reeve the appropriate jib hoist lines to the jib backstay. Hydraulic connections are required for a power luffing jib. The details of these assemblies are beyond the scope of this module.

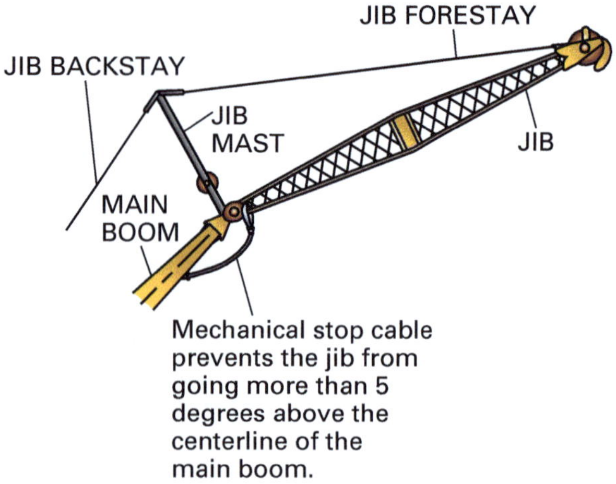

Figure 28 Jib assembly with forestay and backstay pendants.

To complete the jib assembly, erect or install a wire rope roller guide on each jib extension section used, if applicable. Reeve the hoist rope through the jib roller guide(s) and the jib head sheave. Attach the hook block or ball. Attach and connect the anti-two-block sensor to the jib hoist rope, if appropriate. Also make up any electrical connections between the jib and main boom at the main boom head. When the jib assembly is complete, raise the boom as previously described, observing the applicable precautions.

2.4.0 Disassembly of Lattice Booms

This section provides general guidelines for the disassembly of a lattice boom. Always consult the manufacturer's documentation when performing any kind of disassembly. Prepare the crane for the disassembly process, which can impose unusual stress on the crane. Remember that OSHA standards require that an A/D director be present for disassembly as well as assembly.

2.4.1 Crane Preparation

Move the crane to a broad, level, firm area with enough room to lower and land the boom. Set up barricades to minimize personnel traffic during the disassembly operations. Review the previous sections for other A/D site considerations, if necessary. After positioning the crane, fully extend the outriggers and level the base; block the idlers on a crawler, if required. Verify that the counterweight is correct, and set the swing lock.

CAUTION

Operators should make a thorough inspection of the boom hoist cable before booming down. Ropes damaged during crane operations can break under the increased tension imposed at minimum boom angles.

Lower the load block and/or ball to the ground, then lower the boom in the quadrant of maximum stability or as previously determined by the A/D director. For certain crane models, the operator may have to disable certain LMI safety features or switch to a different operating mode to lower the boom for disassembly. Place blocking under the boom tip and under both sides of every joint. Lower the boom onto this blocking until the pendants go slack. The process may require an assist crane to help lower long booms.

Splitting the Load

Some crane manufacturers have developed a special feature for assisting the erection of very long jibs. One or two permanently attached, lightweight wheels support the jib head on the ground. After the jib extension and its rigging are attached, the operator booms up on the main boom while paying out on the boom luffing hoist pendants. The jib head wheels roll toward the crane as the jib base rises. With this feature, the ground supports half the weight of the jib while raising the boom. At the appropriate point permitted by the load chart, the operator hauls in on the jib luffing hoists, raising the far end of the jib into position.

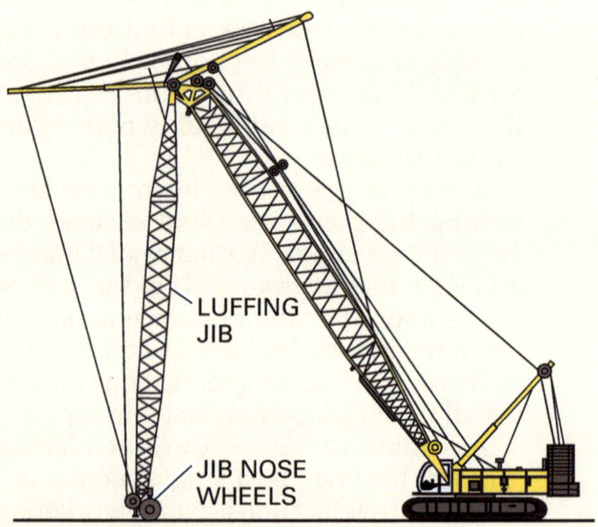

Figure Credit: Liebherr USA, Co.

Disconnect and remove the anti-two-block weights and switches. Disconnect the headache ball and unreeve the hook block according to the documentation. Pay in the load ropes, making sure they don't snag on the boom's roller sheaves. Disconnect all electrical cables along the length of the boom. If applicable, coil up the slack in electrical cables and plug their ends into the dummy socket or connectors provided for this purpose. Remove and stow any removable boom mounted equipment, such as high-voltage proximity sensors, cameras, anemometer, or FAA warning light.

2.4.2 Boom Disassembly

Disassembly of a lattice boom is simply the reverse of the assembly process described earlier. As always, refer to the specific crane manufacturer's documentation for the disassembly procedure. With the head of the boom resting on blocking and the boom pendants slack, shorten the pendants by removing a length and re-anchoring them at a lower boom section. Properly stow and secure the disconnected pendants. Observe the maximum cantilever length limitations during the disassembly process, if applicable. Elevate the boom slightly to minimize stress on the outer section's connector points, but do not elevate it so far that the boom head lifts off its blocking.

CAUTION

Never remove boom connector pins for sections between the pendant support points and the boom base. This will cause the boom to buckle. Always work on sections beyond the point where the pendants support the boom.

WARNING!

Workers must not perform boom disassembly while standing directly under or inside of a boom. Boom collapse can occur during disassembly, resulting in serious injury or death.

After positioning the boom, workers can remove the lower pins on the sections ahead of the pendants. This may require some experimenting with the boom angle to produce the slack in the connectors needed to permit removal of the pins. After properly stowing the removed pins, lower the boom back down onto the blocking. When the blocking for the outer sections bears their weight, workers can remove the upper pins on the sections ahead of the pendants. Store the removed pins in the brackets provided for this purpose (*Figure 29*) if available.

The operator may have to back the crane away to provide some working space between boom sections as the disassembly progresses. Repeat the following process until all the inserts are separated and resting on blocking:

- Disconnect and move the pendants inward to the next insert.
- Raise the boom so that workers can remove the lower connector pins.
- Lower the boom end onto blocking to remove the upper pins.

If required, workers must set up the gantry and/or mast rigging to remove the base section. Many cranes have the capability of self-disassembly. They can stage the boom sections for transport or load them onto trailers using the base section for lifting. At the appropriate point in disassembly, workers will disassemble the crane base itself, which is essentially the reverse of the process described previously.

Figure 29 Boom pin properly stowed.

Additional Resources

ASME P30.1, Planning for Load Handling Activities. Current edition. The American Society of Mechanical Engineers. Current Edition. New York, NY: American Society of Mechanical Engineers.

Mobile Crane Safety Manual. 2014. Milwaukee, WI: Association of Equipment Manufacturers.

29 *CFR* 1926, Subpart CC,*Cranes and Derricks in Construction.* **www.ecfr.gov**

2.0.0 Section Review

1. All the following actions are appropriate when preparing to assemble a lattice boom *except* _____.
   a. extending and locking outriggers to their maximum extent
   b. removing the crawler assemblies
   c. erecting the gantry and boom mast rigging
   d. extending a crane's extendable counterweight

2. The maximum cantilever length of a lattice boom is the maximum _____.
   a. length of a luffing jib
   b. load chart radius of the boom
   c. length of a lowered boom beyond the pendants
   d. boom length permitted by the manufacturer

3. When installing a jib on a lattice boom, attaching the jib may require a _____.
   a. belly line
   b. jib adapter
   c. diagonal strut
   d. back hitch assembly

4. Which boom connector pins are removed first during disassembly?
   a. Lower pins
   b. Upper pins
   c. Upper, then lower pins on the boom head section
   d. Upper, then lower pins only on the intermediate sections

Summary

Lifting operations depend on the proper setup of the crane at the work site. This involves assembly of the crane base, mounting the proper counterweights, and assembly of the lattice boom. Preparation for boom assembly begins with identifying a suitable area for crane assembly.

Every crane has specific manufacturer instructions for boom assembly and disassembly. It is essential that operators and A/D personnel follow these procedures. Like all crane operations, successful crane assembly and disassembly involves planning, coordination among different groups of workers, and supervision by a qualified A/D director, as required by safety and industry standards.

Review Questions

1. The function of a gantry on a crane is to _____.
   a. support a jib extension at various offsets
   b. support the boom base hinge pins
   c. position the crawler assemblies during assembly
   d. provide a mechanical advantage for boom hoisting

2. When lifting lattice boom sections, it is permissible to _____.
   a. attach the hook directly to a boom main chord
   b. use bridles at boom lifting lugs
   c. use bridles attached to boom section connector pins
   d. lift by the boom lattice lacing

3. Which assembly area preparation is intended to minimize hazards to people not involved in the assembly of a crane?
   a. Checking the ground bearing pressure
   b. Posting trip hazard signs near stacks of blocking
   c. Erecting a barricade around the assembly area
   d. Providing a traffic director for trucks transporting boom components

4. Five important steps for assembling a lattice boom are listed below. Which sequence shows the typical order for these procedures?

   (A) install counterweight sections
   (B) install crawler assemblies
   (C) erect gantry and boom mast
   (D) deploy outriggers
   (E) set down on crawlers

   a. A, B, C, D, E
   b. E, D, C, B, A
   c. D, C, B, E, A
   d. A, B, E, C, D

5. Which sequence of steps describes the proper method for continuing the attachment of a boom insert to a previously assembled boom section?
   a. Move pendants out onto previous section, align connectors with next section, insert upper connector pins, boom up, and insert lower connector pins.
   b. Boom up, move pendants out onto next section, boom down, insert lower connector pins to previous section, boom up, and insert upper connector pins.
   c. Insert all lower connector pins for remaining sections, move pendants to boom head, boom up, insert all upper connector pins, boom down to set boom head on blocking.
   d. Insert all upper connector pins, move pendants to boom head, boom up, insert all lower connector pins.

6. Which of the following items refers to the LMI state when checking the crane's readiness to raise the boom following its assembly?
   a. Crawler-crane carbody jacks retracted and pinned in stowed position
   b. Spreader installed (if required) at proper location
   c. Setup mode is turned off
   d. Automatic boom stop installed

7. A long lattice boom is generally considered _____.
   a. one that is longer than 75 feet
   b. one that is longer than 150 feet
   c. any that tends to sag as it comes closer to a horizontal position
   d. one that requires FAA collision avoidance lights to be turned on at night

8. Which of the listed tasks must assembly workers complete first when attaching a jib extension?
   a. Rig the jib mast
   b. Secure the backstay pendant
   c. Reeve the fly sheave
   d. Pin the jib to the boom head

9. When installing a jib extension, one of the final steps is to _____.
   a. install the jib base pins
   b. erect the jib mast
   c. reeve the hoisting line and attach the block or ball
   d. install the forestay and backstay pendants

10. When disassembling a lattice boom, begin the process by _____.
    a. removing the lower connector pins between the boom pendant and base hinge point
    b. disconnecting the boom head and working toward the boom base
    c. disconnecting the boom base and working toward the boom head
    d. removing all the upper boom connector pins first

Trade Terms Introduced in This Module

Back hitch: The mechanism that permits adjusting the position of the boom gantry to maximize its leverage on the boom hoist cables.

Belly lines: Additional pendants attached to the boom hoisting bail and secured at a location near the midpoint of a long boom to minimize sagging.

Boom stops: Devices that mechanically prevent the boom from exceeding a maximum boom angle, reducing the potential for backward tipping or structural damage.

Cantilever length: The length of a horizontal boom that projects beyond the support of the boom pendants, leaving that portion of the boom supported at only one end, like a cantilever.

Diagonal struts: In a lattice boom with a rectangular cross section, pieces of structural pipe or angle iron welded to diagonally opposite chords of the boom to stiffen the assembly.

Gantry: A heavy frame with roller sheaves mounted on the upperworks of a crane, used for providing a mechanical advantage to the boom hoist ropes; also refers to a jib strut or mast.

Hoist kick-out: A term for any safety device that automatically stops a hoist drum when a limit is reached.

Lattice lacing: In a lattice boom, the zig-zag pattern of structural metal supporting the main chords along the sides, top, and bottom of the boom sections.

Live boom mast: A heavy framework or lattice boom with sheaves, hinged at the lower end of its base on the crane's upperworks, whose purpose is to control the angle of the main boom through attached boom pendants. Boom hoist cables reeved through the boom gantry sheaves or hydraulic actuators control its position.

Main chords: The heavy structural pipe or angle iron members running the length of a boom section.

Pendant: A length of heavy wire rope or flexibly connected sections of flat metal bars, equipped with connectors at both ends, used to hoist and/or support sections of a crane's boom.

Suspension rope: The rope reeved through a crane's gantry and mast sheaves to the sheaves suspending the boom pendants.

Manitowoc Model 18000 Product Guide – Boom Combinations

boom combinations

23

No. 55 or No. 55A Boom Combinations

Boom Length m (ft)	Boom Inserts 3,0 m (10 ft)	6,1 m (20 ft)	12,2 m (40 ft)
36,6 (120)	1	1	1
42,7 (140)	1	-	2
48,8 (160)	1	1	2
54,9 (180)	1	2	2
61,0 (200)	1	1	3
67,1 (220)	1	2	3
73,2 (240)	1	1	4
79,2 (260)	1	2	4
85,3 (280)	1	1	5
91,4 (300)	1	2	5
97,5 (320)	1	1	6

NOTE: 36,6 m (120') basic boom consists of 6,1 m (20') butt, 3,0 m (10') insert with drum, 6,1 m (20') insert, 12,2 m (40'), and 9,1 m (30') top.

May use 12,2 m (40') No. 55 insert with or without sheaves.

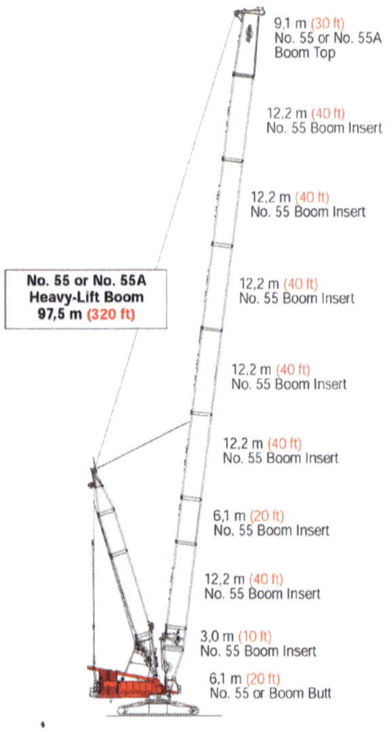

Model 18000
No. 55 or No. 55A
Heavy-Lift Main Boom
97,5 m (320 ft)

No. 55-79A Boom Combinations

Boom Length m (ft)	Boom Inserts 55* 3,0 m (10 ft)	55* 6,1 m (20 ft)	55* 12,2 m (40 ft)	79 6,1 m (20 ft)	79 12,2 m (40 ft)	No. 55-79A Transition 12,2 m (40 ft)
36,6 (120)	1	1	-	-	-	1
42,7 (140)	1	-	1	-	-	1
48,8 (160)	1	1	1	-	-	1
54,9 (180)	1	1	1	1	-	1
61,0 (200)	1	1	1	-	1	1
67,1 (220)	1	1	1	1	1	1
73,2 (240)	1	1	1	-	2	1
79,2 (260)	1	1	1	1	2	1
85,3 (280)	1	1	1	-	3	1
91,4 (300)	1	1	1	1	3	1
97,5 (320)	1	1	1	-	4	1
103,6 (340)	1	1	1	1	4	1
109,7 (360)	1	1	1	-	5	1
115,8 (380)	1	1	1	1	5	1
121,9 (400)	1	1	1	-	6	1

NOTE: 36,6 m (120') basic boom consists of 6,1 m (20') butt, 3,0 m (10') insert with drum, 6,1 m (20') insert, 12,2 m (40'), and 9,1 m (30') top.

May use 12,2 m (40') No. 55 insert with or without sheaves.

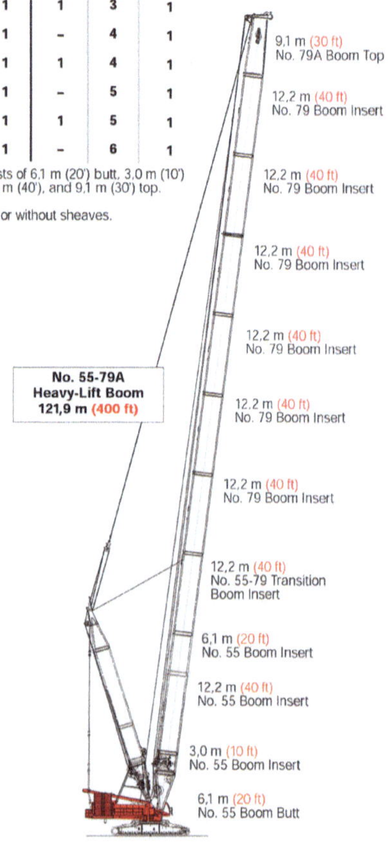

Model 18000
No. 55-79A Heavy-Lift
Main Boom
121,9 m (400 ft)

model 18000

boom combinations

24

No. 79A Fixed Jib Combinations		
	Boom Inserts	
Jib Length m (ft)	6,1 m (20 ft)	12,2 m (40 ft)
27,4 (50)	-	-
33,5 (70)	1	-
39,6 (90)	-	1
45,7 (110)	1	1

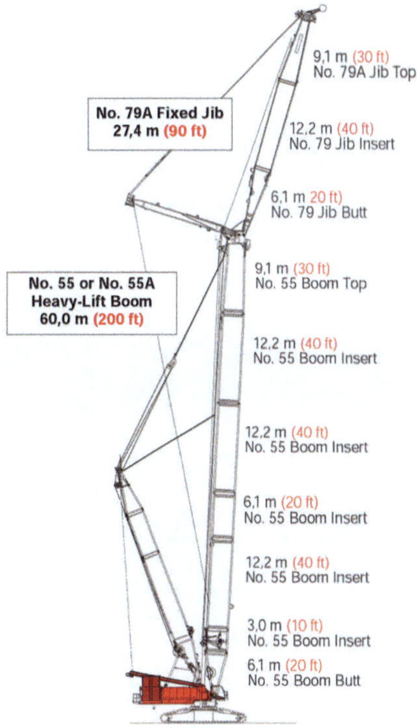

Model 18000
No. 79A Fixed Jib on
No. 55 or No. 55A Heavy-Lift Main Boom
88,0 m (290 ft)

model 18000

Manitowoc

boom combinations

25

No. 44 Luffing Jib Combinations

Luffing Jib Length m (ft)	Boom Inserts 3,0 m (10 ft)	6,1 m (20 ft)	12,2 m (40 ft)
21,3 (70)	-	-	-
24,4 (80)	1	-	-
27,4 (90)	-	1	-
30,5 (100)	1	1	-
33,5 (110)	-	-	1
36,6 (120)	1	-	1
39,6 (130)	-	1	1
42,7 (140)	1	1	1
45,7 (150)	-	-	2
48,8 (160)	1	-	2
51,8 (170)	-	1	2
54,9 (180)	1	1	2
57,9 (190)	-	-	3
61,0 (200)	1	-	3
64,0 (210)	-	1	3
67,1 (220)	1	1	3
70,1 (230)	-	-	4
73,2 (240)	1	-	4

No. 79A Luffing Jib Combinations

Luffing Jib Length m (ft)	Boom Inserts 6,1 m (20 ft)	12,2 m (40 ft)
27,4 (90)	2	-
33,5 (110)	1	1
39,6 (130)	2	1
45,7 (150)	1	2
51,8 (170)	2	2
57,9 (190)	1	3
64,0 (210)	2	3
70,1 (230)	1	4
76,2 (250)	2	4
82,3 (270)	1	5
88,3 (290)	2	5
94,5 (310)	1	6

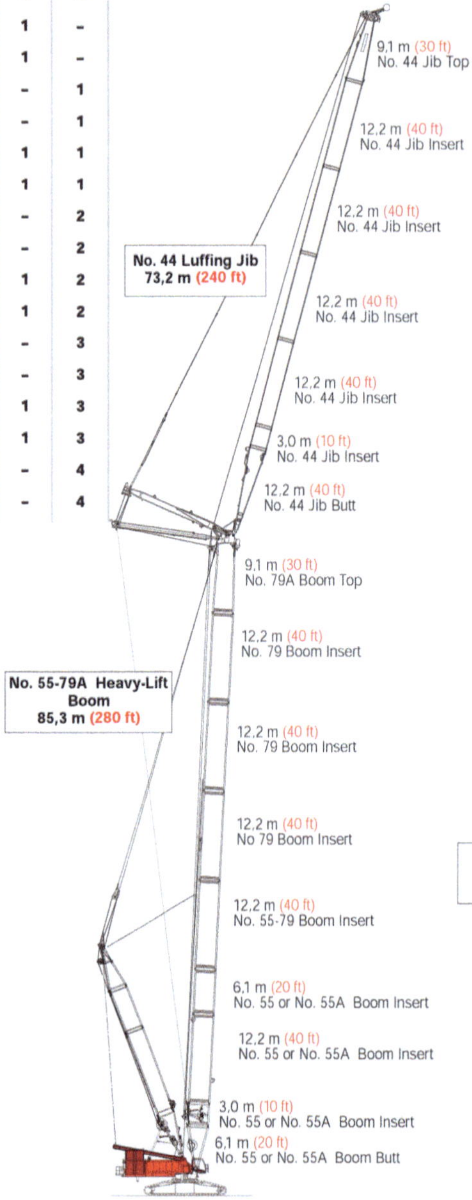

Model 18000
No. 44 Luffing Jib on
No. 55-79A Heavy-Lift Main Boom
158,5 m (520 ft)

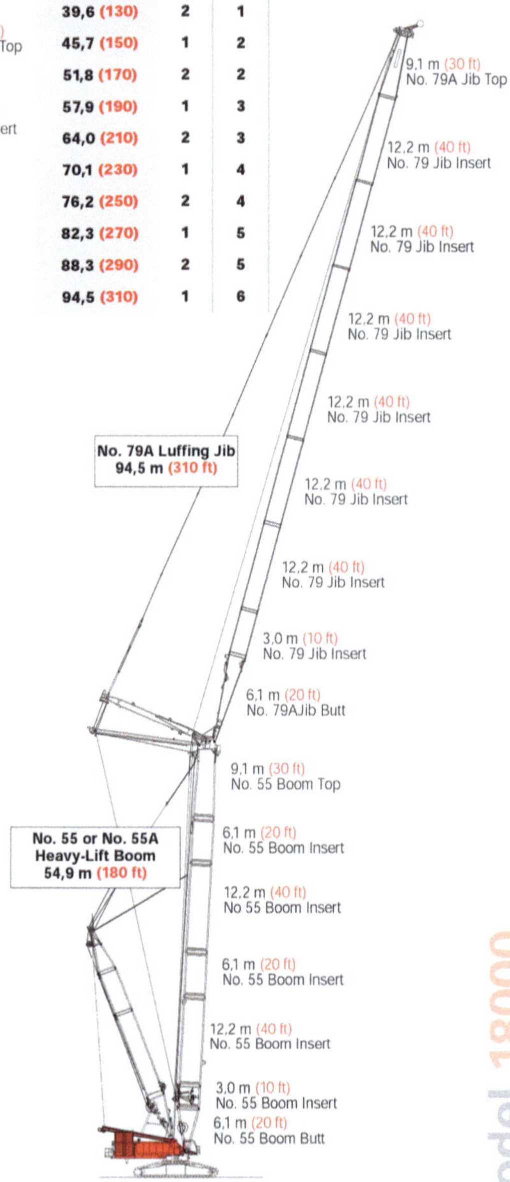

Model 18000
No. 79A Luffing Jib on No. 55
or No. 55A Heavy-Lift Main Boom
149,3 m (490 ft)

model 18000

Manitowoc

Additional Resources

This module is intended as a thorough resource for task training. The following reference materials are recommended for further study.

ASME P30.1, Planning for Load Handling Activities. Current edition. The American Society of Mechanical Engineers. Current Edition. New York, NY: American Society of Mechanical Engineers.

Mobile Crane Safety Manual. 2014. Milwaukee, WI: Association of Equipment Manufacturers.

29 *CFR* 1926, Subpart CC,*Cranes and Derricks in Construction*. **www.ecfr.gov**

Figure Credits

Link-Belt Construction Equipment Company, Module opener, Figures 10, 27

The Manitowoc Company, Inc., Figures 1, 5, 11–14 (line art), 18, Appendix

Mammoet USA South Inc., Figure 9, 26

Liebherr USA, Co., SA01

Section Review Answer Key

Answer	Section Reference	Objective
Section One		
1. d	1.0.0	1a
2. b	1.2.0	1b
Section Two		
1. b	2.1.1	2a
2. c	2.2.2	2b
3. b	2.3.0	2c
4. a	2.4.2	2d

NCCER CURRICULA — USER UPDATE

NCCER makes every effort to keep its textbooks up-to-date and free of technical errors. We appreciate your help in this process. If you find an error, a typographical mistake, or an inaccuracy in NCCER's curricula, please fill out this form (or a photocopy), or complete the online form at **www.nccer.org/olf**. Be sure to include the exact module ID number, page number, a detailed description, and your recommended correction. Your input will be brought to the attention of the Authoring Team. Thank you for your assistance.

Instructors – If you have an idea for improving this textbook, or have found that additional materials were necessary to teach this module effectively, please let us know so that we may present your suggestions to the Authoring Team.

NCCER Product Development and Revision

13614 Progress Blvd., Alachua, FL 32615

Email: curriculum@nccer.org

Online: www.nccer.org/olf

❑ Trainee Guide ❑ Lesson Plans ❑ Exam ❑ PowerPoints Other ______________________

Craft / Level: Copyright Date:

Module ID Number / Title:

Section Number(s):

Description:

Recommended Correction:

Your Name:

Address:

Email: Phone:

Hoisting Personnel

Overview

Using a crane to hoist personnel is potentially hazardous. Loss of control of the crane or the failure of a hoisting component will likely result in the serious injury or death of the workers being hoisted. It was once common practice for workers to ride a ball as a convenient means of moving between the ground and a high point. However, the potential for catastrophic injuries was extreme, and falls and injuries were common. For this reason, the government enacted strict safety regulations pertaining to lifting personnel by crane. This module covers the details of these regulations.

Module 21305

Trainees with successful module completions may be eligible for credentialing through the NCCER Registry. To learn more, go to **www.nccer.org** or contact us at 1.888.622.3720. Our website has information on the latest product releases and training, as well as online versions of our *Cornerstone* magazine and Pearson's product catalog.

Your feedback is welcome. You may email your comments to **curriculum@nccer.org**, send general comments and inquiries to **info@nccer.org**, or fill in the User Update form at the back of this module.

This information is general in nature and intended for training purposes only. Actual performance of activities described in this manual requires compliance with all applicable operating, service, maintenance, and safety procedures under the direction of qualified personnel. References in this manual to patented or proprietary devices do not constitute a recommendation of their use.

21305 V3

21305

Hoisting Personnel

Objective

When you have completed this module, you will be able to do the following:

1. Identify and describe the requirements and operational techniques related to personnel hoisting.
   a. Describe the required characteristics of personnel platforms.
   b. Identify the testing and inspection requirements for personnel platforms.
   c. Identify and discuss the operational requirements involved in personnel lifts.

Performance Task

Under the supervision of your instructor, you should be able to do the following:

1. Complete a personnel lift plan using the form in *Appendix C* and a provided scenario.

Trade Terms

Bridle slings
Flemish eyes
Nondestructive testing (NDT)
Poured socket
Proof test
Trial lift

Industry Recognized Credentials

If you are training through an NCCER-accredited sponsor, you may be eligible for credentials from NCCER's Registry. The ID number for this module is 21305. Note that this module may have been used in other NCCER curricula and may apply to other level completions. Contact NCCER's Registry at 888.622.3720 or go to **www.nccer.org** for more information.

Contents

1.0.0 Hoisting Personnel ... 1
1.1.0 Platform Characteristics ... 1
1.1.1 Design ... 1
1.1.2 Platform Construction ... 3
1.2.0 Testing and Inspections ... 4
1.2.1 Initial Manufacturing Tests ... 4
1.2.2 Proof Testing ... 4
1.2.3 Trial Lifts ... 5
1.2.4 Inspections and Maintenance ... 5
1.3.0 Personnel Hoisting Operations ... 6
1.3.1 Key Individuals ... 6
1.3.2 Fall Protection ... 7
1.3.3 Crane Instrumentation and Configuration ... 8
1.3.4 Personnel Hoisting Near Powerlines ... 9
1.3.5 Planning and Preparations ... 10
1.3.6 Work Practices and Crane Operations ... 10
Appendix A Personnel Lift Platform Pre-Lift Inspection Form ... 17
Appendix B Power Line Planning Form ... 18
Appendix C Personnel Platform Lift Planning and Authorization Form .. 20

Figures and Tables

Figure 1 A personnel lift at a marine work site ... 2
Figure 2 Example of a suspended personnel hoisting platform ... 3
Figure 3 Hoisting rope slings fitted with spliced Flemish eyes and permanent thimbles ... 4
Figure 4 Manbasket with manufacturer supplied test weight ... 5
Figure 5 The lift director conducts the pre-lift meeting ... 7
Figure 6 Fall arrest equipment ... 7
Figure 7 Crane hook with a positive-locking throat latch ... 8
Figure 8 Personnel hoisted near de-energized and grounded powerlines ... 9
Figure 9 Test lift at beginning of a personnel hoisting operation ... 11
Figure 10 Occupants lifted over water must wear approved flotation devices ... 12

Table 1 Energized Powerline Clearances for Mobile Cranes ... 9

SECTION ONE

1.0.0 HOISTING PERSONNEL

Objective

Identify and describe the requirements and operational techniques related to personnel hoisting.

a. Describe the required characteristics of personnel platforms.
b. Identify the testing and inspection requirements for personnel platforms.
c. Identify and discuss the operational requirements involved in personnel lifts.

Performance Task

Complete a personnel lift plan using the form in *Appendix C* and a provided scenario.

Trade Terms

Bridle slings: Lift rigging consisting of multi-leg, wire rope, or chain slings.

Flemish eyes: Spliced loops formed in the end of wire rope rigging slings. For personnel hoist rigging, the eyes must also contain a thimble and be mechanically crimped with a ferrule.

Nondestructive testing (NDT): Testing the integrity of fabricated metal structures, especially welds, by various means to detect hidden flaws without destroying the object or rendering it useless. Common NDT methods include dye penetrant (PT), magnetic particle (MT), X-ray and radiography (RT), and ultrasound (UT).

Poured socket: Referring to wire rope terminations, a method of attaching a fitting by placing the end of the rope inside the fitting socket, then pouring molten zinc, another metal alloy, or a special resin into the socket. After hardening, these terminations can provide 100 percent of the wire rope's minimum breaking strength. This process is also referred to as speltering.

Proof test: A strength test of an object, usually under a load exceeding that expected under normal use, to ensure it is suitable for safe use. In personnel hoisting, a proof test requires loading the platform with a weight equal to 125 percent of its rated load and suspending it for at least 5 minutes.

Trial lift: For a personnel hoisting platform, a lift of the platform and rigging without occupants, but with an equivalent load, under the same conditions and over the same route as the planned personnel lift.

Two key standards address personnel hoisting platforms, often called *manbaskets*. For the design, manufacturing, and operation of such equipment, *ASME B30.23, Personnel Lifting Systems*, provides technical details. 29 *CFR* 1926.1431, *Hoisting Personnel*, provides the safety regulations for hoisting personnel with cranes. The OSHA standard incorporates the relevant parts of the ASME consensus standard and makes them enforceable laws.

29 *CFR* 1926.1431 prohibits the hoisting of personnel except when the employer demonstrates that workers can't complete their tasks by any other means. The OSHA standard notes that Subpart R (rather than 1926.1431) covers hoisting requirements specific to steel erection.

Some form of hoisting platform meeting the requirements of the standard is necessary for all applicable lifts *except*:

- Into or out of drill shafts less than 8 feet (2.44 m) in diameter
- Pile-driving operations
- Personnel transfers at marine worksites (*Figure 1*)
- In storage tanks, shafts, and chimneys

The OSHA standard provides specific requirements applicable to these situations.

1.1.0 Platform Characteristics

ASME B30.23 provides detailed requirements for the design, construction, testing, and inspection of personnel lifting platforms. It also includes operational guidelines to ensure the safe operation of such platforms. 29 *CFR* 1926.1431 incorporates the essential ASME standard's platform characteristics relating to occupant safety.

1.1.1 Design

The first design concern addressed by the ASME standard is the design load factor. For a platform equipped for a single-point lift, the attachment point must have a design factor of 7. This means that it must be able to support seven times the

Figure 1 A personnel lift at a marine work site.

design load of the platform's occupants and materials, including the platform's weight. For platforms with two or more hoisting attachments, any loaded pair of attachment points must have a design factor of 5.

> **NOTE**
> 29 *CFR* 1926.1431 states that if wire rope slings are made of rotation-resistant rope, then the slings must be able to support a minimum of 10 times the designed load.

Besides the requirement for a qualified person familiar with structural design to design platforms, design features for a personnel lift platform must include the following:

> **NOTE**
> For the purposes of establishing design criteria, *ASME B30.23* assumes an occupant weighs 300 pounds.

- Minimum capacity of 300 pounds (136 kg)
- A guardrail consisting of a top rail, intermediate rail, lower barrier, and toe board of specified dimensions. The barrier may be solid, perforated, or of expanded metal mesh with openings no larger than $\frac{1}{2}$ inch (13 mm).
- All structural components of the platform enclosure capable of resisting specified minimum downward and outward forces to ensure security of the platform occupants.
- Each anchor point for attaching fall protection devices capable of stopping the combined weight of all allowed occupants falling through 6 feet (1.8 m).
- Hand railing with a specified standoff along entire perimeter of enclosure except at access doors/gates.
- Slip-resistant decking with drain holes.
- Means to secure loose tools, materials, and other items within the enclosure.
- A permanent identification plate or markings containing manufacture date, testing, standards, maximum occupancy and weight, and equipment ID.
- Access systems to prevent accidental opening. Access doors/gates must not swing outward during the lift operation. OSHA permits outward-swinging gates for small, one-person lift platforms.
- Sufficient headroom for occupants to stand without crouching.
- In addition to hard hats, overhead protection (roof) if occupants are exposed to falling objects. Protection can't obscure visual path between the operator and the occupants.

- A suspension system designed to minimize platform movement due to movement of occupants.
- For platforms that are attachable to crane booms, the attachment mechanism must keep the platform floor within 10 degrees of level at all boom angles.

Figure 2 shows many of the design features on an actual personnel hoisting platform.

1.1.2 Platform Construction

ASME B30.23 and 29 *CFR* 1926.1431 specify minimum construction requirements for personnel lifting platforms. All welding must be in accordance with American Welding Society or ASME procedures.

Welders shall smooth all surfaces to avoid punctures or lacerations to occupants. After completion of fabrication, the manufacturer shall permanently attach an identification plate to the platform documenting the required certification testing, discussed later in this module. The platform must include a weatherproof box or placard for storing or displaying the operating instructions. Placarding shall include a full set of hand signals as described in 29 *CFR* 1926.1422, *Signals—Hand Signal Chart*.

Manufacturers shall design and assemble **bridle slings** or other rigging specifically for suspension of the platform equipment (*Figure 3*). The manufacturer should make these slings and attachments in such a way that it would be impractical to remove them and use the suspension system for any other purpose. It is widely accepted that the intent is to dedicate the necessary suspension system/rigging to hoisting personnel only, and not use the equipment for other tasks. Per*ASME B30.23*, slings shall have the following characteristics whenever each type is used (note that synthetic slings and ropes made from synthetic or natural fibers must not be used):

- Wire rope slings with mechanically spliced **Flemish eyes** must have thimbles in all eyes. Wire rope clips, wedge sockets, or knots shall not be used.
- Wire rope slings with **poured socket** end connections must conform to the manufacturer's or a qualified person's application instructions.
- Chain-sling suspension systems shall be made from a minimum of grade 80 chain.

All slings must use a master link or a bolt-type shackle with cotter pin for attachment to the crane hoist. The master link shall include a tag showing the rated load for the equipment and identification as a part of personnel lifting equipment so

Figure 2 Example of a suspended personnel hoisting platform.

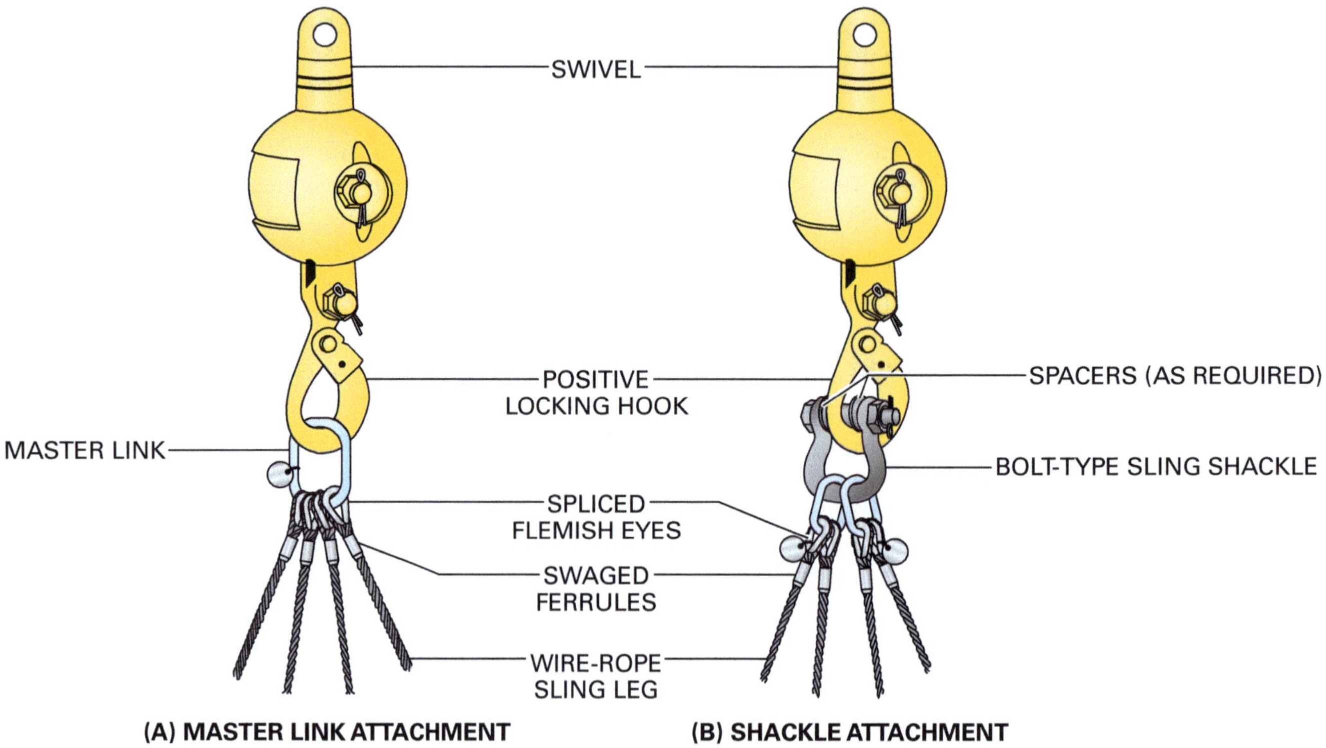

Figure 3 Hoisting rope slings fitted with spliced Flemish eyes and permanent thimbles.

it isn't used for any other purpose. *ASME B30.23* also states that rope and chain slings shall have a permanent tag or label attached to each leg that displays the rated load of the leg.

1.2.0 Testing and Inspections

The safety of lifting platform occupants is the primary purpose of the ASME and OSHA standards. However, the design and construction standards simply establish a set of expectations that manufacturers and users must meet. They can't account for construction flaws or normal wear and tear due to use. To ensure that the equipment meets and maintains the standards, builders, owners, and users must also complete certain tests and inspections.

1.2.1 Initial Manufacturing Tests

ASME B30.23 requires that after building a personnel lift platform, the manufacturer must conduct load tests of the device to ensure it meets the applicable design standards. The tests must include the load suspension attachment points, the occupant safety features, and the platform rating. Quality assurance inspectors must evaluate the attachment point welds with approved **nondestructive testing (NDT)**.

The builder must test the platform suspension system components separately to twice (200 percent) the design load for each leg. For master-link suspension systems, the master link must support twice the sum of the platform's weight and its design load. The manufacturer must document the satisfactory completion of all required certification testing on the platform's identification plate.

1.2.2 Proof Testing

Both *ASME B30.23* and 29 *CFR* 1926.1431 require that users conduct a special test of a lift platform at every new work site before permitting the hoisting of personnel. This **proof test** of the platform and its rigging entails loading the platform evenly with a weight equal to 125 percent of its rated load and suspending it for at least 5 minutes. Most platform manufacturers provide a customized test weight for each model of platform that users can easily attach and detach for this purpose (*Figure 4*).

Before and after completing the test, a competent person as defined in the OSHA standards shall inspect the platform and suspension rigging. Users must correct any defects or problems noted before using the platform for hoisting personnel. If the inspection reveals defects, users must conduct another proof test after correcting the defects.

ASME B30.23 requires that after any structural repair or modification to the platform, users must perform a proof test of 150 percent of the platform's rated load. Operators must raise a

Figure 4 Manbasket with manufacturer supplied test weight.

suspended platform to a height, then lower it at a rate exceeding 100 feet per minute (30 m/min). The operator then applies the hoist drum brake to stop the platform, and keeps it suspended for at least 5 minutes at that point. A competent person must inspect the platform. If there are any defects, they must be corrected and the platform retested before use.

1.2.3 Trial Lifts

In addition to the proof test, which tests the personnel lift hardware, 29 *CFR* 1926.1431 requires a **trial lift** of the platform to evaluate the lift operation itself. Users may combine a trial lift with the initial site proof test, if feasible. The OSHA standard requires that users conduct the trial lift from wherever the hoisting operation will occur, whether from ground level or from other points of origin. If the scheduled tasks require lifting occupants to different locations using a single crane setup, then a trial lift must be over the path to each location. In addition, users must complete trial lifts as follows:

- Prior to each shift during which a personnel hoist will occur
- Any time the crane moves to a new work site or returns to a previous location
- Whenever the lift route changes, unless a competent person determines the new route introduces no new factors affecting safety

> **NOTE**
> Interpretive discussion appended to *ASME B30.23* indicates that a *new work site* is a geographic location to which users must transport the lift platform for use via road transport. The proof test verifies that the platform did not receive any damage during transit. Moving a crane to a location at the same work site does not require another proof test.

Trial lifts for destinations below grade level are particularly important. Crane operators may describe this as *going negative*. There is a real possibility that a crane could run out of hoist rope if the work destination is too far below grade. Planning needs to take this possibility into account and operators must conduct such tests carefully while monitoring the remaining cable on the hoist drum.

29 *CFR* 1926.1431 also specifies other actions during the trial lift. Operators must accurately measure the load radius used for the trial lift and verify that the load will not exceed 50 percent of the crane's capacity at that radius during the lift. After the trial lift, a competent person performs the inspections discussed earlier. These must include examining the ground support, the crane, and the platform for any deficiencies. The individual must also verify that workers remove the platform's test weight.

1.2.4 Inspections and Maintenance

The ASME and OSHA standards provide only the minimum guidance regarding inspections and maintenance. Competent persons must perform inspections prior to and after every lift involving personnel, using a form similar to the *Personnel Lift Platform Pre-Lift Inspection* checklist, found in *Appendix A*. Using the checklist ensures inspection items are not skipped.

ASME B30.23 requires that both frequent and periodic platform inspections be accomplished. Frequent inspections are made at least once each day, prior to use. Periodic inspections must be conducted at least once every 12 months. Since frequent inspections are conducted often, documentation is completed at the point of use and the active checklist remains with the platform. The documents for the most recent periodic inspection must also be kept with the platform.

The platform manufacturer provides a manual for maintaining the equipment. The manual normally includes a parts list as well as guidelines for setting up a maintenance program. *ASME B30.23* requires that owners and users establish and carry out a maintenance program according to the manufacturer's instructions. Such programs will include specific inspections. Instructions also include guidelines for repairs and adjustments. Any repair or replacement of parts must meet or exceed the manufacturer's original specifications. The platform owner is responsible for maintaining records for all repairs to, or replacement of, platform structural components.

For personnel hoisting equipment adjustments and repairs, the following points apply:

- Owners and users must correct hazardous conditions identified during any inspection.
- Only certified welders may make welding repairs.
- Adjustments must only restore the correct functioning of equipment.
- Only qualified persons may repair platform rigging or suspension components.
- If any rope or chain slings show signs of wear, users must take them out of service.
- Owners and users may only make modifications approved in writing by the manufacturer.

1.3.0 Personnel Hoisting Operations

This section addresses the operational requirements and considerations that apply when hoisting personnel using mobile cranes. Due to the hazardous nature of personnel hoisting operations, 29 *CFR* 1926.1431 states that all such operations are categorized as critical lifts as defined in *ASME P30.1, Planning for Load Handling Activities.* The planning and execution of a personnel hoisting operation must be detailed, complete, and safe.

1.3.1 Key Individuals

ASME B30.23 requires that a qualified individual be appointed as personnel lift authorizer and lift director. Among the main duties of a personnel lift authorizer are the following:

- Verify the need for a personnel lift
- Ensure the equipment used meets the technical and legal requirements
- Provide final authorization for the operation to proceed
- Require the lift to follow all applicable requirements

The lift director's responsibilities include the following:

- Review and approve the lift plan
- Hold the pre-lift meeting (*Figure 5*)
- Ensure qualified persons are assigned to each position
- Ensure all administrative and legal requirements are met
- Verify that the crane operator is fit to conduct the lift and meets the rest requirements
- Designate one platform occupant as the signal person
- Supervise the lift

The crane operator is responsible for complying with the technical and operational requirements established in *ASME B30.23* pertaining to the equipment, load, and operations under the individual's direct control. The standard also places the responsibility on the operator to evaluate his or her own physical and mental competency to conduct a personnel lift. One aspect of competency involves working a limited number of hours prior to the lift operation. The ASME standard prohibits an operator from conducting the lift if the following conditions apply:

- Had less than 8 hours off prior to the current shift
- Has worked for more than 10 hours prior to beginning the lift
- Will exceed 12 hours on shift before the personnel lifting has been completed

The crane operator may not engage in any activity that will divert attention from the operation (*e.g.*, taking a call on a cell phone). The operator will respond only to the designated signal person, but must obey a Stop signal issued by anyone participating in the lift. An operator may stop a lift at any time in the interest of safety. The operator must remain at the crane's controls at all times during the lift.

The ground crew must be qualified and trained for their responsibilities, which include the following:

- Attend the pre-lift meeting
- Visually inspect the platform and its rigging for hazards prior to and during the operation
- Assist platform occupants
- Ensure the hoist line does not foul the platform
- Keep persons clear from passing underneath the platform
- Maintain continuous communications with the platform occupants and crane operator
- Must not perform any duties other than those associated with the personnel lifting operation

Figure 5 The lift director conducts the pre-lift meeting.

The platform occupants are also required to have specific training and qualifications. Their responsibilities include the following:

- Recognize hazards associated with personnel lifts
- Understand lift procedures and safety precautions
- Attend the pre-lift meeting
- Maintain platform stability
- Keep all parts of their bodies inside the platform during platform movement (occupants are permitted to extend their arms outside the platform for the purposes of positioning the platform and when serving as a signal person)
- Keep their fall protection device secured at all times unless work requires otherwise
- Know the emergency Stop hand signal
- If the platform has controls, they must be qualified to operate them

1.3.2 Fall Protection

Occupants of personnel hoist platforms shall use personal fall arrest systems (*Figure 6*), except for operations over water. For operations over water, see 29 *CFR* 1926.106, *Working Over or Near Water*. 29 *CFR* 1926.502, *Fall Protection Systems Criteria and Practices*, found in Subpart M, contains fall protection requirements applicable to crane operations. The OSHA standard outlines performance criteria as well as the design and usage of the harnesses, lanyards, and attachment hardware.

Figure 6 Fall arrest equipment.

Some of the more critical fall protection requirements include the following:

- Body belts alone are not acceptable as a personal fall arrest system.
- Harnesses must include D-rings and snap hooks that comply with OSHA and ANSI strength and fabrication specifications. Harnesses and fall arrest components should normally include labeling to this effect.
- Workers must inspect personal fall arrest systems before donning them. Any components showing wear or other defects must be removed from service.
- Unless snap hooks are of the locking type and designed for the following specific uses, platform occupants shall *not* allow snap hooks to be engaged:
  - To each other
  - Directly to webbing or wire rope
  - To D-rings retaining another snap hook
  - To any object that isn't compatible in shape to accept the snap hook
- Platform occupants shall not fasten fall arrest systems to guardrails or handrails or to any platform hoist or suspension point.
- Workers may not use fall protection equipment for any other purpose (*e.g.,* hoisting tools or materials).
- Workers shall immediately remove from service any fall arresting components subjected to impact loading while arresting a fall.

1.3.3 Crane Instrumentation and Configuration

Modern mobile cranes already include most of the hoist and instrumentation features that *ASME B30.23* and 29 *CFR* 1926.1431 require. The following list describes some of these important requirements:

- Operators must disable any feature that allows freewheeling of the hoist drum. The crane must provide the ability to positively control the lowering of the hoist line by a means other than just a friction brake.
- There must be an operable anti-two-block safety switch on the main boom or boom extension.
- A boom-angle indicator must be readily visible to the operator.
- Telescopic-boom cranes must have a boom length indicator or marks readily visible to the operator.
- Hoist hooks must have a positive-locking throat latch that prevents the lift bridle from disengaging (*Figure 7*).
- The crane's hoist controls must have a deadman feature that stops hoist rope movement when released.
- Hoist drums must have a positive method of holding their position upon the loss of hydraulics or other power source (*e.g.,* hydraulic check valves, or pinion pawls or dogs).
- There must be positive-locking devices for outriggers or stabilizers in the extended position to prevent inadvertent retraction.
- 29 *CFR* 1926.1431 permits the use of a power luffing jib for personnel hoisting, but requires that cranes have an operable jib-angle indicator readily visible to the operator, and a separate anti-two-block device for the jib hoist.

Lift planners must establish the crane configuration required for the operation. Crane assembly workers will set up the crane in the specified configuration. For any personnel hoisting operation, the operator must fully extend and lock the crane's outriggers. Critical lift plans also require an evaluation of the ground support (ground bearing pressure) at the lift site. Crane setup must include appropriate blocking, matting, or other supports as required. For land-based operations, the crane must be set up and maintained within 1 percent of level for the entire operation.

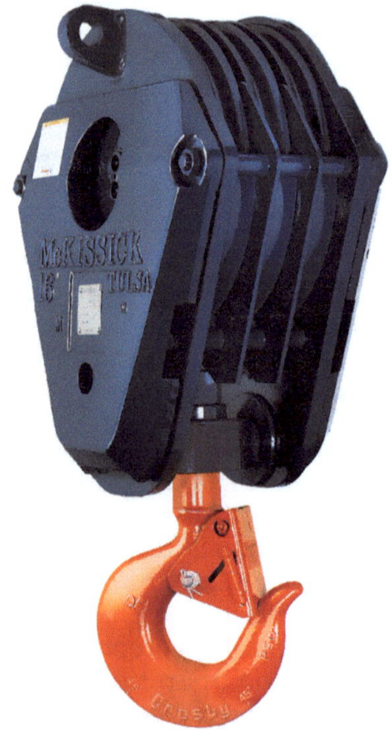

Figure 7 Crane hook with a positive-locking throat latch.

1.3.4 Personnel Hoisting Near Powerlines

Work near overhead powerlines (*Figure 8*) is especially hazardous to cranes and operators. Hoisting personnel near powerlines is an extremely dangerous operation.

The upper section of *Table 1* lists the minimum required standoff distances given in 29 *CFR* 1926.1408 and 1926.1409. OSHA allows three options for work near powerlines:

Option 1 – De-energize and ground all conductors at the work site.

Option 2 – Ensure no part of the crane, rigging, or platform approaches closer than 20 feet (for voltages under 350 kV).

Option 3 – Maintain the standoff distances listed in *Table 1* for voltages higher than 350 kV, regardless of the voltage.

In any case, OSHA prohibits any lift where the crane, rigging, or platform could encroach on the limits specified in the table. Preparations for such work should include coordination with the electrical utility to de-energize and ground the lines, if possible. Otherwise, planners must be able to accomplish the needed work from beyond the permitted standoff distances.

Figure 8 Personnel hoisted near de-energized and grounded powerlines.

Table 1 Energized Powerline Clearances for Mobile Cranes

CRANE IN OPERATION	
POWER LINE VOLTAGE IN	MINIMUM CLEARANCE IN FEET (METERS)
0 to 50	10 (3.05)
50 to 200	15 (4.60)
200 to 350	20 (6.10)
350 to 500	25 (7.62)
500 to 750	35 (10.67)
750 to 1,000	45 (13.72)
Over 1,000	Distance established by the utility owner / operator or registered professional engineer who is a qualified person in power transmission and distribution
CRANE IN TRANSIT (with no load and the boom or mast lowered) (1)	
POWER LINE (kV)	MINIMUM CLEARANCE IN FEET (METERS)
0 to 0.75	4 (1.22)
0.75 to 50	6 (1.83)
50 to 345	10 (3.05)
345 to 750	16 (4.87)
750 to 1,000	20 (6.10)
Over 1,000	Distance established by the utility owner / operator or registered professional engineer who is a qualified person in power transmission and distribution

Note 1: Environmental conditions such as fog, smoke, or precipitation may require increased clearances.

The lower section of *Table 1* lists the clearances permitted for a crane traveling with the boom and platform lowered to the travel position. OSHA allows these standoff distances only if the crane is not carrying a load. This precludes workers occupying a personnel lift platform during travel, even if otherwise permitted.

Large projects often have many contractors requiring crane services. *Appendix B* is an example of a *Power Line Awareness Permit* form. Project policies should require contractors and their crane operators to complete a form similar to *Appendix B* before performing work at the project site. Lift planners for personnel lifts near powerlines should complete the form and submit it with the lift plan for approval.

Robots Handle Live Powerlines Better than People

Safety is always a concern when working with powerlines, but the quick repair of an outage remains essential. Robotic manipulator systems are becoming more popular with major utility companies to tackle the most hazardous and critical live-line chores. An operator can remain in the safety and comfort of a control vehicle while the manipulator arms do the work. Capable of handling shocks and arcs exceeding 1,000,000 volts, the systems are immune to inclement weather or other conditions that often prevent direct intervention by human hands. Specialized equipment such as this prevents human workers in a suspended platform from being exposed to these hazards.

1.3.5 Planning and Preparations

Planning for a personnel hoisting operation must satisfy all the initial requirements described in 29 *CFR* 1926.1431 and *ASME B30.23* as summarized previously. Lift planners must also consider any other requirements established by OSHA and ASME standards relevant to critical lift operations. For example, planners must develop a written lift plan approved by the lift director. Lift planning and critical lift considerations are covered in greater detail in Module 21304 of this curriculum.

After planners develop the lift plan, the lift director reviews and approves it, then provides it to the personnel lift authorizer. The lift plan may include a lift data sheet modified as a *Personnel Platform Lift Planning and Authorization* form (similar to *Appendix* C) to help management focus on the key aspects of the personnel lift. When the personnel lift authorizer approves the plan, it passes back to the lift director for implementation.

OSHA and ASME standards require that the lift director conduct a pre-lift meeting with the lift participants to review the safety requirements, setup, and procedures to be used, as well as the work planned. Participants shall include the crane operator, lift platform occupants, signal person(s), supervisors of the planned work, and essential ground and support persons. The lift director discusses preparations, assigns responsibilities to personnel, and reviews the communications plan, including any special signals, if required. The agenda also includes general and public safety precautions, as well as emergency procedures. Participants should raise any concerns they have so that issues can be resolved before the operation begins.

Following the meeting, work crew supervisors complete the preparations according to the lift plan schedule and requirements. Prior to the lift, some of the key preparations generally include the following:

- Verify completion of all required inspections and certifications of the crane and hoisting equipment
- Perform the platform proof test, if not already completed
- Perform the applicable trial lift, if not already completed
- Establish traffic barriers/controls around the site of the lift
- Test radio communications, if used
- Verify all participants are in position and equipped with required PPE, tools, and materials
- If the lift will be near overhead powerlines, establish the conditions, barriers, and warning signs stipulated by OSHA standards
- Check the existing site weather conditions and the weather forecast
- Obtain permission to begin the lift from the lift authorizer and lift director

1.3.6 Work Practices and Crane Operations

All standard crane safety practices apply to personnel hoisting operations as well. In addition, the following list summarizes work practices that specifically apply to personnel lifts:

- The total load of the platform, occupants, rigging, and load line must not exceed 50 percent of the crane's rated capacity for the configuration and load radius.
- The standards prohibit traveling during a hoist except for cranes on fixed rails or runways, or when the employer can show that no other method exists to accomplish the work.
- The standards prohibit personnel lifts from cranes on rubber under any circumstances.
- Normally, if the platform is not resting on a stable surface for entry or egress of the occupants, it must be secured to the structure where access or the work occurs while suspended from a crane.

- If the platform is secured to any structure, the operator must not move the platform until informed that it is freely suspended.
- At the beginning of the lift, with all occupants, tools, and materials on board, operators shall conduct a test lift of a short distance (*Figure 9*). A competent person shall inspect the platform and suspension rigging to ensure that the platform is secure and properly balanced.
- Before proceeding with the lift, a competent person must verify that there are no hoist line deficiencies and the parts of line are not fouled with each other. That individual must also make sure the master link/shackle is centered above the platform and, if the hoist rope is slack, verify that all parts of line are properly seated in their sheaves.
- The ground crew must use taglines to control the platform as needed.
- The crane operator must move the platform only under the control of the designated signal person.
- The hoisting of personnel must be slow, cautious, and smooth. Raising or lowering speeds shall be less than 100 feet per minute (30 m/min) during lifts involving personnel hoisting.
- Operators may not position any part of the crane or a platform within 20 feet (6 m) of overhead powerlines up to 350 kV, or within 50 feet (15 m) of powerlines over 350 kV, unless the conditions permitted by 29 *CFR* 1926.1408 or 1926.1409 exist.
- The operator shall not hoist on any other lift lines while conducting a personnel lift.
- The operator should avoid simultaneous operation of more than one hoisting control unless doing so increases the safety of the lift operation.
- When the platform is at the work location, the operator shall set all brakes and locks on the crane hoist and slew controls before platform occupants begin working.
- During welding operations from the platform, electrode holders must prevent contact with any portion of the lift platform.
- Lift supervision must terminate hoisting operations if wind speed at the level of the platform exceeds 20 mph (32 kph); for electric storms; or for other adverse weather conditions.

Figure 9 Test lift at beginning of a personnel hoisting operation.

- For cranes on floating barges, stop personnel lifting operations if the barge list exceeds 5 degrees or the limits of the crane's load chart under list conditions. Operators also need to consider wave and wake action.
- For work over water, platform occupants may detach fall arrest devices and wear USCG-approved personal flotation devices; when over land, they must reconnect fall arrest devices. Note that the basket shown in *Figure 10* is used for transferring personnel only in marine environments under 29 *CFR* 1926.1431.
- Occupants of the platform must move slowly to avoid unbalancing the platform, and keep all body parts inside the platform during its movement. They must not intend to increase their height by sitting or standing on any part of the platform other than the floor.

Figure 10 Occupants lifted over water must wear approved flotation devices.

Additional Resources

ASME B30.23, Personnel Lifting Systems. Current edition. The American Society of Mechanical Engineers. 2011. New York, NY: American Society of Mechanical Engineers.

ASME P30.1, Planning for Load Handling Activities. Current edition. The American Society of Mechanical Engineers. New York, NY: American Society of Mechanical Engineers.

29 *CFR* 1926, Subpart CC,*Cranes and Derricks in Construction*. **www.ecfr.gov**

29 *CFR* 1926, Subpart E, *Personal Protective and Life Saving Equipment*. **www.ecfr.gov**

29 *CFR* 1926, Subpart M, *Fall Protection*. **www.ecfr.gov**

29 *CFR* 1926, Subpart R, *Steel Erection*. **www.ecfr.gov**

1.0.0 Section Review

1. For a platform equipped for a single-point lift, the attachment point must have a design factor of _____.
   a. 4
   b. 5
   c. 7
   d. 9

2. During a proof test, operators must raise a suspended platform to a height, then lower it, bringing it to a stop with the hoist drum brake. It must be lowered at a rate exceeding _____.
   a. 50 feet per minute (15 m/min)
   b. 100 feet per minute (30 m/min)
   c. 150 feet per minute (46 m/min)
   d. 200 feet per minute (61 m/min)

3. Which of the following statements about the crane operator's competency for a personnel lift is *true*?
   a. The operator must have had at least 12 hours off prior to the current shift.
   b. The operator cannot have worked for more than 4 hours prior to beginning the lift.
   c. The operator cannot exceed 12 hours on shift before the personnel lifting will be completed.
   d. The operator must have at least 1 hour prior to beginning the lift to rest outside the cab area.

SUMMARY

Consensus and safety standards prohibit the hoisting of personnel by cranes unless the employer demonstrates there is no other feasible method for the task. The design, manufacturing, and operation of personnel hoisting equipment is covered in *ASME B30.23,Personnel Lifting Systems*. 29 *CFR* 1926.1431, *Hoisting Personnel*, provides the safety regulations for these operations. According to OSHA, personnel hoisting operations are categorized as critical lifts.

Consistent testing and inspections ensure that the platform and rigging remains safe to use. Documentation of maintenance and testing helps to ensure this goal. Key personnel must plan, prepare for, and execute personnel lifts safely. Safety considerations are critical for lift plans that involve hoisting personnel near powerlines or over water.

Review Questions

1. The safety regulations for hoisting personnel with mobile cranes are provided in _____.
   a. 29*CFR*1926.1408
   b. *ASME B30.23*
   c. 29*CFR*1926.1431
   d. *ASME B30.5*

2. Which of the statements regarding personnel hoisting platform access doors and gates is *correct*?
   a. For larger platforms, access door and gates must swing inward.
   b. Access door and gates must be tested using a 300-lb (136 kg) weight.
   c. Access doors and gates must open without any effort.
   d. Access doors and gates must have hand- or grabrails.

3. For the purposes of personnel platform design, the ASME standard assumes an occupant weighs _____.
   a. 300 pounds
   b. 250 pounds
   c. 200 pounds
   d. 180 pounds

4. The initial builder's test of a newly constructed personnel lift platform and each leg of its bridle system must be equal to what percentage of the equipment's rated load?
   a. 100 percent
   b. 125 percent
   c. 150 percent
   d. 200 percent

5. 29 *CFR* 1926.1431 requires a proof test after any structural repair to a lifting platform. This involves lifting the platform loaded to what percentage of the platform's rated load?
   a. 200 percent
   b. 175 percent
   c. 150 percent
   d. 100 percent

6. According to 29 *CFR* 1926.1431, personnel platform trial lifts are required for all the following *except* _____.
   a. prior to each shift involving a personnel hoist
   b. when the platform occupant(s) change
   c. any time the crane moves to a new location or returns to a previous one
   d. whenever the lift route changes

7. Of the following, who is responsible for ensuring that the hoist line does not foul the personnel platform during a lift?
   a. Personnel lift authorizer
   b. Lift director
   c. Crane operator
   d. Ground crew

8. Which of the following statements pertaining to fall arrest equipment is *true*?
   a. Body belts alone are acceptable for crane hoisted personnel lifts if they are made from synthetic materials.
   b. Platform occupants may snap fall arrest harnesses to platform handrails or guardrails.
   c. Occupants can inspect and reuse a fall arrest harness again after it has arrested a fall.
   d. Platform occupants may not fasten the fall arrest harness to any platform hoist attachment point.

9. For personnel lifts near overhead powerlines, which of the following statements is *true*?
   a. Standards prohibit entry of the crane boom or suspended platform within the minimum standoff distances of energized powerlines.
   b. Standards permit entry into the prohibited zones near energized powerlines if the crane and platform are both grounded.
   c. Standards allow personnel platforms to approach energized 480 kV powerlines to a minimum distance of 20 feet.
   d. Suspended personnel platforms may not enter the OSHA-specified standoff zone near overhead powerlines under any condition.

10. Hoisting personnel with a mobile crane on rubber is _____.
    a. permitted if travel with a suspended platform is required
    b. permitted when the employer determines it is the only way to accomplish the task
    c. never permitted
    d. permitted only on a firm, stable runway

Trade Terms Introduced in This Module

Bridle slings: Lift rigging consisting of multi-leg, wire rope or chain slings.

Flemish eyes: Spliced loops formed in the end of wire rope rigging slings. For personnel hoist rigging, the eyes must also contain a thimble and be mechanically crimped with a ferrule.

Nondestructive testing (NDT): Testing the integrity of fabricated metal structures, especially welds, by various means to detect hidden flaws without destroying the object or rendering it useless. Common NDT methods include dye penetrant (PT), magnetic particle (MT), X-ray and radiography (RT), and ultrasound (UT).

Poured socket: Referring to wire rope terminations, a method of attaching a fitting by placing the end of the rope inside the fitting socket, then pouring molten zinc, another metal alloy, or a special resin into the socket. After hardening, these terminations can provide 100 percent of the wire rope's minimum breaking strength. This process is also referred to as speltering.

Proof test: A strength test of an object, usually under a load exceeding that expected under normal use, to ensure it is suitable for safe use. In personnel hoisting, a proof test requires loading the platform with a weight equal to 125 percent of its rated load and suspending it for at least 5 minutes.

Trial lift: For a personnel hoisting platform, a lift of the platform and rigging without occupants, but with an equivalent load, under the same conditions and over the same route as the lift with personnel will take.

Personnel Lift Platform Pre-Lift Inspection Form

Personnel Lift Platform Pre-Lift Inspection

Date/Time: ____________________

Platform ID: ____________________

Inspector: ____________________

1. Structure	**SAT**	**UNSAT**
Load-supporting members	______	______
Load-supporting welds/fasteners	______	______
Flooring		
Barrier from toe rail to intermediate rail	______	______
Top rail	______	______
Gate hinge points and locking mechanism	______	______
Grab-/handrail	______	______
Suspension attachment points	______	______
Fall-protection anchor points	______	______

2. Attachment/Rigging Components (Circle applicable items)

Pins/Eyes/Ears/Bolt-ups	______	______
Wire-rope sling/Chain/Rigid, each leg	______	______
Master link/Shackle	______	______

3. Special-Purpose Items (Overhead protection/Floatation/Platform Controls/Boom Attachment)

List:

a. ____________________	______	______
b. ____________________	______	______
c. ____________________	______	______

4. Markings/Documentation

Platform	______	______
Placard/Manual	______	______
Suspension System	______	______

5. Comments: ____

____

____

____

____

____

Lift Director Date/Time

Appendix B

Power Line Planning Form

POWER LINE AWARENESS PERMIT

Today's Date ____________________ **Project** ____

Contractor Name			
Job Address			
Telephone Number		Fax Number	

Emergency Contact Number	

Survey

Before beginning any project, you must first survey your work area to find power lines at the job site. (See job-site sketch on reverse side.)

Identify

After finding all of the power lines at your site, identify the activities you'll be doing that may put you or your workers at risk.

Mark one or more of the following:

- ☐ Cranes (mobile or truck mounted)
- ☐ Drilling rigs
- ☐ Backhoes/excavators
- ☐ Long-handled tools
- ☐ Other tools/high-reaching equipment
- ☐ Concrete pumper
- ☐ Aerial lifts
- ☐ Dump trucks
- ☐ Ladders
- ☐ Material handling & storage
- ☐ Scaffolding
- ☐ Other ______________

Eliminate or Control

After identifying the power line and high-risk activities on the job site, determine how to eliminate or control the risk of electrocution (a successful determination is often reached only after consultation with the utility).

Mark one or more of the following:

- ☐ Move the activity
- ☐ Change the activity
- ☐ Have the utility de-energize the power line
- ☐ Have the utility move the power line
- ☐ Use barrier protection (insulated sleeves)
- ☐ Use an observer
- ☐ Use warning lines with flags
- ☐ Use non-conductive tools
- ☐ Use a protective technology:
  - ☐ Insulated link
  - ☐ Boom cage guard
  - ☐ Proximity device

WARNING!

It is unlawful to operate any piece of equipment near energized power lines, within the distances specified in the table on the reverse side.

Always maintain your minimum safe clearance distance from the power line, except when the utility has de-energized and visibly grounded the power line.

Voltage (nominal, kVAC)	Minimum clearance distance (feet)
up to 50	10
over 50 to 200	15
over 200 to 350	20
over 350 to 500	25
over 500 to 750	35
over 750 to 1,000	45
over 1,000	as estimated by the utility owner/operator or registered professional engineer who is a qualified person with respect to electrical power transmission and distribution

Note: The value that follows "to" is up to and includes that value. For example, over 50 to 200 means up to and including 200kV.

*information contained in this table is from OSHA 29 CFR Sections 1926.1408 and 1926.1409 (as of 2017)

Job-Site Sketch

(Draw in location of power lines and their proximity to construction site, including such things as proposed excavations, location of heavy equipment, scaffolding, material storage areas, etc.)

Completed by ______________________________ **Date** ______________

Title ______________________________

Approved by ______________________________ **Date** ______________

Title ______________________________

Personnel Platform Lift Planning and Authorization Form

Personnel Lift Planning and Authorization

Date:

Lift Platform ID: Lift Description:

Project: Units: U.S. (ft, lb)

Crane Details Manufacturer: Model Number:

Configuration:		Base Mount Type:			Track/Outrigger Ext:		ft
Boom Type:		Boom Length Used:		ft	Boom/Jib Angle:		deg.
Jib Type:		Jib Length Used:		ft	Tail Swing:		ft
Crane Ballast:	lb	Aux Counterweigh		lb			
Block Capacity:	ton	Line Si:		in			
Single Line Pull:	lb	Parts Line Use			Reeved Capacity:		lb
Heavy Lift Attachments:		Load Radius:		ft	Superlift Weight:		ton

Load Details

	Quantity	Wt. ea.	Total Weight
Basic Weight of Platform		lb	lb
Occupants		lb	lb
Equipment/Tools		lb	lb
Materials		lb	lb
		lb	lb
Total Weight of Loaded Platform:			lb

Rigging Data (size, type, and capacity)

		lb	lb
		lb	lb
		lb	lb
		lb	lb
		lb	lb
		lb	lb
		lb	lb
Total Rigging Weight:			lb

Additional Weight Items

Main Hook		lb	lb
Wire Rope		lb	lb
Other Suspended Hooks		lb	lb
Aux Boom Sheaves		lb	lb
Jib		lb	lb
		lb	lb
		lb	lb
Total Additional Weight:			lb
GROSS LOAD (sum of all items above):			lb

Crane Capacities

Condition:			
Total Load to Crane:	lb	lb	lb
Planned Radius:	ft	ft	ft
Chart Radius Used:	ft	ft	ft
Chart Capacity:	lb	lb	lb
% of Chart Capacity:	%	%	%

MAXIMUM PERCENTAGE OF CHART CAPACITY ALLOWED IS 50 %

MAXIMUM PERCENTAGE OF CHART CAPACITY TO BE USED IS %

Total Suspended Load: lb Reeved Capacity: lb

MAXIMUM PERCENTAGE OF REEVED CAPACITY PLANNED TO BE USED IS %

Document Attachments

- ○ Crane Layout
- ○ Rigging Hookup Diagram
- ○ Crane Load Chart
- ○ Foundation Details
- ○ Crane Cribbing Arrangement
- ○ Lift Plan Category
- ○ AHA
- ○ Risk Evaluation
- ○ Weather Forecast
- ○ Load Schematic
- ○ GBP Source/Calculations
- ○ Project Scope
- ○ Load Weight/CG Source
- ○ Load/Crane Clearances

Notes

	Name (Print)	Signature	Title	Date
Prepared by:				
Checked by:				
Approved:				

Additional Resources

This module is intended as a thorough resource for task training. The following reference materials are recommended for further study.

ASME B30.23, Personnel Lifting Systems. Current edition. The American Society of Mechanical Engineers. 2011. New York, NY: American Society of Mechanical Engineers.

ASME P30.1, Planning for Load Handling Activities. Current edition. The American Society of Mechanical Engineers. New York, NY: American Society of Mechanical Engineers.

29 *CFR* 1926, Subpart CC, *Cranes and Derricks in Construction*. **www.ecfr.gov**

29 *CFR* 1926, Subpart E, *Personal Protective and Life Saving Equipment*. **www.ecfr.gov**

29 *CFR* 1926, Subpart M, *Fall Protection*. **www.ecfr.gov**

29 *CFR* 1926, Subpart R, *Steel Erection*. **www.ecfr.gov**

Figure Credits

© iStock.com/BanksPhotos, Module opener

© iStock.com/landbysea, Figure 1

Tway Lifting Products, Figure 2

Lifting Technologies, Inc., Figures 4, 9

© iStock.com/Tashi-Delek, Figure 5

Fall protection materials provided courtesy of Miller Fall Protection, Franklin, PA, Figure 6

©The Crosby Group LLC, Figure 7

© iStock.com/bentaboe, Figure 8

© Brian McDonald/Shutterstock.com, Figure 10

Section Review Answer Key

Answer	Section Reference	Objective
Section One		
1. c	1.1.1	1a
2. b	1.2.2	1b
3. c	1.3.1	1c

NCCER CURRICULA — USER UPDATE

NCCER makes every effort to keep its textbooks up-to-date and free of technical errors. We appreciate your help in this process. If you find an error, a typographical mistake, or an inaccuracy in NCCER's curricula, please fill out this form (or a photocopy), or complete the online form at **www.nccer.org/olf**. Be sure to include the exact module ID number, page number, a detailed description, and your recommended correction. Your input will be brought to the attention of the Authoring Team. Thank you for your assistance.

Instructors – If you have an idea for improving this textbook, or have found that additional materials were necessary to teach this module effectively, please let us know so that we may present your suggestions to the Authoring Team.

NCCER Product Development and Revision

13614 Progress Blvd., Alachua, FL 32615

Email: curriculum@nccer.org
Online: www.nccer.org/olf

❑ Trainee Guide ❑ Lesson Plans ❑ Exam ❑ PowerPoints Other ______________________

Craft / Level: Copyright Date:

Module ID Number / Title:

Section Number(s):

Description:

Recommended Correction:

Your Name:

Address:

Email: Phone:

Advanced Operational Techniques

Overview

Mobile crane operations span a wide variety of applications and circumstances. The focus of the curriculum to this point has been on operating cranes safely while performing common lifting operations. This module provides information on more advanced mobile crane techniques that require additional expertise. Most operators are likely to encounter some of the situations described in this module as they progress in their professional career.

Module 21303

Trainees with successful module completions may be eligible for credentialing through the NCCER Registry. To learn more, go to **www.nccer.org** or contact us at 1.888.622.3720. Our website has information on the latest product releases and training, as well as online versions of our *Cornerstone* magazine and Pearson's product catalog.

Your feedback is welcome. You may email your comments to **curriculum@nccer.org**, send general comments and inquiries to **info@nccer.org**, or fill in the User Update form at the back of this module.

This information is general in nature and intended for training purposes only. Actual performance of activities described in this manual requires compliance with all applicable operating, service, maintenance, and safety procedures under the direction of qualified personnel. References in this manual to patented or proprietary devices do not constitute a recommendation of their use.

21303 V3

21303
Advanced Operational Techniques

Objectives

When you have completed this module, you will be able to do the following:

1. Identify and describe the requirements and techniques related to various advanced operations.
   a. Identify requirements for and techniques used in critical lifts.
   b. Identify requirements for and techniques used in personnel lifting.
   c. Identify requirements for and techniques used in blind and multi-crane lifts.
   d. Identify requirements for and techniques used in operations near radiofrequency transmitters.
   e. Identify requirements for and techniques used in demolition tasks.
   f. Identify requirements for and techniques used in vacuum lifting.
   g. Identify requirements for and techniques used in electromagnet lifting.
   h. Identify requirements for and techniques used in cold weather operations.

Performance Task

Under the supervision of your instructor, you should be able to do the following:

1. Perform a blind lift as a crane operator.

Trade Terms

Alloy
Brittle fracture
Cavitation
Coefficient of friction
Demolition
Duty cycle
Electromagnet
Ether
Ferrous metals
Limit line
Radiofrequency (RF)
Tailing crane
Ultimate pad capacity (UPC)
Vacuum

Industry Recognized Credentials

If you are training through an NCCER-accredited sponsor, you may be eligible for credentials from NCCER's Registry. The ID number for this module is 21303. Note that this module may have been used in other NCCER curricula and may apply to other level completions. Contact NCCER's Registry at 888.622.3720 or go to **www.nccer.org** for more information.

Contents

1.0.0 Advanced Operational Techniques ... 1
1.1.0 Critical Lifts ... 1
1.2.0 Personnel Lifts ... 2
1.3.0 Blind and Multi-Crane Lifts ... 3
1.3.1 Blind Lifts ... 3
1.3.2 Multiple-Crane Lifts ... 4
1.4.0 Operations Near RF Transmitters ... 11
1.5.0 Demolition Operations ... 13
1.5.1 Vertical-Drop Balling ... 14
1.5.2 Swinging Ball: In-Line with Boom ... 14
1.5.3 Swinging Ball: Slewing ... 15
1.6.0 Vacuum Lifting ... 15
1.6.1 Factors Affecting Lifts ... 17
1.6.2 Lifting Guidelines ... 18
1.7.0 Electromagnet Lifting ... 19
1.7.1 Factors Affecting Lifts ... 19
1.7.2 Lifting Guidelines ... 20
1.8.0 Cold Weather Operations ... 21

Figures

Figure 1 Enclosed personnel lift with fall protection ... 3
Figure 2 Blind lift in progress ... 3
Figure 3 A two-crane coordinated heavy lift ... 5
Figure 4 A dual-crane lift of a large structural member ... 5
Figure 5 Dual-crane lift – load's center of gravity location ... 7
Figure 6 Dual-crane lift – calculating distance B, Trial 1 ... 7
Figure 7 Dual-crane lift – calculating distance B, Trial 1 results ... 8
Figure 8 Dual-crane lift – calculating distance B, Trial 2 ... 8
Figure 9 Dual-crane lift – calculating distance B, Trial 2 results ... 9
Figure 10 Dual-crane lift with equalizer beam ... 10
Figure 11 Dual-crane lift with equalizer beam – determining distances B and C ... 11
Figure 12 Dual-crane lift with equalizer beam showing calculated distances B and C ... 12
Figure 13 Cranes working near RF transmitters ... 13
Figure 14 Pear-shaped wrecking ball ... 13
Figure 15 Dynamic earth compactor ... 15
Figure 16 In-line ball demolition showing the pulling rope ... 15
Figure 17 Crane using a vacuum lifter to place a sheet of plate glass ... 16
Figure 18 Relationship between load angle and the friction force acting on the load ... 18
Figure 19 An electromagnet used for lifting scrap metal ... 19
Figure 20 Operating cranes in cold weather requires additional precautions ... 22

SECTION ONE

1.0.0 ADVANCED OPERATIONAL TECHNIQUES

Objective

Identify and describe the requirements and techniques related to various advanced operations.

a. Identify requirements for and techniques used in critical lifts.
b. Identify requirements for and techniques used in personnel lifting.
c. Identify requirements for and techniques used in blind and multi-crane lifts.
d. Identify requirements for and techniques used in operations near radiofrequency transmitters.
e. Identify requirements for and techniques used in demolition tasks.
f. Identify requirements for and techniques used in vacuum lifting.
g. Identify requirements for and techniques used in electromagnet lifting.
h. Identify requirements for and techniques used in cold weather operations.

Performance Task

Perform a blind lift as a crane operator.

Trade Terms

Alloy: A metal that is a combination of two or more pure metals or other chemical elements.

Brittle fracture: A break in a metal object below a certain temperature specific for the metal, caused by a relatively small amount of stress acting on an existing flaw.

Cavitation: The creation of low-pressure regions within the intake area of a pump as the impeller rotates, forming pockets of gas that collapse violently, creating noise and pressure pulsations.

Coefficient of friction: A ratio that expresses a comparison between the force necessary to move an object over the surface of another material, and the pressure between the two materials.

Demolition: The controlled destruction of a structure.

Duty cycle: For repetitive operations, the ratio of the time a device is on or functioning compared to the total time of one cycle (time on plus time off), expressed as a percentage.

Electromagnet: A device consisting of an insulated electrical conductor (wire) coiled around a magnetic metal core. When current flows through the conductor, a strong magnetic field forms within the core.

Ether: A volatile, flammable, organic fluid that operators sometime use to assist in starting diesel engines, especially in cold weather.

Ferrous metals: Iron and its alloys that have relatively strong magnetic characteristics.

Limit line: When engaged in demolition operations using a wrecking ball, a length of wire rope attached to the ball from a point on the crane to prevent it from swinging out too far, providing a safety margin against overturning the crane or damaging the boom.

Radiofrequency (RF): Describes the lower frequencies of the electromagnetic energy spectrum in which radar, TV, AM/FM radio, cell phones, and all other forms of radio-wave communications and detection operate.

Tailing crane: A crane that helps position a load by turning or tilting it during lifting operations, but does not lift a significant part of the load's weight.

Ultimate pad capacity (UPC): Rating of the lifting force generated by a vacuum-lifter pad, based on the vacuum level and pad area.

Vacuum: The reduced presence or near-total absence of air or other gases, creating a low-pressure condition.

As they advance in their profession, most mobile crane operators are likely to encounter a variety of unique situations requiring special expertise. This module describes crane tasks for which an operator needs advanced knowledge and careful preparation. Information is provided for the safe performance of some of these tasks, including personnel lifts, multi-crane lifts, demolition, lifting with specialized load-handling devices, and cold weather operations.

1.1.0 Critical Lifts

The American Society of Mechanical Engineers (ASME) has defined and provided standards for conducting various categories of load handling lifts. This became necessary as government and industry recognized the hazards associated with crane operations and the need to standardize materials and methods to ensure the safety of personal and property. *ASME P30.1, Planning for Load Handling Activities,* is the current version of these standards. It identifies two categories of lifts and explains how to conduct them safely.

A standard lift is one that does not involve any particular hazard or complexity. Operators can perform such a lift using standard load handling methods.

A critical lift involves factors that make the lift particularly complicated or hazardous, or the load itself may require exceptional care in handling. This may be due to its size, weight, close-tolerance installation, high susceptibility to damage, or other unusual factors. OSHA standards classify certain types of lifting operations, such as personnel lifts and multiple-crane lifts, as critical lifts. When lifting such loads, critical lift planning should establish the following conditions (as well as others if needed):

- *Planning* – Planners must prepare a formal, written lift plan approved by the lift director.
- *Supporting surface* – The ground should be firm and level with the required loadbearing characteristics.
- *Blocking and mats* – The ground bearing pressure must be sufficient to support crane operations, or preparations should include appropriate blocking and matting. Blocking and matting helps to distribute the load over a greater area, reducing pressure.
- *Level* – The crane should be level within 1 percent.
- *Load weight* – Planners should know the exact weight of the load or use an approved method to estimate its weight.
- *Center of gravity* – Workers should know the location of the load's center of gravity (CG); the hoist line or point should be directly above the CG.
- *Load radius, boom length, and boom angle* – Planners should directly measure these for every key point during the planned lift.
- *Weather* – Planners should consider the effects of wind, and other weather-related factors, on the lift. The lift director should delay the lift if the wind loads are significant. The operator and lift director must consult the manufacturer's recommendations when lifting in windy conditions.
- *Reeving* – The assembly director should reeve the crane per the manufacturer's instructions.
- *Load rigging* – Planners, riggers, and the operator should know the exact weight of the load rigging and how it attaches to the load.
- *Load movement* – The operator should make all movements as slowly and smoothly as possible, and should complete the operation using the least number of movements.

NOTE

While the use of the term *should* in the list of items above suggests they are optional or suggested actions, OSHA or ASME standards may specify these items as mandatory during critical lifts. Mandatory requirements in these standards use the term *shall.*

29 *CFR* 1926.751 categorizes lifts as critical when the suspended load weight exceeds 75 percent of the crane's load chart capacity. While this standard pertains specifically to steel erection, many organizations apply the same standard to all lifts. A 1-ton load on a 50-ton capacity crane may seem insignificant. However, if the crane's capacity at the load radius is only 2,400 pounds, it is operating at 83 percent of capacity; according to OSHA's definition, this is a critical lift. Company or regulatory policies may identify other conditions that qualify a lift as critical. Refer to Module 21304 of this curriculum for detailed information on how to prepare for and conduct standard and critical lifts.

1.2.0 Personnel Lifts

29 *CFR* 1926.1431(a) prohibits the use of a crane or derrick to hoist employees on a suspended personnel platform unless another means to accomplish the work would be more hazardous or impossible. Personnel lifts are always considered critical lifts. Prior to conducting a personnel lift, planners and the operator should consult the relevant requirements of the OSHA standard. Because of the potential hazards involved, it is common to complete a pre-lift checklist for such operations. *ASME B30.23, Personnel Lifting Systems* provides an example of such a checklist. It also establishes equipment standards and operational guidelines for lifting personnel.

ASME B30.23 identifies some of the important requirements that apply to operators conducting personnel lifts with a crane, as follows:

- The crane must be set up level on outriggers or stabilizers. The OSHA standard prohibits on-rubber personnel lifts.
- All crane indicators, controls, and safety features specified by the standards must be operational.
- Operators must have been off work for at least 8 hours prior to the shift when conducting personnel lifts, or after they have worked continuously for more than 10 hours. They must not conduct the lift if they feel impaired in any way.

- Personnel lifting equipment must be properly designed and fabricated, including platforms, cages, or cradles. Workers often call these *man-baskets* or *lift cages*.
- The design of the equipment must prevent persons or tools from falling out through some means of enclosure, securement, restraints, and fall arrest points (*Figure 1*).
- Lift equipment may not exceed 10 degrees of tilt at any boom angle. Taglines can prevent swinging and spinning, if needed.
- The lift equipment must be clearly and permanently marked with the maximum load that it can safely carry.
- An inspector must check the equipment before use to ensure that it is in a safe condition to carry personnel.
- Workers and an inspector must perform a trial lift supporting the anticipated weight the personnel lift will contain, moving over the exact same path it will take, before conducting the actual lift.
- Special requirements apply for lifting personnel into or out of drill shafts, storage tanks, or chimneys; during pile-driving operations; and over water.

Refer to Module 21305 for more in-depth coverage of suspended personnel platforms and personnel hoisting operations and the specific OSHA and ASME standards that apply.

Figure 1 Enclosed personnel lift with fall protection.

1.3.0 Blind and Multi-Crane Lifts

Common challenges involving mobile cranes are blind lifts and lifts involving multiple cranes. Blind lifts may or may not be in the critical lift category. Multiple-crane lifts, where both cranes bear a proportional part of the load, are always critical lifts. However, *ASME B30.5* states that operations using a second crane only as a **tailing crane** may not fall into the critical lift category if other factors do not apply.

1.3.1 Blind Lifts

A blind lift exists whenever the crane operator loses direct view of the load at any point during the lift. This may be due to obstructions between the load and operator (see *Figure 2*), or an environmental factor, such as the sun's glare, fog, or dust, that prevents the operator from seeing the load.

In all blind lifts, the operator must rely entirely on the signal person's guidance, which requires clear, real-time communications between them. In some situations, one or more additional signal persons may be required to relay information from one signal person to the operator. This might occur when lowering a load into a deep shaft or placing a load on a high structure. While standards permit relay-type hand signal communications during blind lifts, the execution of the lift must take into account the inherent delays between the primary signal person and the operator during relay communications. Depending on the complexity of the operation or the hazards

Figure 2 Blind lift in progress.

presented by the load, such conditions can categorize a blind lift as critical. In these situations, the communication between the signal and/or relay persons and the crane operator must be via two-way radios or hard-wired devices to facilitate immediate responses.

Trainees and less-experienced operators often practice blind lifts in the yard by simply covering up portions of the crane windows to block the operator's view of the load. Either the signal person is seen from another angle, or electronic communication is used to signal the operator.

1.3.2 Multiple-Crane Lifts

Multiple-crane lifts are always critical lifts, requiring detailed written plans and, when required, written procedures. Planners use multiple cranes when the physical dimensions, characteristics, mass, or placement of the load requires the use of more than one crane. Such lifts need detailed planning, because the weight held by any one of the cranes can change as the lift progresses. Lift planners and operators must avoid allowing any one crane to exceed 75 percent of its load chart capacity.

Planners should consider the following factors when planning multiple-crane lifts:

- Qualified, experienced planners must plan a multiple-crane lift, and a qualified lift director must implement the plan.
- If one of the cranes does exceed 75 percent of its net capacity, the operation requires extra engineering evaluations.
- Planners must determine exactly how much of the load's weight each crane will carry at every point during the lift. The rigging design must divide the load as planned.
- During any movement of the load, the line, swing, and boom speeds of the cranes must work together to produce a smooth, even motion.
- Swing and boom motions should be minimized.
- Whenever possible, the cranes should not travel with the load.
- Participants should conduct a dry run without a load or with a test load, if possible, before lifting the actual load.
- Communications by radio are preferred when hand signals are not effective.
- Only one signal person should direct and control the operation unless that is not feasible. In that case, participants must develop and test specific communication procedures involving multiple signal persons.
- Operators must keep hoist lines vertical at all times. When the lines are not vertical, the cranes are transferring load to each other; one crane may side-load the other or exceed its planned capacity.

Figure 3 shows two 750-ton crawler cranes performing a coordinated critical lift to position prefabricated steel structures. This lift required extensive site preparations and a detailed engineering review.

Whenever two or more cranes must lift a load, lift planners must carefully choose the location of each crane's load block so that the cranes safely share the load. Because cranes change their boom position and load radius as the lift progresses, net capacities are constantly changing. The actual load a crane carries may be more or less than the other crane because the load is not symmetrical or its CG is off-center. The following scenario is a practical example of a multiple-crane lift plan.

Planners must determine the weight lifted by two cranes moving a large 180,000-pound structural member (*Figure 4*). Since the load's CG is off-center, the structural member would impose different weights on the two cranes if lifted from its ends. For critical lifts, load planners must document the maximum percentage of net capacity for each crane that will occur during the lift.

Steps 1 through 9 outline the approach for determining the attachment points for lifting an object with two cranes. For this example, assume Crane One's minimum load chart capacity during the lift is 156,000 lb and Crane Two's is 90,500 lb. The load's weight is 180,000 lb. The following procedure determines the hook-block positions on the load and the percent crane capacities.

Step 1 Determine, from Crane One's load chart, the minimum net capacity that it will have during the whole operation.

Minimum net capacity of Crane One:

156,000 lb − 500 lb (beam clamp) = 155,500 lb

NOTE

When lifting from the main boom or an extension, consider all factors: configuration, quadrant of operation, boom length, boom angle, load radius, weight of rigging, and all capacity deductions. Module 21301 of this curriculum presents detailed instructions for the use and interpretation of mobile crane load charts.

Figure 3 A two-crane coordinated heavy lift.

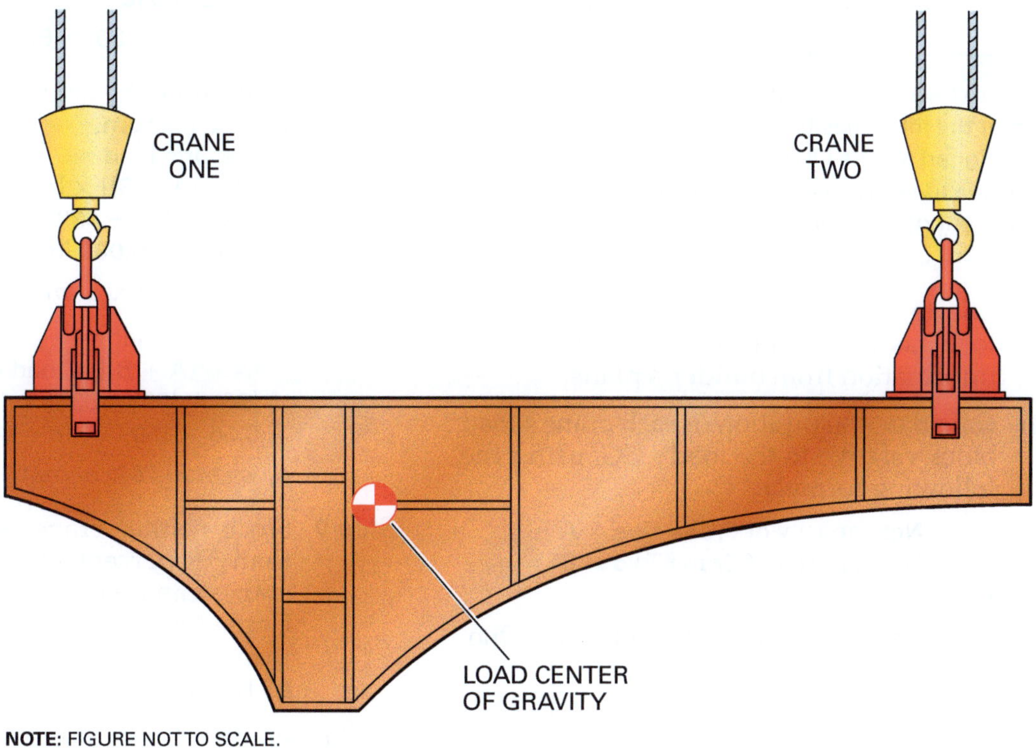

Figure 4 A dual-crane lift of a large structural member.

> **NOTE**
> When using two cranes, measure the load radius of each crane to its attachment point on the load, not to the center of gravity of the load.

Step 2 Repeat Step 1 for Crane Two.

Minimum net capacity of Crane Two:
90,500 lb − 500 lb (beam clamp) = 90,000 lb

Step 3 Determine the total net capacity of the cranes.

Minimum net capacity of the cranes:
155,500 lb + 90,000 lb = 245,500 lb

Step 4 Verify that the load's weight is less than 75 percent of the sum of the two cranes' minimum net capacities.

$$\text{Percent capacity} = \frac{\text{load weight}}{\text{total net capacity}} \times 100$$

$$\text{Percent capacity} = \frac{180{,}000 \text{ lb}}{245{,}500} \times 100$$

$$\text{Percent capacity} = 0.73 \times 100 = 73.00\%$$

> **NOTE**
> This check gives the planner some assurance that the lift is feasible with the two cranes assigned for the lift. However, the planner also needs to verify that neither crane will exceed 75 percent of its capacity.

Step 5 Locate the load's CG from the ends of the load, as shown in *Figure 5*. Obtain this information from builder's plans.

Step 6 Calculate the position of each crane's load block relative to the load's CG, using the following formula:

Net capacity of Crane One × A =
Net capacity of Crane Two × B

Where:

A = distance of load block of Crane One from load CG

B = hook-block distance from load's CG for Crane Two

Step 7 Choose a trial length for A that will be less than the length of C in *Figure 5*. Start with A = 45 ft, for example (see *Figure 6*). Then calculate B:

> **NOTE**
> The equations in *Step 7* are simply the two mathematical solutions for the unknown quantities using the formula given in *Step 6*.

$$B = \frac{\text{Crane One net capacity}}{\text{Crane Two net capacity}} \times A$$

$$B = \frac{155{,}500 \text{ lb}}{90{,}000 \text{ lb}} \times 45'$$

$B = 1.73 \text{ lb} \times 45' = 77.85'$ *(round up to 78')*

This configuration will not work because the load block position for B (78 ft) would be greater than dimension *D* (65 ft) and beyond the end of the right side of the load (see *Figure 7*). To obtain a shorter value of B, choose a shorter length for A— such as, 36 ft (*Figure 8*). This results in a length that fits within the dimensions of the load (*Figure 9*):

$$B = \frac{\text{Crane One net capacity}}{\text{Crane Two net capacity}} \times A$$

$$B = \frac{155{,}500 \text{ lb}}{90{,}000 \text{ lb}} \times 36'$$

$B = 1.73 \text{ lb} \times 36' = 62.28'$ *(round up to 63')*

Step 8 Determine the actual load that each crane will be carrying:

Crane One load =
[B ÷ (A + B)] × load weight =
[63' ÷ (36' + 63')] × 180,000 lb =
0.64 × 180,000 lb = **115,200 lb**

(Less than 75% of min net capacity)

Crane Two load =
[A ÷ (A + B)] × load weight =
[36' ÷ (36' + 63')] × 180,000 lb =
0.36 × 180,000 lb = **64,800 lb**

(Less than 75% of min net capacity)

Step 9 Since neither crane is loaded to greater than 75 percent of its capacity, the two cranes can accomplish the lift as planned. Crane One's load block will be 36 ft from the load's CG and Crane Two's load block 63 ft from it.

Dual-crane lifts can also use an equalizer beam (*Figure 10*) to distribute the weight in proportion to their capacities. By adjusting the crane and/or load attachment points on the beam, it is possible to share the load weight between the cranes in any proportion desired.

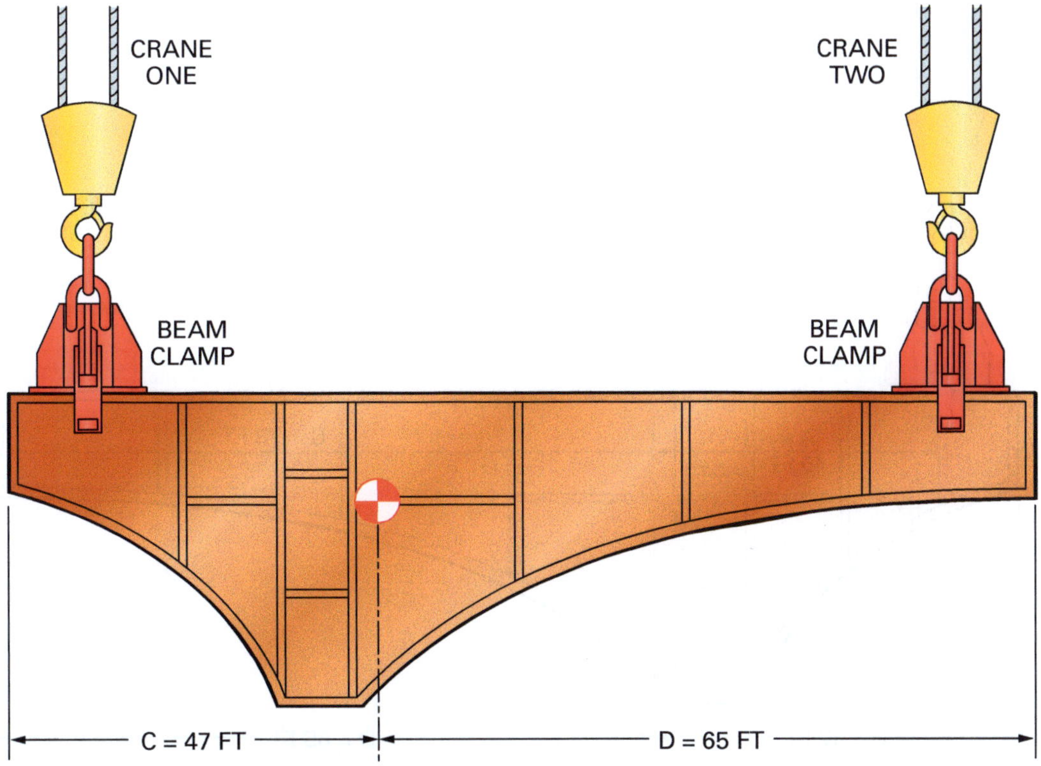

Figure 5 Dual-crane lift – load's center of gravity location.

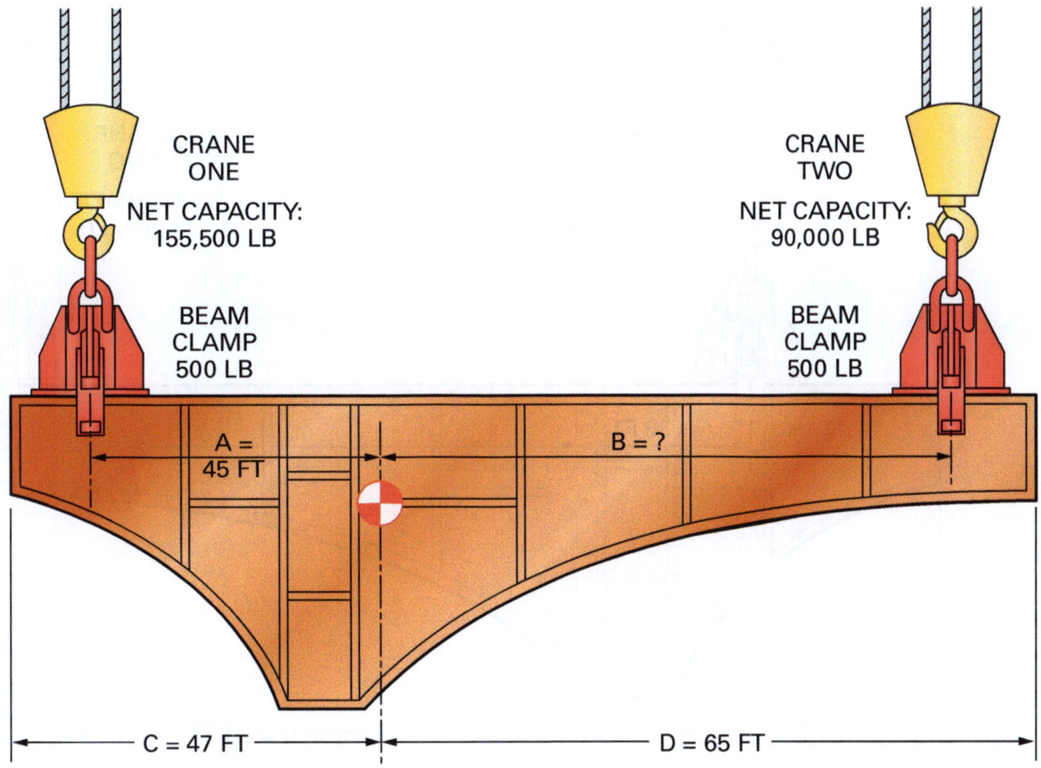

Figure 6 Dual-crane lift – calculating distance B, Trial 1.

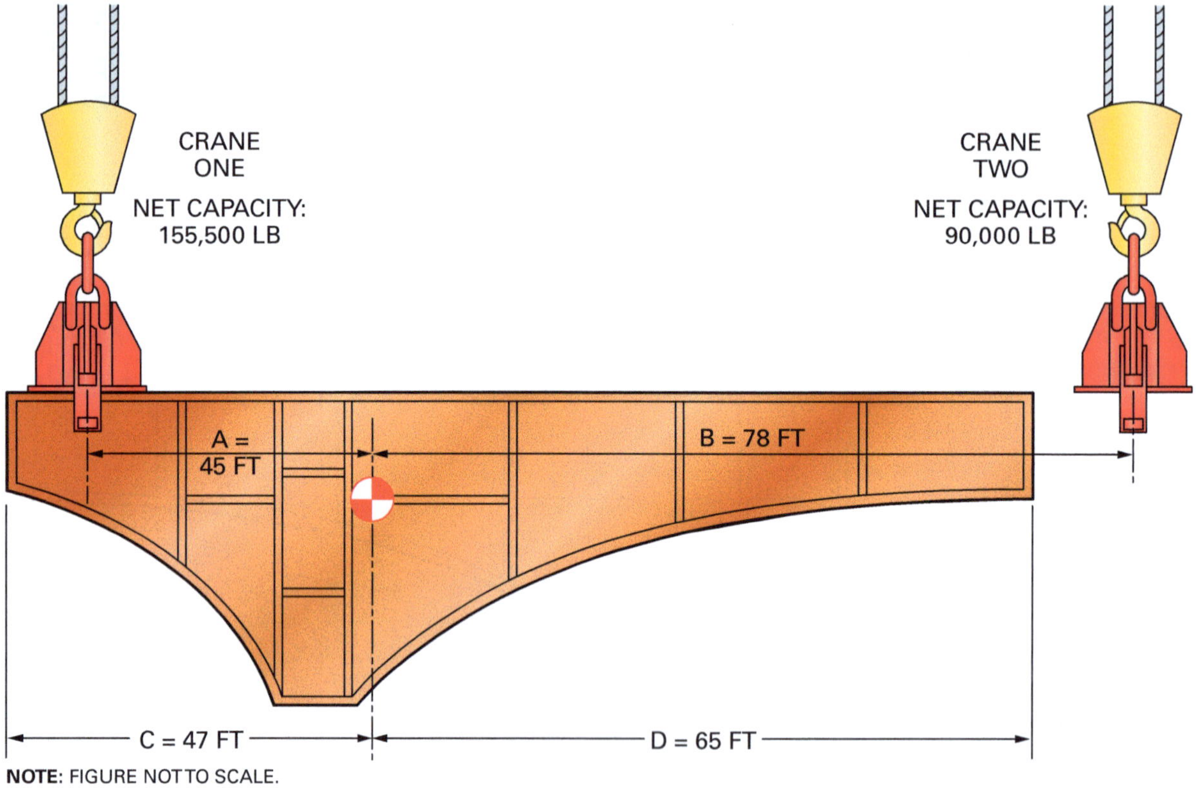

Figure 7 Dual-crane lift – calculating distance B, Trial 1 results.

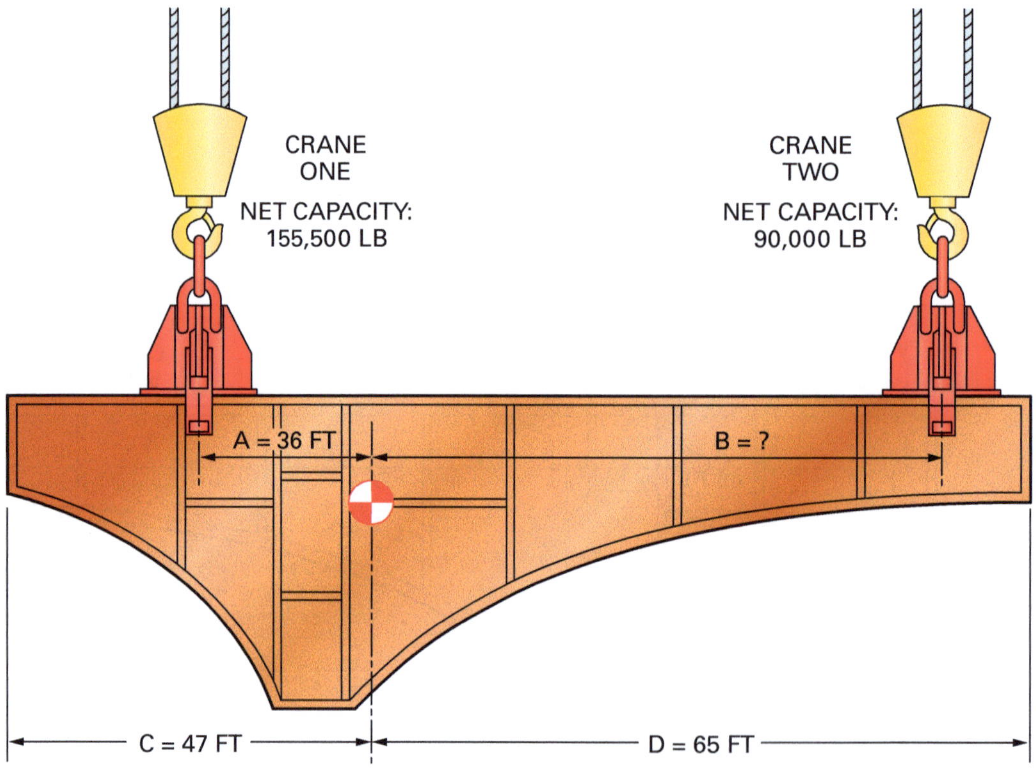

Figure 8 Dual-crane lift – calculating distance B, Trial 2.

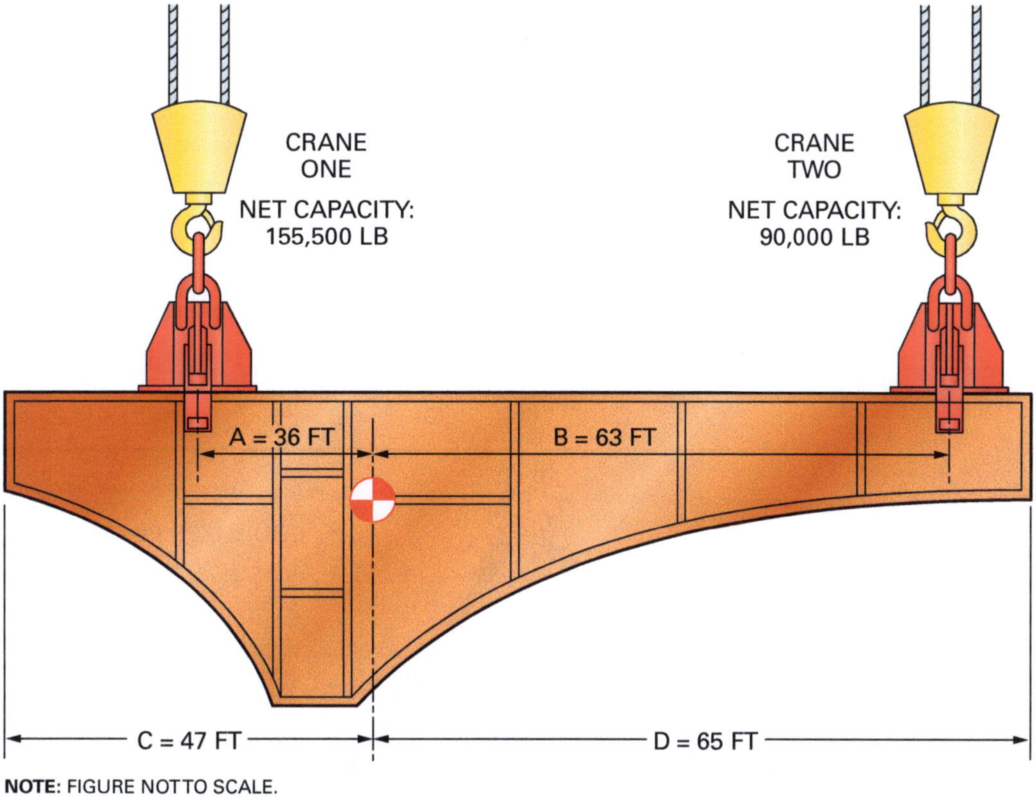

Figure 9 Dual-crane lift – calculating distance B, Trial 2 results.

> **NOTE**
> When using an equalizer beam, measure the load radius of each crane to its pick-up point on the equalizer, not to the center of gravity of the load.

Using a process similar to the one described previously, the lift planner must determine the location on the equalizer beam where the load's hoist point will be, allowing each crane to share the load proportionate to its capacity.

For this example lift plan, the load is a 140,000-lb enclosure. Crane One has a minimum net capacity of 120,000 lb, and Crane Two's minimum net capacity is 80,000 lb. *Figure 11* shows the equalizer bar, the rigging, and the load scheduled for the dual-crane lift.

Step 1 Determine the total capacity of the cranes:

Minimum net capacities:

Crane One = 120,000 lb
Crane Two = 80,000 lb

Total minimum net capacity:

Total capacity = Crane One + Crane Two
Total capacity = 120,000 lb + 80,000 lb

This capacity is greater than the total suspended load weight of 143,000 lb (load + equalizer + rigging). Also, the load weight is less than 75 percent of the cranes' combined capacity (71.5%).

Step 2 Establish the load's hoist point:

> **NOTE**
> For a simple beam problem like this example, the distance of one lifting force from the load's hoist point is proportional to the ratio of the *other* lifting force to the total weight of the load. The equations in the steps that follow illustrate this principle.

$$B = \frac{\text{Crane One net capacity}}{\text{Crane One capacity} + \text{Crane Two capacity}} \times A$$

$$B = \frac{80{,}000\ \text{lb}}{80{,}000\ \text{lb} + 120{,}000\ \text{lb}} \times 25'$$

$$B = \frac{80{,}000\ \text{lb}}{200{,}000\ \text{lb}} \times 25'$$

$$B = 0.40\ \text{lb} \times 25' = \mathbf{10'}$$

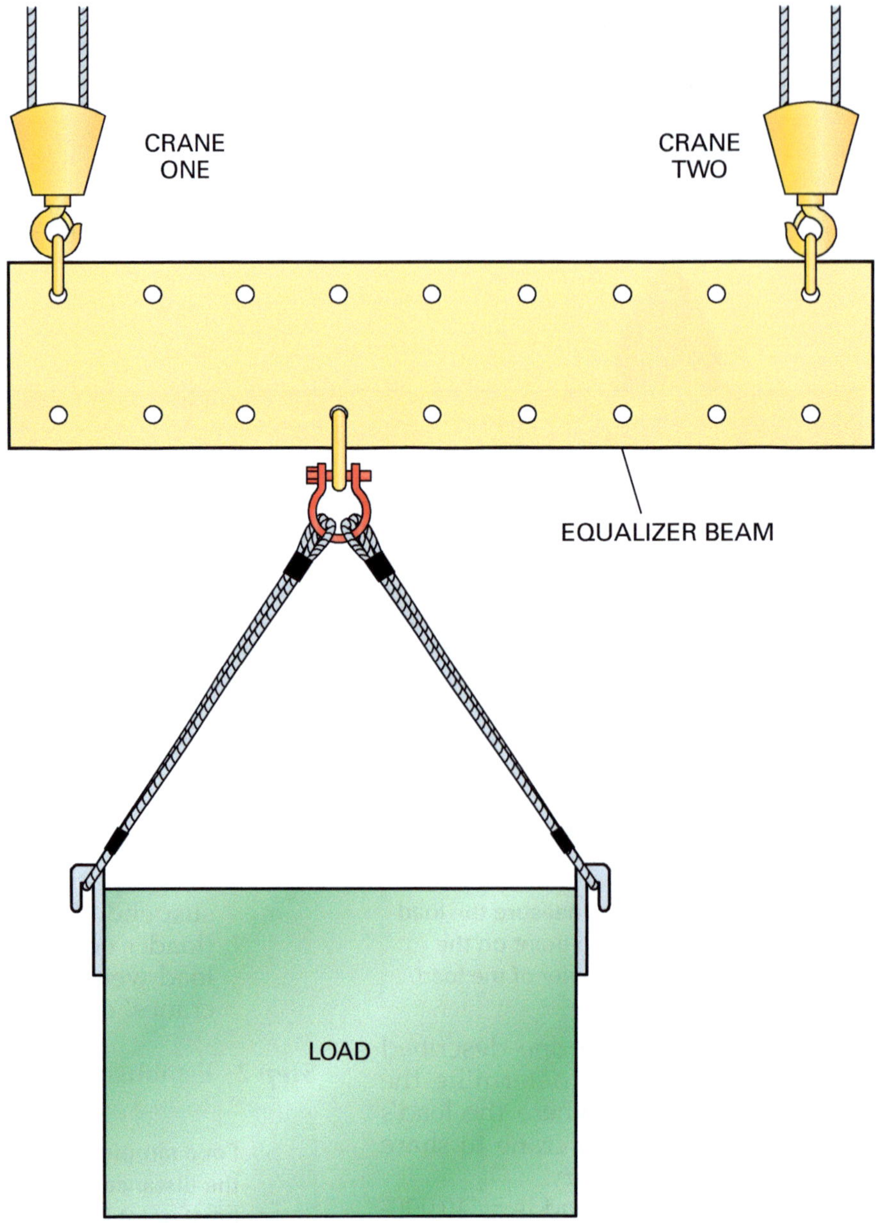

NOTE: FIGURE NOT TO SCALE.

Figure 10 Dual-crane lift with equalizer beam.

On the equalizer beam, if the distance *A* = *B* + *C*, then:

C = A − B
C = 25' − 10'
C = 15'

The distance of Crane Two's load block from the hoist point, then, is 15 feet (shown in *Figure 12*).

C = A − B
C = 25' − 10'
C = 15'

The distance of Crane Two's load block from the hoist point, then, is 15 feet (shown in *Figure 12*).

Step 3 Determine the load that each crane will carry:

Crane One load = (C ÷ A) × load weight
Crane One load = (15' ÷ 25') × 143,000 lb
Crane One load = 0.60 × 143,000 lb
Crane One load = **85,800 lb**

This value is only 71.50 percent of Crane One's minimum net capacity of 120,000 lbs.

Crane Two load = (B ÷ A) × load weight
Crane Two load = (10' ÷ 25') × 143,000 lb
Crane Two load = 0.40 × 143,000 lb
Crane Two load = **57,200 lb**

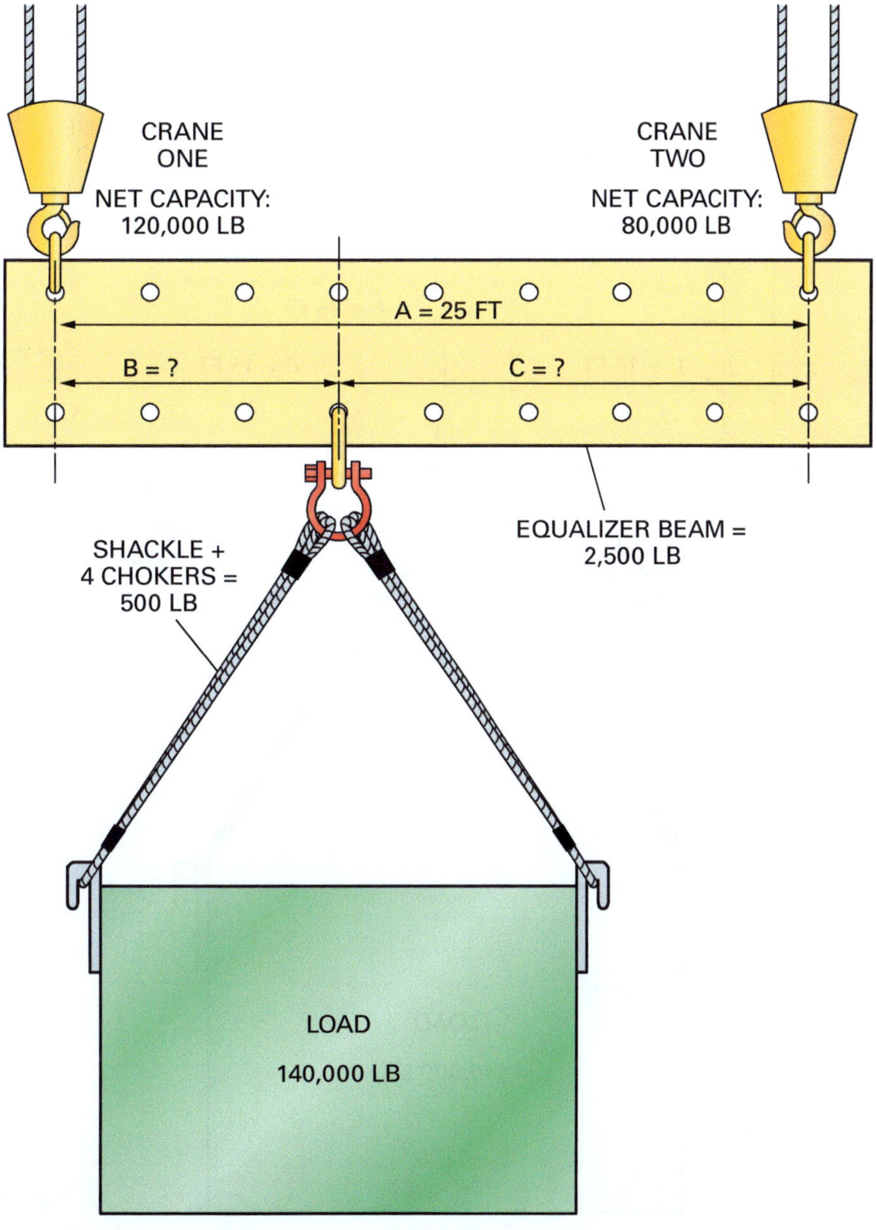

Figure 11 Dual-crane lift with equalizer beam – determining distances B and C.

This value is only 71.50 percent of Crane Two's minimum net capacity of 80,000 lbs. In this example, each crane is carrying the same percentage of their respective net capacity.

As you can see, even these calculations for a simple duo-crane vertical lift are not simple. A far more complex operation, such as slewing a load between cranes or rotating a long horizontal load into an upright position, can involve very complex dynamic calculations to identify the limiting configurations of the cranes during the lift. Crane operators will not always be involved in this part of critical lift planning, but it is important for them to understand how lift planners perform these tasks.

1.4.0 Operations Near RF Transmitters

Radio and TV transmission towers, radars, microwave transceivers, and cellular towers all produce radiofrequency (RF) energy. To transmit RF energy, electronic equipment rapidly oscillates electrical potentials within a transmitter antenna, producing electromagnetic waves radiating at frequencies also found in radio bands. After transmission, when these waves traveling through the air encounter a significant conductive object, the object can act like a receiver antenna. Antennas convert RF energy into other forms of energy, such as electricity or heat. This can be a problem for mobile cranes.

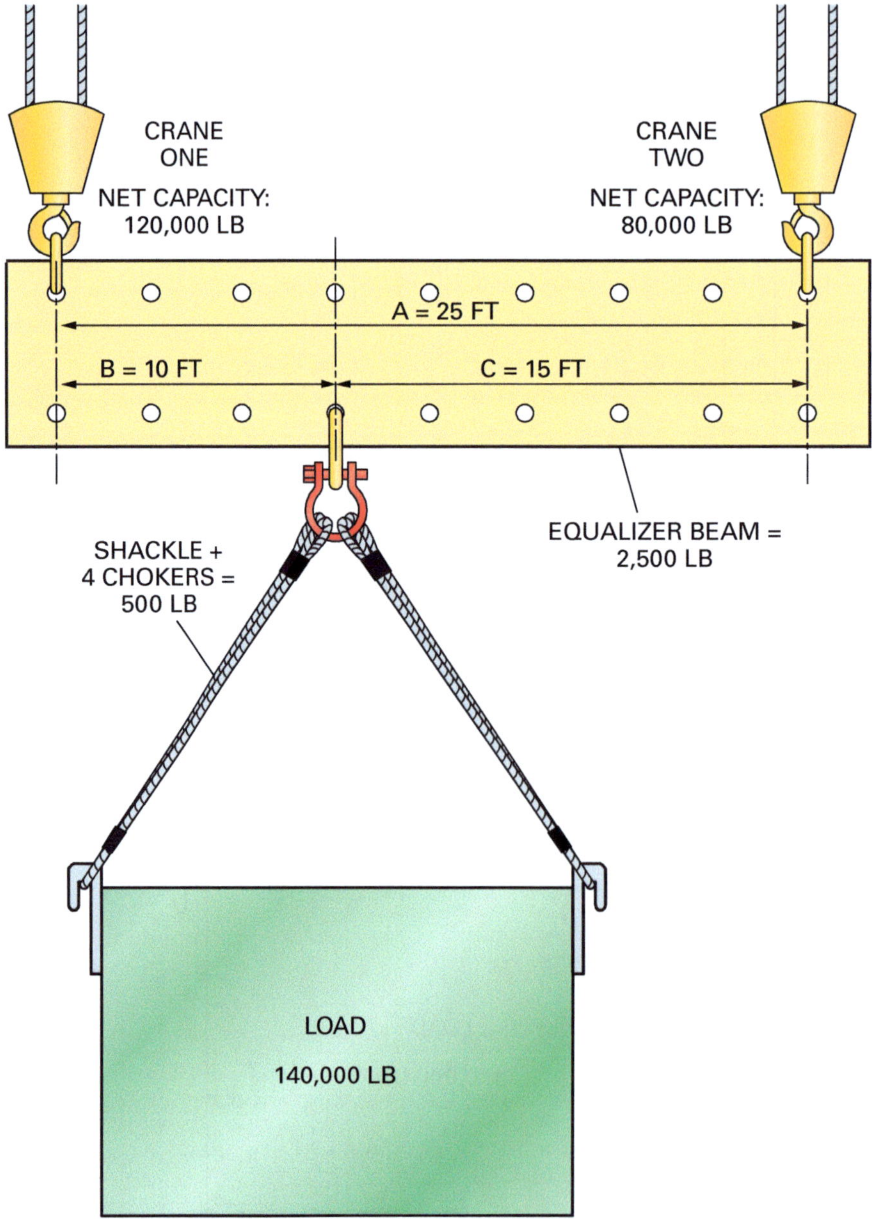

NOTE: FIGURE NOT TO SCALE.

Figure 12 Dual-crane lift with equalizer beam showing calculated distances B and C.

Medium frequency (MF) radio waves, such as those emitted by AM radio stations, have wavelengths between 100 and 1000 meters (328 ft–3,280 ft); high-frequency (HF) RF waves have wavelengths between 10 and 100 meters (32.8 ft–328 ft). High-powered government and private sector radio communications use the HF band. Mobile cranes have the dimensions and structure necessary to unintentionally receive this RF energy. Operators all over the world have reported shocks and burns when working on cranes located within strong RF fields.

NOTE

More common sources of RF energy, such as FM radio, VHF and UHF TV stations, microwave networks, and cellular towers, produce wavelengths that are far smaller than the dimensions of cranes. Normally, these sources don't present shock or thermal hazards to operators. However, there have been reports that interference with crane electronic control systems and LMIs can occur.

Prior to crane operations near transmitter towers (*Figure 13*), where there is risk of the crane generating an RF electrical voltage, the transmitter should be de-energized if possible. Otherwise, engineers must perform tests to determine if an induced electrical voltage is present on the crane. For powerful regional AM radio stations, the transmitter tower can be several miles away from the crane and still cause problems. In most studies, the first indications of a problem were shocks, tingling sensations, or burns. None of the reported injuries was life-threatening.

Studies have shown that there is no single reliable method for eliminating RF induced electrical voltages or solving interference with crane electronics. When RF induced voltages may be present, workers should take the following precautions:

- Ground the entire crane structure with conductors to one or more grounding rods.
- Provide affected equipment with an electrical ground attached directly to the upper rotating structure supporting the boom.
- Attach grounded jumper cables to materials lifted by boom equipment after detecting an electrical voltage. Provide workers with nonconductive poles having large alligator clips, or other similar protection, to attach the ground cable to the load.
- When operators or workers receive shocks or detect RF voltages, remove combustible and flammable materials from the immediate area prior to operation.

Figure 13 Cranes working near RF transmitters.

- For extended operations within an RF field, it may be possible to insert an insulated fitting above the load block or ball to isolate the load from induced voltages in the crane.
- Greasing the hoist sheaves may provide some temporary insulation of the load from the crane.

WARNING!

While induced radiofrequency (RF) voltages from transmitters are not an electrocution hazard, they can create sparks between objects or people. The sparks can initiate fires or explosions if combustible materials are nearby. They can occasionally result in minor burns and shocks from the crane or its load.

1.5.0 Demolition Operations

29 *CFR* Subpart T, 1926.859, *Mechanical Demolition*, provides the safety regulations for **demolition** of structures using wrecking balls or clamshell devices. In balling-demolition operations, workers suspend a round- or pear-shaped wrecking ball (*Figure 14*) from the hoist rope of a crane. The crane operator swings the ball by various means to strike a structure and cause it to collapse. Demolition cranes may use a cube-shaped wrecking block for certain tasks.

The OSHA Mechanical Demolition standard includes the following key requirements:

- The weight of the demolition ball cannot exceed 50 percent of the crane's minimum-rated capacity during operations or 25 percent of the hoist line's nominal breaking strength, whichever is less.
- The boom and hoist line must be as short as possible.

Figure 14 Pear-shaped wrecking ball.

- The ball connection shall be by a swivel to prevent twisting the hoist line, and securely attached.
- Specific precautions are necessary to avoid injury to personnel and uncontrolled collapse of walls or floors during demolition.

The crane's boom and structure receive dynamic loading by the swinging and impact of the wrecking ball. This is the greatest challenge of the job. The magnitude of dynamic loading varies widely depending on wrecking ball usage, the skill of the operator, and the impact resistance of the structure. Because of these factors, operators must work well within the manufacturer's guidelines. Demolition experts state that telescopic cranes should not engage in swinging ball demolition due the stresses placed on hydraulic seals and components. The task is almost exclusively performed by lattice-boom cranes.

Usually, several heavy impacts are required to cause serious cracking. Follow with lighter impacts until part of the object collapses. Repeat the process according to the demolition plan.

WARNING!

Position cranes used for high-structure demolition as far as practical from the structure during demolition. In some cases, a portable concrete barrier between the structure and crane is needed to help prevent debris from striking the crane. Workers must rope off the area around the demolition site and crane to prevent injury or death due to flying debris or sudden structure collapse. Demolition preparations should include compliance with environmental protection requirements, such as removal of toxic materials and/or water suppression of dust.

CAUTION

Certain manufacturers do not recommend the use of their cranes for demolition operations. Always consult the crane manufacturer when planning demolition work.

1.5.1 Vertical-Drop Balling

In this type of operation, demolition occurs by raising the ball vertically and then allowing it to drop by gravity. The ball attaches to the hoist rope of a crane with a free-fall capability. It should be raised a short distance, held by the hoist brake, and then permitted to fall until it lands. Do not apply the brake until the ball has stopped moving; otherwise, structural damage or crane tipping could result. A limit line prevents the ball from swinging beyond the permitted load radius for the operation.

WARNING!

To prevent crane damage or tipping during vertical-drop balling, do not apply the crane brake until the ball has struck and stopped. If the ball is still moving due to a glancing blow, or if it has passed through an object, it may be possible to apply the crane brake slowly to bring the ball to a stop. Make sure that sufficient rope is on the hoist drum to allow for any excessive and unexpected ball travel.

Where the point of impact is above ground level, there is the possibility of the ball missing the intended point of impact. It may pass through the object or bounce off. In this situation, operators should keep the drop distance to a minimum. At the intended point of impact, there should be sufficient length of rope left on the crane hoist drum to allow stopping the ball slowly with the brake to avoid endangering the crane.

Some cranes are designed for the stresses caused by heavy vertical-impact loading. The crane shown in *Figure 15* is a high-cycle dynamic compactor used for earth compaction at depths up to 40 feet. It is also useful for certain types of vertical-drop demolition. It is constructed entirely from high-strength, low-alloy steels. The upperworks are hard-mounted to the carbody and can't rotate. This eliminates damage to swing-circle bearings caused by heavy loading and unloading forces. Front and rear pendants secure the boom to prevent boom whip due to the sudden dynamic loads. A high-speed hoist uses a large-diameter hoist drum and disc-style braking for the drum to eliminate overspooling problems and increase cycle speeds. These cranes can use drop weights up to 20 tons from heights exceeding 100 feet. Regardless of whether a crane is technically a high-cycle dynamic compactor, the characteristics of a lattice boom with the upperworks hard-mounted to the carbody are preferred for this type of work.

1.5.2 Swinging Ball: In-Line with Boom

This demolition procedure uses a pulling rope attached to the ball from a second drum on the crane to pull the ball in towards the crane boom. Upon releasing the pulling rope, the ball swings outward, in the plane of the boom, to strike the object (*Figure 16*). The pulling rope serves a second purpose as a limit line.

This method is the most advisable for demolishing low-level objects because it produces the least strain on the crane. However, operators should always take precautions to avoid allowing the ball to swing outward beyond the

Figure 15 Dynamic earth compactor.

Figure 16 In-line ball demolition showing the pulling rope.

maximum allowed operating radius. With a swinging ball, the operating radius is the distance from the crane's center to the ball's center of gravity at its farthest point from the crane. This means the operating radius can change on each swing.

1.5.3 Swinging Ball: Slewing

Demolition by slewing is necessary for taller structures when little else is practical. Setup suspends the ball about ten feet below the boom head. Operators engage the slew motion, causing the ball to swing in an arc and strike the object. Slewing should be disengaged and the slew brake applied to stop the boom at a point in line with the point of impact. As with the in-line swing method, the operator should use a limit line to prevent the ball from swinging out of the safe radius. Professionals recognize that extended demolition operations by slewing is not advisable because it results in large torsion (twisting) loads on the boom and superstructure. However, experienced operators can reduce stress and wear though coordinated control of the boom and hoist.

1.6.0 Vacuum Lifting

Operations using a **vacuum** lifter (*Figure 17*) are common when lifting loads that have a smooth, rigid surface suitable for vacuum-lifting pads. The load surface must be clean, flat, dry, and nonporous, or nearly so. Operators may use a vacuum-lifting device in place of a standard hook or as a below-the-hook device. Regardless of the attachment method, *ASME B30.20, Chapter 20-2, Vacuum Lifting Devices*, provides the relevant requirements for preparing and using such devices.

Because there is no physical attachment, vacuum lifting is less secure than hoisting using physically attached hooks and rigging. Vacuum equipment generates a lifting force by creating an airtight seal between the lift pads and the surface of the load. Any surface irregularity, pad leakage, or failure of the source of the vacuum results in loss of the lifting force—and potentially, a dropped load. This obviously has dangerous consequences to people below the lift as well as to the load and other objects below it. For this reason, standards specify detailed requirements for the design, construction, testing, and operation of vacuum-lifting devices.

Figure 17 Crane using a vacuum lifter to place a sheet of plate glass.

CAUTION

Loads hoisted by vacuum-lift systems are often large, flat sheets of material. If the load is dropped from a significant height, it could develop aerodynamic lift, which causes it to glide sideways in an unpredictable way (like a falling leaf). This increases the hazard to people and structures outside the boundaries of the lifting area. Operators and planners should consider this potential when planning high lifts with vacuum-lifting devices.

The following requirements apply to vacuum lifting:

- The standard provides detailed requirements for label plate and other markings that must be visible on vacuum-lifting equipment.
- Workers and operators must complete periodic and in-use inspections of vacuum-lifting equipment.
- For horizontal-surface lifts, the load weight must not be greater than 50 percent of the **ultimate pad capacity (UPC)**.
- For vertical-surface lifts, the load weight must not be greater than 25 percent of the UPC.
- The crane operator should be fully trained in the use of the equipment. The operator must also demonstrate proper use of the equipment before assuming responsibility for lift operations.

Precast Concrete Construction

Precast building construction uses concrete panels that are tilted up and supported until multiple panels are positioned, each supporting the other. Workers must exercise extreme care in the rigging and lifting of precast panels. Panels must often be lifted from multiple points on the panel to minimize the stress. Spreader bars are commonly used. Sling angles should never be less than 45 degrees, due to the inherent weakness of the lifting anchors embedded in the concrete. These panels must often be rotated in the air, which sometimes requires two cranes working together.

Figure Credit: Robert I. Carr, Ph.D., P.E.

Did You Know?

Getting Something from Nothing

A pure vacuum consists of nothing—it's a volume of space in which nothing exists. Such a vacuum is a theoretical idea. Not even deep space is a true, perfect vacuum. Astronomers estimate that cool space between the stars have a million molecules per cubic centimeter (cm^3), which, compared to earth's atmosphere at sea level (10^{19} molecules per cm^3), is still a pretty good vacuum!

Vacuum lifters work by using a vacuum to develop a force that presses against the load strong enough to lift it. This is possible by pumping out most of the air between a specially constructed pad and the load's surface. The edges of the pad form a tight seal against the surface. The outside air pushes against the lifting pad because there is little air pressure underneath it. The force that air pressure exerts on the pad is equal to the area of the pad times the difference between the atmospheric pressure and the remaining air pressure under the pad.

If a perfect vacuum existed under the pad, the pressure difference would be 14.7 pounds per square inch (psi) minus 0 psi, or 14.7 psi. Assuming a lifting pad has an effective area of 100 square inches (in^2), then the maximum vacuum force it could generate is:

$$\text{Force} = \text{area} \times \text{pressure}$$
$$\text{Force} = 100\ \text{in}^2 \times 14.70\ \text{lb/in}^2$$
$$\text{Force} = 1{,}470\ \text{lb}$$

Typical vacuum lifting pads operate at about 24 inches of mercury (in Hg) vacuum, or 2.9 psi. So the example lifting pad operating at this vacuum would produce about 1,180 pounds of force. That's something from basically nothing!

- In use, the operator is responsible for all aspects of the vacuum lifter and its support systems. The standard notes that the operator is not responsible for hazards or conditions adversely affecting the equipment's operations that are not under his/her direct control.

1.6.1 Factors Affecting Lifts

ASME B30.20 identifies the following factors that can affect the efficiency and load capacity of a vacuum lifter:

- *Load rigidity and strength* – Depending on the nature of the materials lifted, they may deform under the pressure difference created by the vacuum pads. This can lead to uneven loading of the lifter. More rigid materials are less subject to this condition. Load rigidity can vary depending on its orientation to the lifter. Lifting can possibly fracture materials that cannot handle the stress at the vacuum pads. A loss of the vacuum seal because of a material failure will release the load.
- *Load surface conditions* – The vacuum pad requires a tight seal against the load's surface to maintain the vacuum. Any roughness, unevenness, dirt, oil, or other flaw that reduces the effectiveness of the seal can significantly lower lifting capacity. Similarly, when load surfaces are vertical, the effectiveness of the lifter depends significantly on frictional force between the pad and the surface. Anything that decreases that friction will increase the likelihood of slippage.
- *Load overhang* – A lifter contacts the load only within the surface area of the lifter pads. If the load extends beyond the pads, there is a risk the load can flex or bend at the perimeter. Depending on the strength of the load, it could break due to deflection stress, or the load could peel away from the lifting pad. Vacuum lifters should have sufficiently large dimensions and enough lifting pads to prevent excessive bending of the load due to overhang.
- *Angle of the load* – Cranes often lift sheet-like loads, (*e.g.*, large plate glass windows or prefabricated wall panels) to the point of installation. Positioning these types of loads requires a vertical or near-vertical orientation. When using vacuum lifters, the weight of the load is acting perpendicular to the direction of the force applied by the vacuum lifting pads, and, and as noted above, friction is the only thing keeping the load from slipping from the lifter.

For horizontal lifts, the weight of the load acts in line with the vacuum force exerted by the lifting pad. Friction between the pad and the load has no influence on the lifting force in this orientation. As the angle of the load changes from the horizontal, an increasing portion of the load's weight acts perpendicular to the vacuum force (*Figure 18*), and friction becomes more important in preventing the load from slipping. When the lift is vertical, all the load's weight is perpendicular to the vacuum force and only friction holds the load in place. As long as the friction between the lifting pad and the load is equal to the load's weight plus any other dynamic forces, the load is secure.

If the pad seal is some form of organic polymer and the load is a steel plate, the **coefficient of friction** between the two materials could be as high as 0.6. This means that a lifter pad could lift a plate of steel in the vertical orientation weighing no more than 6/10 of the vacuum force of the pad when attached to the plate. Frictional limitations are the reason standards restrict vertical lifts to 25 percent of the UPC to guide manufacturers in setting their equipment capacities.

Other factors related to determining lifting capacity are identified as follows:

- *Number of pads* – The capacity of a lifter is proportional to the number of lifting pads with which it is equipped. A lifter having eight pads rated with a certain UPC will have twice the capacity of a lifter equipped with only four of the same pads. Operators should be aware that the malfunction of even one pad on a multi-pad vacuum lifter can seriously reduce load security.
- *Load temperature* – Handling very hot or very cold loads with a vacuum lifter is not common. However, a steel plate exposed to the noon sun can become very hot, and handling the hot plate with a vacuum lifter can be a problem. Extremes of temperature can seriously affect the performance of the vacuum pad seals. Cold temperatures can reduce the flexibility of the materials, preventing them from forming a good seal against the load. High temperatures can damage the seals, degrading or even melting them. A change in composition can also affect the coefficient of friction between the lifting pad and load material.
- *Elevation and vacuum available* – The vacuum force is created as the equipment removes air from under the lifting pad. Atmospheric pressure presses in on the pad's surface in an attempt to equalize the pressure difference. The vacuum force is the difference between the inside and outside pressures, multiplied by the area of the pad. At higher elevations the atmospheric pressure is reduced, which lowers the lifting capacity.

Manufacturers typically rate the capacities of their vacuum lifters for sea-level operation unless otherwise stated in the equipment manual. Atmospheric pressure is maximum at sea level. At higher elevations, air pressure gradually decreases. For example, in Denver, Colorado, the average atmospheric pressure is only 82 percent of sea level. Therefore, operators should derate a vacuum lifter's sea-level capacity at least 18 percent for operations at this location.

Every vacuum lifter is fitted with a device that gives a visual indication of the state of the vacuum at any time. If the system begins to fail and the vacuum force falls, the operator must immediately warn everyone in the vicinity and land the load as quickly as possible.

1.6.2 Lifting Guidelines

Prior to the first use of a vacuum lifter, or before the device is placed back into service after repairs, a load test must be conducted. The pad contact surface must be clean and free of any loose particles. The load rating should not be more than 80 percent of the maximum load sustained during the test. The test load must not exceed 125 percent of the rated load of the system unless the manufacturer or a qualified individual recommends otherwise.

For normal operations, the operator must make sure that the pad surface is clean, the vacuum level has reached the required level, and there are no vacuum line kinks, twists, or loops wrapped

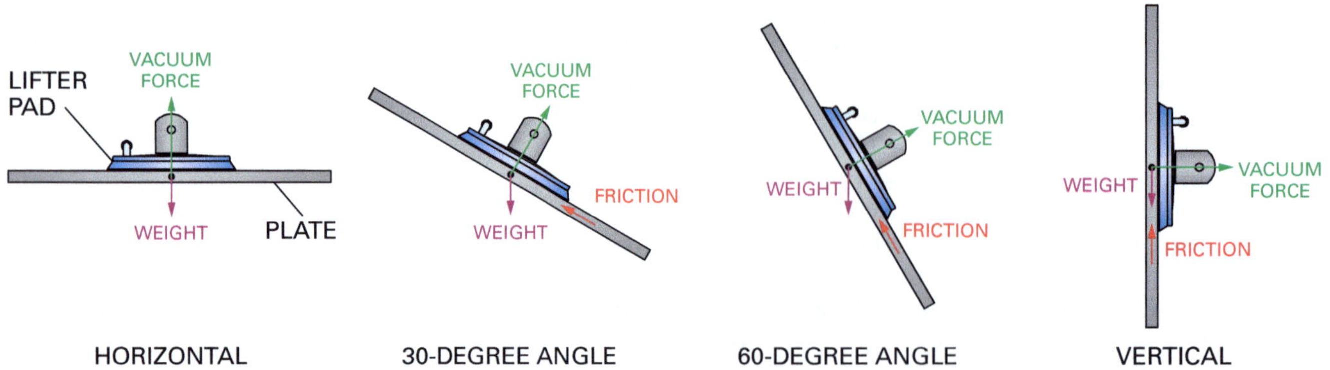

Figure 18 Relationship between load angle and the friction force acting on the load.

around portions of the lifter. Operators must ensure the load is evenly distributed and that the lifter adequately supports the load. The operator is required to lift the load a few inches to make sure the vacuum-lifting device is working correctly and the vacuum is stable. The operator must be gentle with the load, without jerking or snatching it, and must never carry a load over people.

WARNING!

Due to possibility of a sudden vacuum failure, personnel should never position themselves under or near a vacuum lifter load.

1.7.0 Electromagnet Lifting

Another below-the-hook device used for lifting and handling **ferrous metals** and other magnetic materials is the **electromagnet** lifter (*Figure 19*). These devices use 250–500 VDC applied to a large solenoid (coil) of wire housed inside the lifting device. An external AC power supply converts the current to DC and delivers it through cables rigged to the electromagnet. Some magnets are powerful enough to lift thick, solid, flat, or block-shaped ferrous loads of nearly 60 tons. Load characteristics, including the type and shape of the metal, greatly affect the amount of load that an electromagnet will lift.

WARNING!

Electromagnet lifting poses an electrocution hazard. Before connecting or disconnecting a crane electromagnet, secure the power source or open the supply breaker. Use appropriate safety precautions, such as lockout/tagout procedures.

As with vacuum lifters, electromagnets have no direct physical connection to the load. They depend on a magnetic field interacting with ferrous metals. The magnetic field exists as long as an electrical current is flowing through the coil. If any interruption of the current occurs, the magnetic field is quickly lost, and the load drops. A backup power supply can give the operator time to land the load, in some cases.

ASME B30.20, Chapter 20-4, Remotely Operated Lifting Magnets, provides the requirements for the design, construction, testing, and operations for electromagnets. Many of the principles are similar to those for vacuum lifters, differing only in the details. Some of the requirements found in the standard are listed as follows:

Figure 19 An electromagnet used for lifting scrap metal.

- Lifting magnets should be weather resistant
- Cab status indicators should have backup bulbs to avoid false indications from burned out lamps.
- Control voltages for electromagnets shall not exceed 150 VAC or 300 VDC.
- If radio-controlled, loss of the radio signal will not result in de-energizing the magnet.
- Initial, periodic, and in-use inspections and testing will be performed by qualified individuals and documented in maintenance records.
- Crane operators must receive training in all aspects of electromagnet use. Operators must complete a practical demonstration before operating independently.
- All persons must be clear of the magnet, and controls must be off or in neutral before energizing the lifter.
- Operational practices for electromagnet lifters are similar to those for vacuum lifters.

1.7.1 Factors Affecting Lifts

ASME B30.20 identifies a number of factors that affect operations with an electromagnet lifter that parallels those for vacuum lifters. These factors are identified as follows:

Draglines

Figure Credit: © iStock.com/sakakawea7

Draglines are used to excavate soil and similar materials and deposit it in trucks, on a nearby spoil pile for later loading, or into a pile to construct a levee. The operator can dig a pit significantly deeper than the soil level surrounding it without the crane entering the pit. Crawler cranes with lattice booms are typically the best choice for this task. Longer boom lengths require smaller buckets. Dragline operations can be somewhat tricky due to the added resistance of digging through the ground in addition to the weight of the material in the bucket.

The crane operator must swing the bucket to cast it out ahead of the boom. Working a dragline requires multiple drums. The operator uses the hoist line to control the depth of the dig. The dragline is released to dump the bucket contents. Operators must consult several different resources to determine the optimum depth of cut for the material being excavated to maximize productivity.

- *Load thickness* – The measure of the strength of any magnet is the number of magnetic lines of force passing through an area of space. The greater the number of lines of force, the stronger the magnetic force being generated. Similarly, the magnetic force depends on the number of lines of force from the magnet passing through its metal. The number of lines of force within the load cannot be greater than that generated by the magnet itself. The thicker the ferrous load, the greater the number of lines of force it can contain, up to the number of field lines produced by the magnet. Further increases in load thickness don't result in any additional magnetic force. In practice, the electromagnet lifter's rated capacity is greater for thicker loads than thinner loads.
- *Load alloys and metals* – Low-carbon steels, such as SAE 1020 steel, are nearly as effective magnetically as pure iron. However, with exception of cobalt and nickel, most pure nonferrous metals and their alloys are nonmagnetic, and many other alloys of iron, such as stainless steel, contain nonmagnetic materials in a form that produces a lower affinity for magnetism. Nonferrous metals include copper, aluminum, brass/bronze, lead, zinc, manganese, and tin.
- *Load surface conditions* – Anything that creates an air gap between the magnet and the load disrupts the magnetic lines of force and thus reduces the lifting capacity of a magnet. Paper, dirt, rust, paint, and scale act the same as air. A rough surface finish is the same as dirt because it creates an air gap between the magnet and the load.
- *Load length or width* – As the length or width of the load increases, it ceases to remain flat when lifted and the edges droop. This drooping or sagging of the load, called *peel*, can create an air gap between the load and the magnet, reducing the lifting capacity. Multiple magnets attached to one or more spreader beams are available for lifting long or wide loads.
- *Attitude of load* – As the angle of the lifting surface changes from horizontal to vertical, the lifting capacity of the magnet reaches a minimum and becomes dependent upon the coefficient of friction for the lifting surface (0.5–0.8 for unpolished steel to steel).
- *Load temperature* – The temperature of the load can cause damage to the magnet and, if high enough, can even reduce the magnetic affinity of the load.

1.7.2 Lifting Guidelines

The lifting charts supplied by the magnet manufacturer show the effects of air gap, load thickness, and load overhang/peel on lifting capacity. As the thickness of the load or the density of metallic pieces decreases, so does the capacity of the magnet. Guidelines show the maximum weight or load size the magnet can lift for each thickness under varying air-gap conditions. Do not exceed either the maximum weight or size for each thickness.

When first lifting materials with a magnet, lift the load several inches to confirm that the magnet has a secure grip on the load. Adjust the control power to maximum before raising the load higher than the trial lift. After making these

Did You Know?

Magnetic Metals

Iron is fourth most common element in the earth's crust and the second most common metal after aluminum. Many alloys containing iron are also magnetic. However, iron isn't the only metal that a magnet can attract. There are two other elements that are magnetic in their natural state—cobalt and nickel. Together, these three elements make up the ferromagnetic elements and exhibit the property of ferromagnetism. Ferromagnetism can also occur when certain metal and nonmetal elements combine into compounds. These can produce very strong permanent magnets as well as efficient electromagnet cores.

checks, raise the load to the desired height, and follow through on the lift. The adjustable input control should remain in the Full or Full On position until removing the load from the magnet. All movements are to be smooth, free of jerks, and clear of any obstructions. The operator must not move the load over people.

When using a magnet to pick up scrap metal, work from deep piles. Allow the magnet's weight to settle on and compress the pile before energizing it. If this is not done, two things happen. First, smaller pieces will jump on the magnet and block the rest of the metal, preventing a full load. Second, unwanted materials around the perimeter can be attracted to the magnet. After you turn the power on, let the magnet sit for 5–10 seconds to establish a steady lifting current through the magnetic coil.

WARNING! Due to possibility of a sudden electrical failure, personnel should never be under or near an electromagnet load.

Electromagnets, like some other devices that develop a significant amount of heat, have a rating related to their design **duty cycle**. Do not exceed the duty-cycle rating of the magnet. Exceeding the rating will reduce the lifting capability because of the higher temperatures that develop in the magnet. A hotter magnet increases the electrical resistance in the electromagnet coil, reducing current flow and the magnetic field it produces. The higher temperatures also can reduce the operating life of many of the materials in the lifter.

The duty-cycle rating is calculated in percent by the following equation:

$$\text{Duty cycle} = \frac{\text{Time on}}{\text{Time off} + \text{time on}} \times 100$$

Time is measured in minutes for electromagnets. The following is an example calculation for determining the duty cycle (3 minutes on, 1 minute off):

$$\text{Duty cycle} = \frac{\text{3 minutes}}{\text{1 minute} + \text{3 minutes}} \times 100$$

$$\text{Duty cycle} = \tfrac{3}{4} \times 100 = \mathbf{75\%}$$

NOTE From an ambient starting temperature, it is normal for an electromagnet to heat up to an operating temperature of 220°F to 270°F (104–132°C) in use. However, at normal operating temperatures the magnet can lose up to 25 percent of its cold capacity. Operators should be aware of this normal change in load capacity.

To avoid damage to the magnet, consult the manufacturer's manual for the recommended maximum continuous energized operation. To prolong the life of a lifting electromagnet, follow these user guidelines:

- Minimize the operating duty cycle.
- Avoid exposure to high temperatures.
- Do not exceed design control current voltages.
- Keep the magnetic lifter dry. Never cool it with water, which could short out the magnet and shock someone.
- Never disconnect the magnet from its power source while energized. Electrical arcing may occur, damaging hardware and causing flash burns.

1.8.0 Cold Weather Operations

Cold weather (*Figure 20*) places additional stress on cranes. Lubricants within the crane become more viscous, reducing their lubricating properties. The engine takes longer to warm up. Dimensional tolerances are tighter, so wear increases when attempting to operate cold equipment. It is

Figure 20 Operating cranes in cold weather requires additional precautions.

important to consult the manufacturer's manual for operating in cold weather, and monitor the hydraulic fluid temperature gauge prior to operating when weather conditions warrant.

Crane operators must approach operations in extreme cold (ambient temperatures below 10°F or –12°C) with caution. Some metals become prone to **brittle fracture** when cold, although the materials in cranes generally have good fracture toughness. Even so, operators should minimize shock loading, which could lead to catastrophic failure of a part or wire rope. Manufacturers may also require derating a crane for very low temperatures.

The following are example guidelines for restricting crane operations in cold temperatures as suggested by one construction company:

- *Ambient temperatures below –40°F (–40°C):* Prohibit all lifts except for emergency response. Lift category is automatically a critical lift.
- *Ambient temperatures –22°F to –40°F (–30°C to –40°C):* Derate cranes by 40 percent for any lift. Supervisors and operators should consider suspending lifting activities.
- *Ambient temperatures –5°F to –22°F (–15°C to –30°C):* Avoid shock loading cranes and rigging. Monitor hydraulic systems closely. Derate cranes 25 percent for critical lifts.

CAUTION

Work in extreme cold is hard on people as well as equipment. Extreme wind chill affects the ability of operators, riggers, and signal persons to do their jobs. Stop operations if these individuals can't perform their tasks safely under the current conditions.

Operators of cranes, especially those equipped with hydraulic systems, should follow these guidelines when working in cold temperatures:

- Do not start or operate a crane when the air temperature is below the minimum specified by the manufacturer.
- Do not touch metal surfaces that could freeze to your skin.
- Clean the crane, especially the boom, of all ice and snow. Snow accumulation can add undocumented weight to the boom, derating it's load chart capacity by some unknown amount.
- Allow ample time for hydraulic oil to warm up.
- Keep the engine at average speed until the hydraulic oil has reached 50°F (10°C).
- Carry out all crane functions at least twice once the oil reaches operating temperature in order to displace the cold oil in all parts of the hydraulic system.

Wind Chill

Wind chill is something that living organisms experience when exposed to moving air at temperatures below normal body temperature. The wind chill effect has two aspects: the organisms (workers, in this case) feel colder than they would if there were no breeze, and the moving air removes heat from the skin surface, cooling the body and placing more stress on the body's temperature regulatory mechanisms. The wind chill effect increases with colder temperatures and higher wind speeds. Safety supervisors should make wind chill charts (such as the one shown here) available when operating in cold weather.

It is important to note that wind chill cannot cool a person or object colder than the temperature of the air. For example, even though the wind chill chart says that –15°F air blowing at 30 mph has a wind chill of –46°F, the temperature of hydraulic oil systems exposed to such a wind will be no lower than –15°F.

WIND CHILL CHART

Wind (mph) \ Temperature (°F)	40	35	30	25	20	15	10	5	0	-5	-10	-15	-20	-25	-30	-35	-40	-45
Calm	40	35	30	25	20	15	10	5	0	-5	-10	-15	-20	-25	-30	-35	-40	-45
5	36	31	25	19	13	7	1	-5	-11	-16	-22	-28	-34	-40	-46	-52	-57	-63
10	34	27	21	15	9	3	-4	-10	-16	-22	-28	-35	-41	-47	-53	-59	-66	-72
15	32	25	19	13	6	0	-7	-13	-19	-26	-32	-39	-45	-51	-58	-64	-71	-77
20	30	24	17	11	4	-2	-9	-15	-22	-29	-35	-42	-48	-55	-61	-68	-74	-81
25	29	23	16	9	3	-4	-11	-17	-24	-31	-37	-44	-51	-58	-64	-71	-78	-84
30	28	22	15	8	1	-5	-12	-19	-26	-33	-39	-46	-53	-60	-67	-73	-80	-87
35	28	21	14	7	0	-7	-14	-21	-27	-34	-41	-48	-55	-62	-69	-76	-82	-89
40	27	20	13	6	-1	-8	-15	-22	-29	-36	-43	-50	-57	-64	-71	-78	-84	-91
45	26	19	12	5	-2	-9	-16	-23	-30	-37	-44	-51	-58	-65	-72	-79	-86	-93
50	26	19	12	4	-3	-10	-17	-24	-31	-38	-45	-52	-60	-67	-74	-81	-88	-95
55	25	18	11	4	-3	-11	-18	-25	-32	-39	-46	-54	-61	-68	-75	-82	-89	-97
60	25	17	10	3	-4	-11	-19	-26	-33	-40	-48	-55	-62	-69	-76	-84	-91	-98

Frostbite Times: 30 minutes; 10 minutes; 5 minutes

Figure Credit: NOAA

- During freezing weather, frequently check all air tanks for water and ice buildup at air compressor unloader valves.
- Always handle propane tanks according to the supplier's instructions.
- Before lifting a load, perform a test hoist to determine if it is frozen to the ground.
- Be extremely careful when using aerosol spray or starting fluids containing ether or other volatile components that can cause explosions or fire. Note that the use of such products may be prohibited by the manufacturer or the employer.

In addition, truck-crane operations should follow these guidelines:

- Start the truck and run at the manufacturer's recommended idle speed for proper warmup.
- After a sufficient warmup time, depress the clutch pedal to engage the power take-off (PTO).
- At engine idle, release the clutch pedal slowly until fully engaging the PTO.
- With the PTO fully engaged and the truck engine running at idle speed, let the hydraulic system oil circulate.

During the warmup, do not race the engine and overspeed the hydraulic pumps. Pump cavitation and permanent damage may result. If at any time you hear any hydraulic pump noise, such as metal grinding or popping, shut down the unit immediately. Check for clogging of the hydraulic line leading to the suction port of the pump. Hydraulic fluid can also become too thick to pump easily in extreme cold. Either condition can prevent the flow of hydraulic oil through the pump.

After the crane's fluid systems reach normal operating temperatures, prepare the crane for operation by completing the following steps, as applicable:

Step 1 Slowly extend horizontal outrigger cylinders approximately 6 inches and retract; extend again approximately halfway and retract; then extend full stroke.

Step 2 Follow the procedure on crane deployment, and extend and retract boom actuator cylinders.

Step 3 When completed, begin crane-slewing operation by slowly rotating approximately $^1/_8$ revolution one way, return to the previous position, and rotate in the opposite direction. Do this several times, and then rotate 90 degrees and return.

Step 4 If equipped with a hydraulic winch, begin winch operation by slowly opening the control valve, allowing several revolutions in one direction. Then, reverse the control valve, and operate several revolutions in the opposite direction. Repeat this several times, allowing longer operation intervals.

CAUTION

Engage valves slowly during the warmup cycle to prevent sudden hydraulic pressure spikes or hammer, which may damage hydraulic components.

Additional Resources

ASME Standard B30.5, Mobile and Locomotive Cranes. Current edition. New York, NY: American Society of Mechanical Engineers.

ASME Standard P30.1, Planning for Load Handling Activities. Current edition. New York, NY: American Society of Mechanical Engineers.

Guide to Mobile Crane Safety Management. 2015. Specialized Carriers and Rigging Association (SC&RA). Centreville, VA: SC&RA.

29 *CFR* Subpart CC, 1926.1431,*Hoisting Personnel.* **www.ecfr.gov**

29 *CFR* Subpart T, 1926.859,*Mechanical Demolition.* **www.ecfr.gov**

1.0.0 Section Review

1. Which of the following characteristics does *not* alone require categorizing a lift as critical?
   a. Complicated, close-tolerance installation
   b. Hazardous load
   c. Extremely heavy load
   d. Repetitive lift

2. Which of the following statements about personnel lifts is *true*?
   a. Personnel lifting equipment must be specially designed and tested before use.
   b. Lifts of people into tanks is prohibited.
   c. Personnel lifts are routine enough to be considered standard lifts.
   d. LMI instrumentation does not need to be operational, since personnel lifts are low-weight.

3. A blind lift occurs when the _____.
   a. operator needs corrective lenses for operating a crane
   b. signal person can't see the lift director's hand signals
   c. signal person can't see the crane
   d. operator loses direct view of the load

4. The radiofrequency (RF) transmitter with the greatest potential for RF-induced voltages is _____.
   a. FM radio
   b. AM radio
   c. Cellular networks
   d. UHF TV

5. The swinging and impact of a wrecking ball affects the crane due to _____.
   a. dynamic loading
   b. a short duty cycle
   c. the coefficient of friction
   d. arcing

6. Operators moving loads with a vacuum lifter should pay attention to the device that provides a visual indication of _____.
   a. the state of the vacuum force
   b. the horizontal or vertical orientation of the lifter
   c. the temperature of the load
   d. the amount of load overhang

7. The force used by an electromagnet lifter to hoist and move loads is _____.
   a. electrical force
   b. vacuum force
   c. magnetic force
   d. adhesive force

8. What do some crane companies consider an appropriate action when ambient temperatures are below –40°F (–40°C)?
   a. Add thinning agents to the hydraulic fluid.
   b. Prohibit all crane lifts except for emergency response.
   c. Derate critical lifts by 25 percent.
   d. Derate all lifts by 40 percent.

SUMMARY

This module covers the standards, regulations and safety guidelines for advanced crane operations that many operators are likely to encounter on a work site. OSHA regulations define critical lifts, which include personnel and multiple-crane lifts. Cranes operating near transmitter towers must take precautions to prevent the crane from unintentionally receiving RF energy. Operators must adhere to OSHA safety regulations for demolition operations, and ASME standards that govern vacuum lifting and electromagnetic lifts. Operations in extremely cold weather involve special precautions to prevent damage to equipment and injury to personnel.

Review Questions

1. Which of the statements regarding critical lifts is *true*?
   a. ASME standards identify certain operations as critical lifts by definition.
   b. The standards categorize repetitive lifts as critical lifts.
   c. Companies may identify conditions that qualify a lift as critical.
   d. Crane operations from barges are critical lifts by definition.

2. Which of the following statements about personnel lifts is *not* true?
   a. Personnel lifting equipment must use some means of enclosure to keep passengers and objects from falling out.
   b. Inspectors must perform a trial lift of the lift equipment at full weight.
   c. Equipment tilt during the lift must always be less than 10 degrees.
   d. Operators must not perform personnel lifts over water.

3. Which condition would *not* require categorizing a lift as a blind lift?
   a. Fog or dust
   b. Smudges on the windshield
   c. An obstruction between the load and the operator
   d. Load momentarily lifted above the operator's line of sight

4. The most reliable method for minimizing the generation of RF-induced voltages in a crane or its load is to _____.
   a. de-energize the applicable RF transmitter(s)
   b. grease the sheaves with an insulating lubricant
   c. ground the crane to the earth
   d. move the crane farther from the transmitter

5. Cranes are likely to pick up certain strong RF signals because _____.
   a. the crawlers provide a direct path to ground
   b. they contain materials such as glass and rubber
   c. they have the dimensions and structure necessary to unintentionally receive RF energy
   d. of their hoist cables

6. The demolition operation that involves swinging a wrecking ball by rotating the crane upperworks is _____.
   a. drop-balling
   b. swinging ball in-line with boom
   c. compacting
   d. swinging ball by slewing

7. When vacuum lifting flat objects oriented vertically, the load weight cannot exceed _____.
   a. 25 percent of the UPC
   b. 50 percent of the UPC
   c. 75 percent of the UPC
   d. 80 percent of the UPC

8. The capacity of a vacuum lifter can be reduced by _____.
   a. lower load temperatures
   b. more rigid load materials
   c. higher work site elevations
   d. smaller load angles

9. For a given amount of current supplied to an electromagnet lifter, what factor does *not* affect its load capacity?
   a. The thickness of the material
   b. Atmospheric pressure
   c. Load alloys and metals
   d. Attitude of the load

10. How does the wind chill affect crane operations in bitterly cold weather?
    a. It can be dangerous to operators, riggers, and signal persons.
    b. It causes out-of-vertical lifts.
    c. It can cool hydraulic oil below ambient temperature so it gels.
    d. It makes diesel engines harder to start than if there were no wind.

Trade Terms Introduced in This Module

Alloy: A metal that is a combination of two or more pure metals or other chemical elements.

Brittle fracture: A break in a metal object below a certain temperature specific for the metal, caused by a relatively small amount of stress acting on an existing flaw.

Cavitation: The creation of low-pressure regions within the intake area of a pump as the impeller rotates, forming pockets of gas that collapse violently, creating noise and pressure pulsations.

Coefficient of friction: A ratio that expresses a comparison between the force necessary to move an object over the surface of another material, and the pressure between the two materials.

Demolition: The controlled destruction of a structure.

Duty cycle: For repetitive operations, the ratio of the time a device is on or functioning compared to the total time of one cycle (time on plus time off), expressed as a percentage.

Electromagnet: A device consisting of an insulated electrical conductor (wire) coiled around a magnetic metal core. When current flows through the conductor, a strong magnetic field forms within the core.

Ether: A volatile, flammable, organic fluid that operators sometime use to assist in starting diesel engines, especially in cold weather.

Ferrous metals: Iron and its alloys that have relatively strong magnetic characteristics.

Limit line: When engaged in demolition operations using a wrecking ball, a length of wire rope attached to the ball from a point on the crane to prevent it from swinging out too far, providing a safety margin against overturning the crane or damaging the boom.

Radiofrequency (RF): Describes the lower frequencies of the electromagnetic energy spectrum in which radar, TV, AM/FM radio, cell phones, and all other forms of radio-wave communications and detection operate.

Tailing crane: A crane that helps position a load by turning or tilting it during lifting operations, but does not lift a significant part of the load's weight.

Ultimate pad capacity (UPC): Rating of the lifting force generated by a vacuum-lifter pad, based on the vacuum level and pad area.

Vacuum: The reduced presence or near-total absence of air or other gases, creating a low-pressure condition.

Additional Resources

This module is intended as a thorough resource for task training. The following reference materials are recommended for further study.

ASME Standard B30.5, Mobile and Locomotive Cranes. Current edition. New York, NY: American Society of Mechanical Engineers.

ASME Standard P30.1, Planning for Load Handling Activities. Current edition. New York, NY: American Society of Mechanical Engineers.

Guide to Mobile Crane Safety Management. Specialized Carriers and Rigging Association (SC&RA). 2015. VA: SC&RA.

29 *CFR* Subpart CC, 1926.1431,*Hoisting Personnel*. **www.ecfr.gov**

29 *CFR* Subpart T, 1926.859,*Mechanical Demolition*. **www.ecfr.gov**

Figure Credits

© iStock.com/Art Wager, Module Opener

© iStock.com/imamember, Figure 1

Link-Belt Construction Equipment Company, Figure 2

Mammoet USA South Inc., Figure 3

© iStock.com/Nikada, Figure 13

© iStock.com/ShooteRhys, Figure 14

Lampson International LLC, Figure 15

© Dtfoxfoto/Dreamstime.com, Figure 16

Robert I. Carr, Ph.D., P.E., SA01

© Bildagentur Zoonar GmbH/Shutterstock.com, Figure 17

© iStock.com/sakakawea7, SA02

© AngelPet/Shutterstock.com, Figure 19

© iStock.com/ewg3D, Figure 20

NOAA, SA03

Section Review Answer Key

Answer	Section Reference	Objective
Section One		
1. d	1.1.0	1a
2. a	1.2.0	1b
3. d	1.3.1	1c
4. b	1.4.0	1d
5. a	1.5.0	1e
6. a	1.6.1	1f
7. c	1.7.0	1g
8. b	1.8.0	1h

NCCER CURRICULA — USER UPDATE

NCCER makes every effort to keep its textbooks up-to-date and free of technical errors. We appreciate your help in this process. If you find an error, a typographical mistake, or an inaccuracy in NCCER's curricula, please fill out this form (or a photocopy), or complete the on-line form at **www.nccer.org/olf**. Be sure to include the exact module ID number, page number, a detailed description, and your recommended correction. Your input will be brought to the attention of the Authoring Team. Thank you for your assistance.

Instructors – If you have an idea for improving this textbook, or have found that additional materials were necessary to teach this module effectively, please let us know so that we may present your suggestions to the Authoring Team.

NCCER Product Development and Revision

13614 Progress Blvd., Alachua, FL 32615

Email: curriculum@nccer.org

Online: www.nccer.org/olf

❏ Trainee Guide ❏ Lesson Plans ❏ Exam ❏ PowerPoints Other ________________

Craft / Level: Copyright Date:

Module ID Number / Title:

Section Number(s):

Description:

Recommended Correction:

Your Name:

Address:

Email: Phone:

Glossary

A/D director: The individual specified by OSHA standards to be responsible for the assembly and disassembly of cranes.

Alloy: A metal that is a combination of two or more pure metals or other chemical elements.

Back hitch: The mechanism that permits adjusting the position of the boom gantry to maximize its leverage on the boom hoist cables.

Belly lines: Additional pendants attached to the boom-hoisting bail and secured at a location near the midpoint of a long boom to minimize sagging.

Boom angle: In load charts, the angle between the boom relative to the horizontal plane. On telescopic-boom cranes, boom angle instruments indicate the angle between the bottom of the base section and the horizontal with the boom loaded. On lattice-boom cranes, it is the angle between the center line of the boom and a horizontal line.

Boom length: Measured from the boom foot pins to the center of the boom-point sheaves. It does not include the length of an attached jib.

Boom stops: Devices that mechanically prevent the boom from exceeding a maximum boom angle, reducing the potential for backward tipping or structural damage.

Boom-extend mode: Refers to the order, and percentage, of extension telescopic-boom sections. The standard mode is to extend all segments the same percentage. Other modes may selectively extend certain sections to provide different capacities for a given load radius.

Boom-point elevation: Distance from the ground to the center of the boom-point sheaves.

Bridle slings: Lift rigging consisting of multileg, wire-rope or chain slings.

Brittle fracture: A break in a metal object below a certain temperature specific for the metal, caused by a relatively small amount of stress acting on an existing flaw.

Cantilever length: The length of a horizontal boom that projects beyond the support of the boom pendants, leaving that portion of the boom supported at only one end, like a cantilever.

Cavitation: The creation of low-pressure regions within the intake area of a pump as the impeller rotates, forming pockets of gas that collapse violently, creating noise and pressure pulsations.

Coefficient of friction: A ratio that expresses a comparison between the force necessary to move an object over the surface of another material, and the pressure between the two materials.

Configuration: A term that addresses all aspects of a crane's condition for a given lift that can include boom assembly and position, hoisting line and accessories, base arrangement, counterweights, and other factors addressed in a load chart.

Connector pin: Any solid, cylindrical, metal part used to fasten, support, or attach the various components and members of a crane's boom and other hoisting gear.

Demolition: The controlled destruction of a structure.

Diagonal struts: In a lattice boom with a rectangular cross section, pieces of structural pipe or angle iron welded to diagonally-opposite chords of the boom to stiffen the assembly.

Duty cycle: For repetitive operations, the ratio of the time a device is on or functioning compared to the total time of one cycle (time on plus time off), expressed as a percentage.

Effective weight: With reference to jib/boom extensions, the weight calculated by the manufacturer which, when applied at the boom tip, has the same effect on the crane capacity as the jib and/or boom extension itself.

Electromagnet: A device consisting of an insulated electrical conductor (wire) coiled around a magnetic metal core. When current flows through the conductor, a strong magnetic field forms within the core.

Ether: A volatile, flammable, organic fluid that operators sometime use to assist in starting diesel engines, especially in cold weather.

Ferrous metals: Iron and its alloys that have relatively strong magnetic characteristics.

Flemish eyes: Spliced loops formed in the end of wire-rope rigging slings. For personnel hoist rigging, the eyes must also contain a thimble and be mechanically crimped with a ferrule.

Fly: With reference to mobile cranes, a boom extension carried on the side of the main boom when not in use, that can swing into position to extend the main boom; also called a *swing-away*.

Gantry: A heavy frame with roller sheaves mounted on the upperworks of a crane, used for providing a mechanical advantage to the boom-hoist ropes. Also refers to a jib strut or mast.

Gross capacity: A crane capacity value as displayed in the crane's load chart for a specified configuration.

Gross load: The total weight of all the items lifted by the crane. The gross capacity of the crane minus any deductions for the crane configuration results in the gross load that can be lifted. Gross load is directly related to net capacity.

Ground bearing pressure (GBP): The maximum force per unit area (pressure), measured in pounds per square foot, that can be exerted on the ground's surface without causing deformation of the surface. Can only be determined by testing.

Hoist kick-out: A term for any safety device that automatically stops a hoist drum when a limit is reached.

Interpolate: To estimate or calculate an unknown intermediate value that falls between two known values.

Lattice lacing: In a lattice boom, the zigzag pattern of structural metal supporting the main chords along the sides, top, and bottom of the boom sections.

Level luffing: The process of luffing the boom of a crane while adjusting the hoist cable and boom extension offset so that the load travels in a horizontal direction.

Lift data sheet (LDS): The essential page or pages of a written lift plan that organize and summarize the important information required to develop a lift plan.

Limit line: When engaged in demolition operations using a wrecking ball, a length of wire rope attached to the ball from a point on the crane to prevent it from swinging out too far, providing a safety margin against overturning the crane or damaging the boom.

Line pull: The maximum pulling force a crane hoist drum can apply to a hoist cable, accounted for by the torque that the motor can apply to the drum and the working radius of the rope as it leaves the drum.

Live boom mast: A heavy framework or lattice boom with sheaves, hinged at the lower end of its base on the crane's upperworks, whose purpose is to control the angle of the main boom through attached boom pendants. Boom hoist cables reeved through the boom gantry sheaves or hydraulic actuators control its position.

Load operating radius: The horizontal distance from a vertical line through the axis of crane rotation to the center of the vertical hoist line or tackle, with a load applied. Also called *load radius, lifting radius*, or *operating radius.*

Luffing: Raising or lowering a crane's boom. A luffing jib can luff independently of the main boom.

Main chords: The heavy structural pipe or angle-iron members running the length of a boom section.

Manual luffing jib: A semi-fixed jib extension whose offset must be manually set between lifts.

Net capacity: The crane's hoisting force available after taking into account the crane's and/or load's configurations relating to a lift. Equal to load chart gross capacity minus weight deductions.

Nondestructive testing (NDT): Testing the integrity of fabricated metal structures, especially welds, by various means to detect hidden flaws without destroying the object or rendering it useless. Common NDT methods include dye penetrant (PT), magnetic particle (MT), X-ray and radiography (RT), and ultrasound (UT).

Offset: The angle formed by an erected boom, measured from an imaginary line extending through the center line of the main boom. The crane's load chart provides capacities for allowable jib or boom-extension offset angles.

Pendant: A length of heavy wire rope or flexibly-connected sections of flat metal bars, equipped with connectors at both ends, used to hoist and/or support sections of a crane's boom.

Poured socket: Referring to wire-rope terminations, a method of attaching a fitting by placing the end of the rope inside the fitting socket, then pouring molten zinc, another metal alloy, or a special resin into the socket. After hardening, these terminations can provide 100 percent of the wire rope's minimum breaking strength. This process is also referred to as *speltering*.

Power luffing jib: A jib extension that has the means to change its offset during a lift by remote control of the operator.

Proof test: A strength test of an object, usually under a load exceeding that expected under normal use, to ensure it is suitable for safe use. In personnel hoisting, a proof test requires loading the platform with a weight equal to 125 percent of its rated load and suspending it for at least 5 minutes.

Radio frequency (RF): Describes the lower frequencies of the electromagnetic energy spectrum in which radar, TV, AM/FM radio, cell phones, and all other forms of radio-wave communications and detection operate.

Ringer crane: A large, high-capacity crane erected on a circular track (the ring, usually installed directly on the ground) that supports the entire crane base.

Stop rail: A heavy metal guiderail used by some crane manufacturers to help align a swing-away to its transport location as it is being stowed.

Suspension rope: The rope reeved through a crane's gantry and mast sheaves to the sheaves suspending the boom pendants.

Tailing crane: A crane that helps position a load by turning or tilting it during lifting operations, but does not lift a significant part of the load's weight.

Tipping axis: In crane operations, an imaginary horizontal line around which a crane's center of gravity would rotate as it tips.

Trial lift: For a personnel hoisting platform, a lift of the platform and rigging without occupants, but with an equivalent load, under the same conditions and over the same route as the lift with personnel will take.

Ultimate pad capacity (UPC): Rating of the lifting force generated by a vacuum-lifter pad, based on the vacuum level and pad area.

Vacuum: The reduced presence or near-total absence of air or other gases, creating a low-pressure condition.

Index

A

A/D director, (21302):1, 9, 19
Alloy, (21303):1, 14, 28
All-purpose cranes. *See* Industrial/all-purpose cranes
Approvals, lift data sheets, (21304):15
ASME standards
  Critical Lift Plan, (21304):4
  Example Lift Plan, (21304):6
  hoisting operations, (21305):6, 8
  lift planning
    critical lifts, (21304):4–5, 15
    executing the lift, (21304):16
    personnel hoisting, (21305):10
    post-lift review, (21304):17
    standard lifts, (21304):5
  Mobile and Locomotive Cranes, (21304):4
  Personnel and Responsibilities, (21304):1
  Personnel Lifting Systems, (21303):2–3, (21305):1
  Planning for Load Handling Activities, (21303):1, (21304):1, (21305):6
  platforms
    construction, (21305):3
    inspections, (21305):4
    testing, (21305):4
  Remotely Operated Lifting Magnets, (21303):19
  Standard Lift Plan, (21304):4
  tipping, (21301):5
Auxiliary boom heads, (21302):8–9

B

Back hitch, (21306):9, 23
Barges, personnel hoisting on floating, (21305):11
Belly lines, (21306):9, 16, 23
Blind lifts, (21303):3–4
Blocking in critical lift planning, (21303):2
Boom angle
  configuration and, (21301):13
  in critical lift planning, (21303):2
  defined, (21301):1, 49
  illustrated, (21301):5
  load chart capacities and, (21301):3
Boom-extend mode, (21302):1, 10, 19
Boom length
  configuration and, (21301):13
  in critical lift planning, (21303):2
  defined, (21301):1, 49
  illustrated, (21301):5
  load chart capacities and, (21301):3
Boom-point elevation, (21301):1, 3, 5, 49
Boom stops, (21306):9, 10, 23
Boom trucks
  capacity, calculating, (21301):38, 43–45
  configuration, (21301):10
Bridle slings, (21305):1, 3, 16
Brittle fracture, (21303):1, 22, 28

C

Cantilever length, (21306):9, 14, 23
Capacity
  calculating
    boom truck example, (21301):38, 43–45
    industrial/all-purpose crane example, (21301):32–33, 34
    lattice-boom crane example, (21301):33, 37–42
    lift data sheets for, (21304):10–12
    operational notes, (21301):31, 35
    telescopic-boom crane example, (21301):33, 35–37
  factors in, (21301):9
  gross, (21301):1, 2, 49
  net, (21301):1, 49
  planning for, (21304):4
  range diagram in determining, (21301):29–30
Carry-deck cranes, (21301):9
Cavitation, (21303):1, 23, 28
Center of gravity (CG), (21301):1–2, 14, (21303):2
CG. *See* Center of gravity (CG)
Cobalt, (21303):21
Coefficient of friction, (21303):1, 18, 28
Commercial impacts, planning for, (21304):2
Complexity, planning for, (21304):2
Configuration
  center of gravity in, (21301):2
  defined, (21301):1, 49
  lift data sheets, (21304):6–10, 21, 22, 23–28
Connector pins, (21302):1, 3, 19, (21306):3, 4
Construction, precast concrete, (21303):16
Cotter pins, (21306):3
Counterweights
  assembly, (21306):7, 9
  configurations, (21301):23–24
  requirements, (21302):10
Crane assembly/disassembly director, responsibilities, (21304):1
Crane base configuration
  counterweights, (21301):23–24
  on crawlers, (21301):18
  on outriggers, (21301):15, 18, 20–22
  ring attachments, (21301):23
  on rubber, (21301):15, 19
  tower attachments, (21301):23
Crane configuration
  boom trucks, (21301):10
  capacity, factors influencing, (21301):10–11
  examples, (21301):10
  industrial/all-purpose cranes, (21301):9
  lattice-boom crane, (21301):10
  telescopic-boom crane, (21301):10
Crane operations, cold weather effects on, (21303):21–24
Crane operators
  assembly/disassembly responsibilities, (21306):1, 3
  fatalities, (21306):2, 4
  fatigue, (21304):1
  responsibilities, (21301):5, (21303):2–3, (21304):1, (21305):6, 8
Crane owner, responsibilities, (21304):1
Cranes. *See also specific types of*
  on floating barges, personnel hoisting, (21305):11
  self-assembling, (21302):9
  wind farms and, (21302):14, 15
Crane user, responsibilities, (21304):1

Crawler cranes
configurations, (21301):10
initial setup, (21306):9
load chart capacities, (21301):18
self-assembling, (21302):9
Critical lifts, (21303):1–2, 4, (21304):4–5, 15, 30–41, (21305):8

D

Demolition
defined, (21303):1, 28
mechanical, (21303):13–14
safety, (21303):13
swinging ball
in-line with boom, (21303):14–15
slewing, (21303):14–15
vertical-drop balling, (21303):14
Diagonal struts, (21306):1, 3, 23
Documents and Attachments section, (21304):13, 22
Draglines, (21303):20
Dual-crane lifts, (21303):4–12
Duty cycle, (21303):1, 28

E

Effective weight, (21301):25, 26, 49
Electromagnet, (21303):1, 19, 28
Electromagnet lifting
alloys and metals and, (21303):20
attitude and, (21303):20
factors affecting lifts, (21303):19–20
guidelines, (21303):20–21
length and, (21303):20
surface conditions and, (21303):20
temperature and, (21303):20
thickness and, (21303):20
width and, (21303):20
Elevation, boom-point, (21301):1, 3, 5, 49
Engineer, responsibilities, (21304):1
Environmental conditions, planning for, (21304):4
Ether, (21303):1, 23, 28

F

Fall protection, (21305):2, 7–8, 11, 12
Ferromagnetism, (21303):21
Ferrous metals, (21303):1, 19, 28
Flemish eyes, (21305):1, 3, 16
Fly, (21302):1, 19

G

Gantry, (21306):1, 3, 23
GBP. *See* Ground bearing pressure (GBP)
General contractor/construction manager, responsibilities, (21304):1
Gross capacity, (21301):1, 2, 49
Gross load, (21301):1, 3, 49
Ground bearing pressure (GBP), (21304):1, 13, 20
Ground crew, qualifications and training, (21305):6
Grove TM1500 carrierier-mounted hydraulic crane, Load Chart, (21304):23–24
Grove TM-1500 truck-mounted telescopic-boom crane, capacity, calculating, (21301):33, 35–37
Grove TM-1500 truck-mounted telescopic-boom crane, load charts, (21301):56–57
Grove YB4409 industrial crane, capacity, calculating, (21301):32–33, 34
Guardrails, (21305):2

H

Hand rails, (21305):2
Hand signals, (21305):3
Hazards, planning for, (21304):2
HF. *See* High frequency (HF) radio waves
High frequency (HF) radio waves, (21303):12
Hoisting. *See also* Lifts; Personnel hoisting
hoist rope, (21301):26, 28
line pull, (21301):28
parts of line, (21301):28–29, 30
range diagram, (21301):29–30, 31
Hoist kick-out, (21306):9, 13, 23
Hoist rope, (21301):26, 28
Hoist sheave bearings, (21301):28

I

Industrial/all-purpose cranes
capacity, calculating, (21301):32–33, 34
configuration, (21301):9
In-line ball demolition, (21303):14–15
Interpolate, (21301):9, 13, 49
Iron, (21303):21
ISO standards, tipping, (21301):5

J

Jib charts, (21301):25–26, 27
Jib extensions, setup and assembly, (21302):1–8
Jibs, assembly, (21306):17–18

L

Lattice-boom cranes
assembly
connector pins, (21306):3, 6
considerations pre- and post-, (21306):6–7
counterweights, (21306):7, 9
hazards, (21306):2
inspections post-, (21306):3
location, determining, (21306):6
logistics, (21306):6–7, 9
procedure table of contents, (21306):5
process, (21306):2
standards, (21306):1, 3
capacity, calculating, (21301):33, 37–38, 39–42
configuration, (21301):10
crawler cranes, load chart, (21301):11, 12
deductions, (21301):4
disassembly
collapse, (21306):4
for transport, (21306):1
inspections, pre- and post-assembly, (21306):3
transport, (21306):1
Lattice booms
assembly
hoist line components, (21306):15–16
initial crane setup, (21306):9
location, determining, (21306):6–7
long booms, (21306):14–17, 18
pre-assembly review, (21306):9
process, (21306):10–12
raising checklist, (21306):12–13
disassembly, (21306):18–19
sections required, determining, (21306):3
transportation inspection and repair, (21306):3
Lattice lacing, (21306):1, 3, 23
LDS. *See* Lift data sheet (LDS)
Length
boom length
configuration and, (21301):13
in critical lift planning, (21303):2
defined, (21301):1, 49
illustrated, (21301):5
load chart capacities and, (21301):3

cantilever, (21306):9, 14, 23
in electromagnet lifting and load, (21303):20
Level luffing, (21302):1, 14, 19
Leverage, (21301):2
Liebherr Lr 1250, (21302):9
Lift data sheet (LDS)
defined, (21304):1, 20
HVAC condenser example
approvals, (21304):15
capacity calculations, (21304):10–12
configuration, (21304):6–10, 21, 22, 23–28
developing, (21304):5–6
Documents and Attachments section, (21304):13, 22
example, (21304):29
load and rigging, (21304):10–13, 22, 23–28
Notes section, (21304):15
review, (21304):15
Lift director, responsibilities, (21304):2, 16–17
Lift personnel
hazards to, (21304):2
responsibilities
crane assembly/disassembly director, (21304):1
crane operator, (21304):1
crane owner, (21304):1
crane user, (21304):1
engineer, (21304):1
general contractor/construction manager, (21304):1
lift director, (21304):2
lift planner, (21304):2
rigger, (21304):2
signal person, (21304):2
site safety officer, (21304):2
site supervisor, (21304):2
spotter, (21304):2
transport operator, (21304):2
Lift planner, responsibilities, (21304):2
Lift plans
considerations
capacity, (21304):4
commercial impacts, (21304):2
complexity, (21304):2
environmental conditions, (21304):4
hazards, (21304):2
personnel responsibilities, (21304):1–2
purposes, (21304):1
repetitive operations, (21304):4
standards, (21304):1
for critical lifts, (21303):2, (21304):4–5, 15, 30–41
for multiple-crane lifts, (21303):4–12
range diagrams in, (21301):29
software for, (21304):7
for standard lifts, (21304):4–5
standards, (21304):1
Lift plans, HVAC condenser example
critical lifts, (21304):15
lift data sheets
approvals, (21304):15
crane capacity calculations, (21304):10–12
crane configuration, (21304):6–10, 21, 22, 23–28
developing, (21304):5–6
Documents and Attachments section, (21304):13, 22
example, (21304):29
load and rigging, (21304):10–13, 22, 23–28
Notes section, (21304):15
review, (21304):15
Lifts
critical, (21303):1–2, 4, (21304):4–5, 15, 30–41, (21305):8
electromagnet
alloys and metals and, (21303):20
attitude and, (21303):20
factors affecting lifts, (21303):19–20
guidelines, (21303):20–21
length and, (21303):20
surface conditions and, (21303):20
temperature and, (21303):20
thickness and, (21303):20
width and, (21303):20
executing, (21304):16
hoist rope, (21301):26, 28
line pull, (21301):28
multiple-crane, (21303):4–12
parts of line, (21301):28–29, 30
range diagram, (21301):29–30, 31
reviews
post-lift, (21304):17
pre-lift, (21304):16
stopping, (21304):16
vacuum
angle and, (21303):17
factors affecting lifts, (21303):17–18
guidelines, (21303):18–19
operations, (21303):15–17
overhang and, (21303):17
rigidity and strength and, (21303):17
surface conditions and, (21303):17
temperature and, (21303):17
Limit line, (21303):1, 14, 28
Line pull, (21301):25, 28, 49
Live boom mast, (21306):9, 23
LMI. *See* Load moment indicator (LMI); Load moment indicators (LMI)
Load
in critical lift planning
movement, (21303):2
radius, (21303):2
rigging, (21303):2
weight, (21303):2
electromagnet lifting
alloys and metals and, (21303):20
attitude and, (21303):20
length and, (21303):20
surface conditions and, (21303):20
temperature and, (21303):20
thickness and, (21303):20
width and, (21303):20
gross, (21301):1, 3, 49
operating radius, (21301):1, 5, 49
radius, (21301):13
vacuum lifting
angle and, (21303):17
overhang and, (21303):17
rigidity and strength and, (21303):17
surface conditions and, (21303):17
temperature and, (21303):17
Load charts
basis for, (21301):3, 5, 10
on crawlers, (21301):18
electronic, (21301):3
examples, (21301):6–7, 13, 14, 15
importance of, (21301):1, 5
information displayed on, (21301):3
interpolating, (21301):13
interpreting correctly, factors in, (21301):10–11
jib boom extensions, (21301):25–26, 27
notes, (21301):25, 27, 30, 35
on outriggers, (21301):15, 18, 20–22

Load charts (*continued*)
quadrants of operation, (21301):14–15, 17
replacing, (21301):2–3
on rubber, (21301):15, 19
specificity of, (21301):2, 11
weight deductions, (21301):25, 26
Load moment indicator (LMI), (21301):13, (21302):1
Load operating radius, (21301):1, 5, 49
Load radius, (21301):13
Location, assembly, (21306):6
Luffing, (21302):1, 2, 19
Luffing jib
manual, (21302):1, 13, 19
power, (21302):1, 14, 19

M

Main chords, (21306):1, 3, 23
Manitowoc 4100W-2 lattice-boom crawler crane
capacity, calculating, (21301):37–42, 54–55
load charts, (21301):52–53
Manitowoc Model 999 Crane Assembly and Disassembly Guide, (21306):5
Manitowoc Model 18000 Boom Combinations, (21306):24–26
Manual luffing jib, (21302):1, 13, 19
Mats, in critical lift planning, (21303):2
Mechanical demolition, (21303):13–14
Medium frequency (MF) radio waves, (21303):12
MF. *See* Medium frequency (MF) radio waves
Multiple-crane lifts, (21303):4–12

N

National Crane 18103 Series truck-mounted, telescopic-boom truck, capacity, calculating, (21301):38, 43–45
NDT. *See* Nondestructive testing (NDT)
Net capacity, (21301):1, 49
Nickel, (21303):21
Nondestructive testing (NDT), (21305):1, 4, 16
Notes
load charts, (21301):25, 27, 30, 35
Terex RT555-1, (21301):51
Notes section, lift data sheet, (21304):15

O

Offset, (21301):25, 49
On-outriggers load charts, (21301):15
Operational quadrants, (21301):14–15, 17
OSHA standards
assembly/disassembly, (21306):2, 4
Cranes and Derricks in Construction, (21304):1
critical lifts, (21303):2
Fall Protection Systems Criteria and Practices, (21305):7
hoisting operations, (21305):8
Hoisting Personnel, (21305):1
Inspections, (21306):3
lift planning, (21305):10
Mechanical Demolition, (21303):13–14
personnel lifts, (21303):2
platforms
construction, (21305):3
testing, (21305):4
Signals- Hand Signal Chart, (21305):3
telescopic-boom attachments
assembly/disassembly, (21302):1, 9
counterweight considerations, (21302):10
tipping, (21301):5
working near powerlines, (21305):9
Working Over or Near Water, (21305):7
Outriggers, personnel lifts and, (21303):2–3
On-outriggers load charts, (21301):18, 20–22

P

Parts of line, (21301):28–29
Pendant, (21306):1, 3, 23
Pendant pins, (21306):3
Personnel hoisting
lift planning, (21305):8, 10, 20
near powerlines, (21305):9–10, 11
platforms
access systems, (21305):2
attachment mechanisms, (21305):3
construction, (21305):3–4
decking, (21305):2
design, (21305):1–3
fall protection anchor points, (21305):2
headroom, (21305):2
identification requirements, (21305):2, 3
inspections, (21305):4, 5–6, 11, 17
maintenance, (21305):5–6
minimum capacity, (21305):2
overhead protection, (21305):2
railings, (21305):2
slings and attachments, (21305):3–4
standards, (21305):1
suspension systems, (21305):3
testing, (21305):4–5, 11
restrictions, (21305):1
standards, (21305):1
weather conditions and, (21305):11
Personnel hoisting operations
instrumentation and configuration, (21305):8
personnel
fall protection requirements, (21305):7–8
key individuals, (21305):6–7
work practices, (21305):10–11
Personnel lift authorizer, (21305):6, 10
Personnel Lift Platform Pre-Lift Inspection Form, (21305):17
Personnel lifts
design, (21303):3
markings, (21303):3
near RF transmitters, (21303):11–13
standards, (21303):2–3
trial lifts, (21303):3
Personnel Platform Lift Planning and Authorization form, (21305):20
Personnel responsibilities, planning for, (21304):1–2
Pick-and-carry, (21301):9
Pins
connector pins, (21302):1, 3, 19, (21306):3, 4
cotter pins, (21306):3
Platform occupants
fall protection requirements, (21305):7–8, 11, 12
movement of, (21305):11
qualifications and training, (21305):7
total load, (21305):10
Platforms
access systems, (21305):2
attachment mechanisms, (21305):3
construction, (21305):3–4
decking, (21305):2
design, (21305):1–3
fall protection anchor points, (21305):2
headroom, (21305):2
identification requirements, (21305):2, 3
inspections, (21305):4, 5–6, 17
maintenance, (21305):5–6
minimum capacity, (21305):2
overhead protection, (21305):2
railings

guardrails, (21305):2
hand rails, (21305):2
slings and attachments, (21305):3–4
suspension systems, (21305):3
testing
initial manufacturing tests, (21305):4
proof testing, (21305):4–5
trial lifts, (21305):5, 11
Poured socket, (21305):1, 3, 16
Power Line Awareness Permit, (21305):18–19
Powerlines
personnel hoisting near, (21305):9–10, 11
Power Line Awareness Permit, (21305):18–19
Power luffing jib, (21302):1, 14, 19
Proof test, (21305):1, 4, 16
Purposes, planning for, (21304):1

Q
Quadrants of operation, (21301):14–15, 17
Qualified persons, (21306):3

R
Radiofrequency (RF), (21303):1, 11–12, 28
Radio transmissions, (21303):11–13
Railings, personnel hoists
guardrails, (21305):2
hand rails, (21305):2
Range diagram, (21301):29–30, 31
Reeving in critical lift planning, (21303):2
Repetitive operations, planning for, (21304):4
RF. *See* Radiofrequency (RF)
Riggers, (21304):2, (21306):3
Rigging
in critical lift planning, (21303):2
lift plans, HVAC condenser example, (21304):10–13, 22, 23–28
planning for, (21304):4
Ringer crane, (21301):9, 23, 49
Rope
hoist rope, (21301):26, 28
suspension, (21306):1, 3, 23
wire rope
capacity, (21301):30
rotation resistant, (21301):26
Rough/all-terrain (RT/AT) cranes, (21301):10
RT/AT. *See* Rough/all-terrain (RT/AT) cranes
On-rubber load charts, (21301):15, 19

S
Sheaves, (21301):28, 30
Signal person, responsibilities, (21304):2
Site safety officer, responsibilities, (21304):2
Site supervisor, responsibilities, (21304):2
Slewing, demolition by, (21303):15
Slings, (21305):3–4
Spotter, responsibilities, (21304):2
Stability
quadrants of operation, (21301):14–15
standards, (21301):5
Standard lifts, (21303):1–2
Stop rail, (21302):1, 3, 19
Struts, diagonal, (21306):1, 3, 23
Suspension rope, (21306):1, 3, 23
Swing-away lattice extension, (21302):1–8
Swinging ball demolition
in-line with boom, (21303):14–15
slewing, (21303):14–15

T
Tailing crane, (21303):1, 3, 28
Telescopic-boom attachments
assembly/disassembly
intermediate boom sections, (21302):10–13
luffing extensions, (21302):13–14
OSHA requirements, (21302):1, 9
pre- and post- considerations, (21302):9
setup
auxiliary boom heads, (21302):8–9
swing-away lattice extension, (21302):1–8
Telescopic-boom cranes
capacity, calculating, (21301):33, 35–37
configuration, (21301):10
deductions, (21301):4
load charts, (21301):11, 18
multiple boom-extension modes, (21301):13–14, 16
Temperature
cold, effects of, (21303):21–24
load
electromagnet lifting and, (21303):17
vacuum lifting and, (21303):17
Terex RT555-1, (21301):50–51
Testing, nondestructive (NDT), (21305):1, 4, 16
Tipping
avoiding, (21301):1
depending on, (21301):5
structural failure before, (21301):34
Tipping axis, (21301):1, 2, 14, 49
Tire-mounted truck cranes, (21301):10
Tower luffing-jib attachments, (21301):23
Transit during hoists, (21305):10
Transport operator, responsibilities, (21304):2
Trial lift, (21305):1, 5, 16
Truck-mounted crane
center of gravity, (21301):2
leverage distances, (21301):2

U
Ultimate pad capacity (UPC), (21303):1, 16, 28
UPC. *See* Ultimate pad capacity (UPC)

V
Vacuum, (21303):1, 15, 28
Vacuum lifting
angle and, (21303):17
factors affecting lifts, (21303):17–18
guidelines, (21303):18–19
operations, (21303):15–17
overhang and, (21303):17
rigidity and strength and, (21303):17
surface conditions and, (21303):17
temperature and, (21303):17
Vertical-drop balling demolition, (21303):14

W
Weather
cold, effects of, (21303):21–24
in critical lift planning, (21303):2
Weather condition, (21305):11
Weight, effective, (21301):25, 26, 49
Wind chill, (21303):23
Wind farms, cranes and, (21302):14, 15
Wire rope
capacity, (21301):30
rotation resistant, (21301):26